U0186272

中 外 物 理 学 精 品 书 系

本 书 出 版 得 到 " 国 家 出 版 基 金 " 资 助

国家出版基金项目
NATIONAL PUBLICATION FOUNDATION

中 外 物 理 学 精 品 书 系

高 瞻 系 列 · 1 5

d-Wave Superconductivity
d 波超导体

向 涛 著
吴从军 译

北京大学出版社
PEKING UNIVERSITY PRESS

图书在版编目 (CIP) 数据

d 波超导体 = d-Wave Superconductivity / 向涛著；吴从军译.
— 北京：北京大学出版社，2020.8
（中外物理学精品书系）
ISBN 978-7-301-31334-3

Ⅰ. ① d… Ⅱ. ① 向… Ⅲ. ① D 波 - 超导体 Ⅳ. ① TM26

中国版本图书馆 CIP 数据核字 (2020) 第 104181 号

书　　　名	_d_-Wave Superconductivity（_d_ 波超导体）
著作责任者	向涛　著，吴从军　译
责 任 编 辑	刘啸
标 准 书 号	ISBN 978-7-301-31334-3
出 版 发 行	北京大学出版社
地　　　址	北京市海淀区成府路 205 号　100871
网　　　址	http://www.pup.cn　新浪微博：@ 北京大学出版社
电 子 信 箱	zpup@pup.cn
电　　　话	邮购部 010-62752015　发行部 010-62750672　编辑部 010-62754271
印 　刷　 者	北京中科印刷有限公司
经 　销　 者	新华书店

730 毫米 × 980 毫米　16 开本　25.5 印张　413 千字
2020 年 8 月第 1 版　2020 年 8 月第 1 次印刷

定　　　价　118.00 元

序　言

　　物理学是研究物质、能量以及它们之间相互作用的科学。她不仅是化学、生命、材料、信息、能源和环境等相关学科的基础,同时还与许多新兴学科和交叉学科的前沿紧密相关。在科技发展日新月异和国际竞争日趋激烈的今天,物理学不再囿于基础科学和技术应用研究的范畴,而是在国家发展与人类进步的历史进程中发挥着越来越关键的作用。

　　我们欣喜地看到,改革开放四十年来,随着中国政治、经济、科技、教育等各项事业的蓬勃发展,我国物理学取得了跨越式的进步,成长出一批具有国际影响力的学者,做出了很多为世界所瞩目的研究成果。今日的中国物理,正在经历一个历史上少有的黄金时代。

　　在我国物理学科快速发展的背景下,近年来物理学相关书籍也呈现百花齐放的良好态势,在知识传承、学术交流、人才培养等方面发挥着无可替代的作用。然而从另一方面看,尽管国内各出版社相继推出了一些质量很高的物理教材和图书,但系统总结物理学各门类知识和发展,深入浅出地介绍其与现代科学技术之间的渊源,并针对不同层次的读者提供有价值的学习和研究参考,仍是我国科学传播与出版领域面临的一个富有挑战性的课题。

　　为积极推动我国物理学研究、加快相关学科的建设与发展,特别是集中展现近年来中国物理学者的研究水平和成果,北京大学出版社在国家出版基金的支持下于 2009 年推出了"中外物理学精品书系",并于 2018 年启动了书系的二期项目,试图对以上难题进行大胆的探索。书系编委会集结了数十位来自内地和香港顶尖高校及科研院所的知名学者。他们都是目前各领域十分活跃的知名专家,从而确保了整套丛书的权威性和前瞻性。

　　这套书系内容丰富、涵盖面广、可读性强,其中既有对我国物理学发展的梳理和总结,也有对国际物理学前沿的全面展示。可以说,"中外物理学精品书系"力图完整呈现近现代世界和中国物理科学发展的全貌,是一套目前国内为数不多的兼具学术价值和阅读乐趣的经典物理丛书。

"中外物理学精品书系"的另一个突出特点是,在把西方物理的精华要义"请进来"的同时,也将我国近现代物理的优秀成果"送出去"。物理学在世界范围内的重要性不言而喻。引进和翻译世界物理的经典著作和前沿动态,可以满足当前国内物理教学和科研工作的迫切需求。与此同时,我国的物理学研究数十年来取得了长足发展,一大批具有较高学术价值的著作相继问世。这套丛书首次成规模地将中国物理学者的优秀论著以英文版的形式直接推向国际相关研究的主流领域,使世界对中国物理学的过去和现状有更多、更深入的了解,不仅充分展示出中国物理学研究和积累的"硬实力",也向世界主动传播我国科技文化领域不断创新发展的"软实力",对全面提升中国科学教育领域的国际形象起到一定的促进作用。

习近平总书记在 2018 年两院院士大会开幕会上的讲话强调,"中国要强盛、要复兴,就一定要大力发展科学技术,努力成为世界主要科学中心和创新高地"。中国未来的发展在于创新,而基础研究正是一切创新的根本和源泉。我相信,在第一期的基础上,第二期"中外物理学精品书系"会努力做得更好,不仅可以使所有热爱和研究物理学的人们从中获取思想的启迪、智力的挑战和阅读的乐趣,也将进一步推动其他相关基础科学更好更快地发展,为我国的科技创新和社会进步做出应有的贡献。

"中外物理学精品书系"编委会主任

中国科学院院士,北京大学教授

王恩哥

2018 年 7 月于燕园

Preface

After the discovery of high temperature superconductors in 1986, the establishment of the theoretical framework of d-wave superconductivity has been an important advance. Supported by a great deal of experimental evidence, this framework demonstrates that, unlike most conventional metallic superconductors, high temperature superconductors are the so-called d-wave superconductors, hence their Cooper pairs possess d-wave pairing symmetry. In other words, two electrons inside each pair carry a relative angular momentum of two. In comparison, most metallic superconductors exhibit s-wave pairing symmetry, where the relative angular momentum of each pair is zero. The superconducting energy gap of s-wave superconductor is finite over the entire Fermi surface, and the low energy excitations are thermally activated. In contrast, the energy gap of d-wave superconductor is zero at certain points on the Fermi surface, and the low-lying excitations can be activated through non-thermal ways. Therefore, physical properties of these two types of superconductors are qualitatively different. Accurate determination of the pairing symmetry is a prerequisite of a comprehensive study of high temperature superconductivity. To reach this point, a thorough understanding of physical properties in the superconducting phase is desired.

In order to determine unambiguously the pairing symmetry, one needs to carry out a comprehensive investigation by comparing the experimental results obtained from optical, electric, magnetic, and thermodynamic measurements with theoretical predictions. Traditional reference books or textbooks on superconductivity consider mostly conventional metallic superconductors, and concentrate on the introduction of s-wave superconductors. A few monographs and review articles, that have been published in recent years on high temperature superconductivity, have described physical properties of d-wave superconductors, but emphasized more on experimental observations. A systematical and comprehensive introduction to the theory of *d*-wave superconductivity is still not available. The main purpose for me to write this book is to fill this gap by giving a comprehensive review on the progress in the theoretical study of high temperature superconductors in the superconducting state. Many of the problems, methods, and ideas discussed in this book have already become essential to the understanding of high temperature superconductivity. I have made a considerable effort to make this book largely self-contained. The materials introduced in this book may be useful to graduate students and research scientists or engineers to become familiar with the field and the methods.

The *d*-wave superconductor is one kind of non-*s*-wave superconductors (also known as unconventional superconductors). It is also one of the representatives of the entire family of unconventional superconductors. Loosely speaking, the theoretical results of *d*-wave superconductors introduced in this book can be readily transformed or generalized to apply to other unconventional superconductors, including, for example, *p*-wave superconductors. Therefore, this book is not only useful for studying high temperature superconductors, but also valuable for studying other unconventional superconductors.

This book is written on the background of high temperature superconductivity. It focuses on the discussion of physical properties of *d*-wave superconductors in the superconducting state. For comparison with theoretical results, some experimental results of high temperature superconductors are analyzed and summarized. This book does not aim at providing a comprehensive review of all physical properties of cuprate superconductors. It discusses neither the microscopic origins of *d*-wave pairing mechanism of high temperature superconductivity, nor the rich properties in the normal state. The reason for doing this is owing to the limitation of time and my expertise. More importantly, it is because the mechanism of high temperature superconductivity remains an issue unsolved. There are quite many abnormal physical phenomena discovered in the normal state of high temperature cuprates. These include, for example, the pseudogap effect, linear resistivity, spin-charge separation, intrinsic charge inhomogeneity and stripes. A coherent and unified theoretical description to these phenomena has not been established. It is still premature to review and summarize these results from a theoretical perspective.

The book is organized into 13 chapters as follows. Chapter 1 introduces the fundamental concepts of superconductivity, and briefly reviews the basic ideas and the framework of the Bardeen-Cooper-Schrieffer (BCS) theory of super-conductivity. This chapter also includes some concepts that are important to the understanding of superconductivity but not well introduced in the standard textbooks, such as the difference between the probability current density of su-perconducting quasiparticles and the electric current density, the off-diagonal long range order and the BCS mean-field theory, spontaneous symmetry break-ing and the Meissner effect, etc. Chapter 2 introduces some microscopic models of high temperature superconductivity. This chapter is relatively independent from other chapters. The aim is to provide a background and starting point for studying d-wave superconductors, and to facilitate the readers towards a more in-depth study on the microscopic mechanism of d-wave superconducti-vity. Chapters 3 through 13 introduce systematically the physical properties of various thermodynamic and electrodynamic response functions of d-wave su-perconductors. These include, for example, the temperature dependences of superconducting energy gap functions and specific heat, excitation spectra and photoelectric response of superconducting quasiparticles, Andreev reflection, single-electron tunnelling, Josephson effect and phase sensitivity experiments, impurity scattering, superfluid density, optical conductivity, thermal conduc-tivity, Raman spectroscopy, nuclear magnetic resonance, and mixed states, etc. These chapters highlight various universal behaviors of d-wave superconductors. Detailed derivations of theoretical formula, together with a comparison with the measurement results of high temperature cuprates, are given. In addition, there are five appendices at the end of the book, which provide useful theorems and mathematical formulas in the analysis of superconducting properties. With regard to citations, I include as many as I can the original references which are important to the historical development and a selection of other references that are crucial to the understanding of the methods or ideas. But the list is surely not a comprehensive bibliography.

In writing this book, I received warm encouragements and kind supports from Lu Yu, Zhongxian Zhao, Changde Gong, and Zhaobing Su. I have also benefited from the discussions with C. Panagopoulos, J. M. Wheatley, W. N. Hardy, Haihu Wen, Nanlin Wang, Jianlin Luo, Guangming Zhang, Zhengyu Weng, Rushan Han, Shiping Feng, Jianxin Li, Qianghua Wang, Qingming Zhang, Xianhui Chen, Honggang Luo, Chengshi Liu, Yuehua Su, Jun Chang, and many other colleagues. I would like to express my heartfelt thanks to them.

Abbreviations Used in This Monograph

1. BCS theory: Bardeen-Cooper-Schrieffer theory
2. BdG equation: Bogoliubov-de Gennes equation
3. ODLRO: off-diagonal long range order
4. GL: Ginzburg-Landau
5. ARPES: angular resolved photo-emission spectroscopy
6. NMR: nuclear magnetic resonance

Contents

Chapter 1

Introduction to Superconductivity

1.1 Basic Properties of Superconductivity

Superconductivity, as an emergent macroscopic quantum phenomenon, is one of the most important subjects of contemporary condensed matter physics. It was first discovered by Dutch physicist Heike Kamerlingh Onnes on April 8, 1911 [1, 2, 3]. In 1908, Onnes and his assistants successfully liquefied helium and for the first time reached low temperatures below 4.25K. This was a historical breakthrough of low temperature physics. When they applied this technique and measured the resistance of mercury, they found that its resistance drops abruptly to zero around 4.2K. This important discovery opened up the whole field of superconductivity and related applications. It also greatly stimulated the study of quantum emergent phenomena in condensed matter physics.

There are two characteristic electromagnetic features of superconductivity: zero direct current resistance and perfect diamagnetism. Zero resistance means that superconductors are ideal conductors, and thus there is no energy loss during electric energy transport using superconducting transmission lines. Moreover, superconductors are more than just ideal conductors. More fundamentally, superconductors exhibit perfect diamagnetism which expels magnetic flux lines from the interior of superconductor. The external magnetic field can only penetrate into superconductors within a short length scale near the surface called the *penetration length*. The perfect diamagnetism of superconductivity was discovered by W. Meissner and R. Ochsenfeld in 1933, which is called the Meissner effect[4]. The Meissner effect is *not* a consequence of zero resistance but an independent fundamental property due to superconducting phase stiffness. A qualified superconductor must exhibit both zero resistance and the

Meissner effect, which are two fundamental criteria to determine experimentally whether a sample is superconducting or not.

The superconducting state is a distinct thermodynamic phase. At zero external magnetic field, the superconducting phase transition is second-order, which occurs at a critical temperature denoted as T_c. The phase transition from a normal metallic or insulating state to a superconducting state is an establishment of macroscopic quantum coherence characterized by the superconducting order parameter. Different from ferromagnetism, the superconducting order is an off-diagonal long range order (ODLRO) which does not have a classical correspondence. This point will be further elaborated later. As lowering temperatures, there exists a critical region from the normal state to the zero resistance superconducting state. The width of this critical region is determined by the fluctuation of superconducting order parameters. In conventional metal-based superconductors, the temperature range of the critical region is very narrow, and the resistance drops to zero abruptly. However, in high-T_c cuprate superconductors or dirty superconductors of metals and alloys, fluctuations are strong. The corresponding critical regions are broad and the resistance drop is relatively slow.

A superconductor has exactly zero direct-current resistivity and is able to maintain an electric current without generating an external voltage in the superconducting state. But it losses the superconducting phase coherence and exhibits a small but finite resistance in the presence of an alternative current. One can also turn a superconductor into a normal conductor by applying a strong magnetic field or a direct electric current. For a given temperature, the highest applied magnetic field or electric currect under which a material remains superconducting is called the upper critical field or the critical current.

1.2 Two Characteristic Length Scales

There exist two characteristic length scales in superconductors: the coherence length ξ and the penetration depth λ. The coherence length is the length scale over which the superconducting order parameter can exhibit significant spatial variations. Because of the Meissner effect, an external weak magnetic field cannot penetrate into the interior of superconductor. But it can survive in a narrow length scale near the surface. The penetration depth describes the length scale over which the external magnetic field decays to zero. Both the coherence length and the penetration depth are temperature-dependent, and they diverge at the superconducting transition temperature. The competition between these two length scales has important consequences on the physical properties of superconductors. Particularly, superconductors are classified into two types according to the ratio of ξ/λ. They are called type-I or type-II superconductors when ξ/λ is larger or smaller than $\sqrt{2}$, respectively. Magnetic fluxes are expelled from the interior of type-I superconductors, but can penetrate into the interior of type-II superconductors, forming quantized vortex lines.

1.3 Two-Fluid Model and London Equations

Historically, an important phenomenological theory of superconductivity is the two-fluid model first proposed by Groter and Casimir [5]. The key assumption of this model is the existence of two different types of electrons in superconductors: normal and superconducting electrons. The density of normal electrons is called the normal fluid density and that of superconducting electrons is called the superfluid density. The sum of these two kinds of densities gives the total density of electrons. Normal electrons carry entropy and behave similarly as in ordinary metals. Their states are changed by scattering with phonons and impurities. In contrast, superconducting electrons are resistance free. They do not carry entropy and have no contribution to thermodynamic quantities such as the specific heat. A static electric field cannot exist in an equilibrium superconducting state. Otherwise, superconducting electrons would be accelerated without attenuation, leading to a divergent electric current. The existence of superconducting electrons with zero electric field explains why the resistance is zero. However, the two-fluid model does not answer the question how superconducting electrons are formed, neither can it explain the Meissner effect.

In order to explain the Meissner effect, London brothers proposed an electromagnetic equation [6] to describe the superconducting current. This equation connects the superconducting current density J_s with the electromagnetic vector potential A. Under the Coulomb gauge, it can be expressed as

$$J_s = -\frac{n_s e^2}{m} A, \tag{1.1}$$

where n_s is the superfluid density of electrons. This equation is called the London equation. It cannot be deduced from the Maxwell equations and should be viewed as an independent electromagnetic equation by treating superconductors as a special class of electromagnetic media.

The equation that the magnetic field satisfies in the superconducting state can be derived by combining Eq. (1.1) with the Maxwell equations. It is dubbed as the first London equation and given by the expression

$$\nabla^2 H = \frac{\mu_0 n_s e^2}{m} H. \tag{1.2}$$

In a semi-infinite plate of superconductor with its surface perpendicular to the x-direction, the solution of Eq. (1.2) is simply given by

$$H(x) = H_0 e^{-(x-x_0)/\lambda}, \tag{1.3}$$

$$\lambda = \sqrt{\frac{m}{\mu_0 n_s e^2}}, \tag{1.4}$$

where x_0 is the x-coordinate of the superconductor-vacuum interface, λ is the penetration depth describing the decay length of the external magnetic field. In the limit $x - x_0 \gg \lambda$, the magnetic field decays to zero. This gives a phenomenological explanation to the Meissner effect.

In spite of its simplicity, the two-fluid model captures the key features of superconductors. The key concepts, i.e. the normal and superconducting electrons, were broadly used in the construction of the microscopic theory of superconductivity. The normal and superconducting electrons correspond to the quasiparticle excitations, and the superconducting paired electrons (called *Cooper pairs*), respectively. The two-fluid model has played an important role in the study of superconductivity, although it does not explain the microscopic mechanism of superconductivity. Even after the establishment of microscopic theory of superconductivity, it is still useful to apply the two-fluid model to understand qualitatively experimental results of superconductors.

1.4 Cooper Pairing

Superconductivity is a quantum many-body effect and cannot be understood based on the single-electron theory and its perturbative expansion. In 1956, Cooper considered a two-electron problem which turned out to be one of the most crucial steps towards microscopic understanding of superconductivity [7]. He showed that if there exists an effective attraction interaction, no matter how weak it is, between two electrons in a background of Fermi sea, the free Fermi surface is no longer stable. Electrons on the Fermi surface will pair each other to form bound states. This reduces the energy of the ground state. The bound state of paired electrons is called a Cooper pair. The Cooper instability results from the interplay between the weak attractive interaction and the Fermi sea. The appearance of the Fermi sea is crucial. Otherwise, the Cooper pairing instability would not happen in an arbitrarily weak attractive potential. In free space, two electrons can form a bound state only if the attractive interaction between them is sufficiently strong (above a finite threshold) in three dimensions.

The proof given by Cooper is based on a simple variational calculation. He considered how the ground state energy is changed by adding two extra electrons with opposite momenta and spins to a filled Fermi sea at zero temperature. Due to the Pauli exclusion principle, these two electrons can only be put outside the Fermi sea. For simplicity in the calculation, he assumed that the attractive potential is nonzero only when both electrons lie between the Fermi energy E_F and $E_F + \omega_c$, and the amplitude of the potential V_0 is momentum independent. Here the cutoff ω_c is a characteristic energy scale determined by the mechanism or resource from which the attraction is induced. If the effective attraction is induced by the electron-phonon interaction, ω_c is just the characteristic frequency of phonons, namely the Debye frequency. After a simple variational calculation, Cooper found that the lowest energy for these two electrons is given by

$$E = 2E_F - 2\hbar\omega_c e^{-2/N_F V_0}, \tag{1.5}$$

where N_F is the electron density of states on the Fermi surface. Eq. (1.5) indicates that the two electrons are bounded and their binding energy is

$$2\Delta = 2\hbar\omega_c e^{-2/N_F V_0}. \tag{1.6}$$

It is also the energy needed to break a Cooper pair. This result shows that the Fermi surface is unstable against an infinitesimally small attractive interaction. It also reveals two important parameters in describing a superconducting state. One is the characteristic attraction energy scale ω_c, and the other is the dimensionless coupling constant defined by product of the density of states at the Fermi level and the depth of the attractive interaction. As discussed later, these two parameters also determine the superconducting transition temperature T_c. The calculation made by Cooper is simple, but it captures the main character of superconductivity.

Eq. (1.6) shows that the dependence of the binding energy on the interaction strength V_0 is singular. Thus the microscopic theory of superconductivity cannot be established through perturbative calculations based on normal conducting states. This is actually the major difficulty in the study of superconducting mechanism, which obstructed the development of microscopic theory of superconductivity for nearly fifty years after its discovery.

1.5 Mean-Field Theory of Superconductivity

In 1957, John Bardeen, Leon Cooper, and John Robert Schrieffer (BCS) proposed the microscopic theory of superconductivity based on the concept of Cooper pairing[8]. Their work established a fundamental theory of superconductivity. It is also a tremendous progress towards the understanding of microscopic quantum world.

In the framework of BCS theory, there are two preconditions for the formation of superconducting condensation. The first is the formation of Cooper pairs through effective attraction. The second is the development of phase coherence among Cooper pairs. Cooper pairing refers to the process that electrons near the Fermi surface form bound states. It is a prerequisite of superconductivity because Cooper pairs carry the feature of bosonic statistics, which eliminates the effective repulsive interaction induced by the Fermi statistics of electrons, and can condense into a superfluid state by forming phase coherence. Cooper pairs are found to exist in all superconductors discovered by far. This is a strong support to the BCS theory. It also indicates the importance of the pairing in the formation of superconductivity.

The BCS work is a variational theory. It is based on the BCS variational wavefunction first proposed by Schrieffer. This wavefunction generalizes the solution of Cooper pair to a many-body system. It captures the main picture of Cooper for the superconducting condensation of paired electrons. The BCS theory is equivalent to the mean-field theory later developed based on the Bogoliubov transformation. This mean-field theory is to take the Gaussian or saddle-point approximation in the framework of quantum field theory. It handles the thermal average of operators, rather than the variational wavefunction of the ground state. Fluctuations of Cooper pairs around the saddle point can be included, for example, by taking the one-loop expansion in the path-integral formulism.

The mean-field theory starts by considering the following reduced BCS pairing Hamiltonian

$$H = \sum_{k\sigma}(\varepsilon_k - \mu)c_{k\sigma}^{\dagger}c_{k\sigma} - \sum_{k,k'}V_{k,k'}c_{k\uparrow}^{\dagger}c_{-k\downarrow}^{\dagger}c_{-k'\downarrow}c_{k'\uparrow}, \qquad (1.7)$$

where ε_k is the energy dispersion of electrons, μ is the chemical potential, and $V_{k,k'}$ is the scattering potential between two Cooper pairs with momenta $(k\uparrow; -k\downarrow)$ and $(k'\uparrow; -k'\downarrow)$, respectively. This Hamiltonian is a simplification to the complex interactions of electrons. It highlights the interaction in the pairing channel and neglects interactions in other channels.

Eq. (1.7) is applicable to superconductors with spin singlet pairing. It can be extended to describe spin triplet superconductors with slight modifications. This Hamiltonian considers the Cooper pairs with zero center-of-mass momen-

tum, and neglects the pairing with finite center-of-mass momentum. It is equivalent to setting the real space pairing interaction range to infinite, consistent
with the spirit of mean-field approximation. The zero momentum pairing is
physically reasonable because the phase space for the finite momentum pairing
is strongly constrained by the Fermi surface geometry and by the momentum conservation [9]. In an external magnetic field, the Fermi surfaces of up-
and down-spin electrons are split, and the pairing with finite center-of-mass
momentum is believed to be favored. Cooper pairs in a current-carrying superconducting state have finite pairing momenta. But the pairing energy is
suppressed and becomes zero when the current exceeds the critical current.

   In real calculation, it is commonly assumed that $V_{k,k'}$ is factorizable with
respect to the momenta k and k'

$$V_{k,k'} = \frac{g}{V}\phi_k\phi_{k'}, \tag{1.8}$$

where V is the system volume and g is the coupling constant. ϕ_k is a symmetry
function that describes the internal structure of a Cooper pair. Substituting Eq.
(1.8) into Eq. (1.7), the Hamiltonian can be reexpressed as

$$H = \sum_{k\sigma}(\varepsilon_k - \mu)c_{k\sigma}^\dagger c_{k\sigma} - \frac{g}{V}\hat{A}^\dagger\hat{A}, \tag{1.9}$$

where $\hat{A} \equiv \sum_k \phi_k c_{-k\downarrow}c_{k\uparrow}$. Taking the mean-field approximation for the second
term on the right hand side of Eq. (1.9),

$$-\hat{A}^\dagger\hat{A} = -\langle\hat{A}^\dagger\rangle\hat{A} - \langle A\rangle A^\dagger + \langle A^\dagger\rangle\langle A\rangle, \tag{1.10}$$

we obtain the mean-field Hamiltonian

$$H_{MF} = \sum_k \left(\sum_\sigma \xi_k c_{k\sigma}^\dagger c_{k\sigma} + \Delta_k c_{k\uparrow}^\dagger c_{-k\downarrow}^\dagger + \Delta_k c_{-k\downarrow}c_{k\uparrow}\right) + \frac{V}{g}\Delta^2, \tag{1.11}$$

where $\xi_k = \varepsilon_k - \mu$, $\langle A\rangle$ represents the expectation value of operator A, and

$$\Delta_k = \Delta\phi_k \tag{1.12}$$

is the superconducting order parameter determined by the equation

$$\Delta = -\frac{g}{V}\langle\hat{A}\rangle = -\frac{g}{V}\sum_k \phi_k\langle c_{-k\downarrow}c_{k\uparrow}\rangle. \tag{1.13}$$

$\langle c_{-k\downarrow} c_{k\uparrow} \rangle$ depends on the value of Δ. Eq. (1.13) is just the celebrated BCS gap equation. It determines completely the low energy elementary excitation spectra of superconductors together with Eq. (1.11). By solving these two equations self-consistently, one can calculate all thermodynamic quantities.

H_{MF} does not conserve the particle number. But the total spin, $\sum_k \sigma c_{k\sigma}^\dagger c_{k\sigma}$, and the total momentum of Cooper pairs remain conserved. H_{MF} can be diagonalized by an unitary matrix using the Bogoliubov transformation introduced in Appendix A:

$$
\begin{pmatrix} c_{k\uparrow} \\ c_{-k\downarrow}^\dagger \end{pmatrix} = \begin{pmatrix} u_k & v_k \\ -v_k^* & u_k^* \end{pmatrix} \begin{pmatrix} \alpha_k \\ \beta_k^\dagger \end{pmatrix}.
$$

After the diagonalization, the Hamiltonian becomes

$$
H_{MF} = \sum_k E_k \left(\alpha_k^\dagger \alpha_k + \beta_k^\dagger \beta_k \right) + \sum_k (\xi_k - E_k) + \frac{V}{g} \Delta^2. \tag{1.14}
$$

$\alpha_k^\dagger$ and $\beta_k^\dagger$ are the creation operators of the Bogoliubov quasiparticles. They describe the single-particle excitations above the superconducting gap, corresponding to the normal electrons in the two-fluid model. The quasiparticle excitation energy is given by

$$
E_k = \sqrt{\xi_k^2 + \Delta_k^2}. \tag{1.15}
$$

On the Fermi surface, $\xi_k = 0$ and $E_k = |\Delta_k|$. Thus Δ_k is the gap function of quasiparticles in momentum space. The matrix elements u_k and v_k satisfy the normalization condition, $u_k^2 + v_k^2 = 1$, and are determined by

$$
u_k = \sqrt{\frac{1}{2} + \frac{\xi_k}{2E_k}}, \tag{1.16}
$$

$$
v_k = -\text{sgn}(\Delta_k)\sqrt{\frac{1}{2} - \frac{\xi_k}{2E_k}}. \tag{1.17}
$$

By calculating the pairing correlation function using the above solution, we can express explicitly the gap equation as

$$
1 = \frac{g}{V} \sum_k \frac{\phi_k^2}{2E_k} \tanh \frac{\beta E_k}{2}. \tag{1.18}
$$

The temperature dependence of the gap function $\Delta(k)$ can be determined by solving this equation self-consistently. Moreover, the superconducting transition temperature T_c can be solved from this equation by setting $\Delta = 0$ in E_k.

At zero temperature, there are no quasiparticle excitations, and both $\langle \alpha_k^\dagger \alpha_k \rangle$ and $\langle \beta_k^\dagger \beta_k \rangle$ are zero. The ground state wavefunction can be obtained by projecting out both α- and β-types of quasiparticles from an arbitrary initial state $|\Psi_0\rangle$ not orthogonal to the ground state

$$|\Psi\rangle = \prod_k \left(1 - \alpha_k^\dagger \alpha_k\right) \left(1 - \beta_k^\dagger \beta_k\right) |\Psi_0\rangle. \tag{1.19}$$

To set $|\Psi_0\rangle$ as the vacuum state $|0\rangle$, the above wave-function after renormalization then becomes

$$|\Psi\rangle = \prod_k \left(u_k + v_k c_{k\uparrow}^\dagger c_{-k\downarrow}^\dagger\right) |0\rangle = \prod_k u_k \exp\left(\frac{v_k}{u_k} c_{k\uparrow}^\dagger c_{-k\downarrow}^\dagger\right) |0\rangle, \tag{1.20}$$

which is just the BCS variational wavefunction with v_k^2 the pairing probability. In the above expression, the states inside and outside the Fermi surface can be separated

$$|\Psi\rangle = \prod_{|k|>k_F} \left(u_k + v_k c_{k\uparrow}^\dagger c_{-k\downarrow}^\dagger\right) \prod_{|k|<k_F} \left(u_k c_{-k\downarrow} c_{k\uparrow} + v_k\right) |\text{Fermi Sea}\rangle. \tag{1.21}$$

Based on this expression, it is clear that quasiparticle excitations with momenta outside and inside the Fermi surface are electron- and hole-like, respectively. Here the definition of electrons and holes is the same as in normal conductors.

The quasiparticle operators α_k and β_k contain both the creation and annihilation operators of electrons. Clearly, they are not particle-number eigenoperators. However, real physical process should preserve the electric charge or the particle-number conservation. Does this imply that it is improper to use these operators to describe physical observables? The answer is affirmative. To gain an intuitive understanding, let us introduce the creation and annihilation operators of Cooper pairs, $\hat{B}^\dagger$ and $\hat{B}$, and redefine the quasiparticle operators

α_k and β_k as

$$\alpha_k = u_k c_{k\uparrow} - v_k \hat{B} c^\dagger_{-k\downarrow},$$
$$\beta^\dagger_k = v_k \hat{B}^\dagger c_{k\uparrow} + u_k c^\dagger_{-k\downarrow}.$$

The pair operators $\hat{B}^\dagger$ and $\hat{B}$ create and annihilate two electrons, respectively. Operators α_k and β_k such defined maintain the charge conservation. They change the particle number by -1 and 1, respectively. This gives a more rigorous definition for the creation and annihilation operators of Bogoliubov quasiparticles. In the superconducting state, Cooper pairs condense, and $\hat{B}$ and $\hat{B}^\dagger$ can be replaced by their expectation values, i.e. $\hat{B} = \hat{B}^\dagger \approx \langle B \rangle$. Thus we can set $\hat{B}$ and $\hat{B}^\dagger$ as constants and eliminate them from the above expressions by absorbing them into the redefinition of v_k. The above equations then return back to the original expressions of α_k and β_k. This implies that the charge conservation is still preserved in the BCS theory, although it is formally broken in the definition of quasiparticle operators α_k and β_k. Thus quasiparticle excitations and related physical quantities described by these operators are physically observable. This is in fact confirmed by enormous experimental measurements.

1.6　Bogoliubov-de Gennes Self-Consistent Equations

The BCS gap equation and other formulas introduced in the preceding section are derived based on the translation invariance. They need to be modified in a system with impurities or magnetic vortices where the translation symmetry is broken. In order to describe the spatial variations of superconducting order parameters and other physical quantities, it is more convenient to work directly in coordinate space rather than in momentum space.

In a spatially inhomogeneous system, if there are no magnetic impurities or other sources of interactions that break time-reversal symmetry, the BCS mean-field Hamiltonian can be generally expressed as

$$H_{MF} = \int \mathrm{d}r\mathrm{d}r' (c^\dagger_{r\uparrow}, c_{r\downarrow}) \begin{pmatrix} H_0(r)\delta(r-r') & \Delta(r,r') \\ \Delta^*(r,r') & -H_0(r)\delta(r-r') \end{pmatrix} \begin{pmatrix} c_{r'\uparrow} \\ c^\dagger_{r'\downarrow} \end{pmatrix},$$
$$(1.22)$$

where

$$H_0 = -\frac{\hbar^2}{2m}\nabla^2 + U(r) - \mu,$$

and $U(r)$ is a scalar scattering potential. In real space, the gap function $\Delta(r, r')$ is defined as the pairing order parameter for the two electrons at r and r',

$$\Delta(r, r') = -g\langle c_{r\uparrow}c_{r'\downarrow}\rangle, \tag{1.23}$$

which can be calculated self-consistently.

The Hamiltonian defined in Eq. (1.22) is quadratic. Its trace is zero, i.e. $\mathrm{Tr}H_{MF} = 0$. From the particle-hole symmetry, it can be shown that if E_n is an eigenvalue of H_{MF}, so is $-E_n$. H_{MF} can be diagonalized using an unitary matrix through the Bogoliubov transformation,

$$\begin{pmatrix} c_{r\uparrow} \\ c_{r\downarrow}^{\dagger} \end{pmatrix} = \sum_n \begin{pmatrix} u_n(r) & -v_n^*(r) \\ v_n(r) & u_n^*(r) \end{pmatrix} \begin{pmatrix} \alpha_n \\ \beta_n^{\dagger} \end{pmatrix}. \tag{1.24}$$

For superconducting systems without time-reversal symmetry, for example, in the presence of an external magnetic field where up- and down-spin electrons are mixed by the Zeeman interaction, a similar Bogoliubov transformation can be defined. But the above 2×2 transformation matrix needs to be generalized and replaced by a 4×4 matrix.

$u_n(r)$ and $v_n(r)$ define the wavefunction of Bogoliubov quasiparticles. They are determined by the eigen equation of H_{MF}:

$$\int dr' \begin{pmatrix} H_0(r)\delta(r - r') & \Delta(r, r') \\ \Delta^*(r, r') & -H_0(r)\delta(r - r') \end{pmatrix} \begin{pmatrix} u_n(r') \\ v_n(r') \end{pmatrix} = E_n \begin{pmatrix} u_n(r) \\ v_n(r) \end{pmatrix}. \tag{1.25}$$

In the quasiparticle representation, the gap function $\Delta(r, r')$ can be expressed using $u_n(r)$ and $v_n(r)$ as

$$\Delta(r, r') = -\frac{g}{2}\sum_n [u_n(r)v_n^*(r') + u_n(r')v_n^*(r)]\tanh\frac{\beta E_n}{2}. \tag{1.26}$$

Eqs. (1.25) and (1.26) are the Bogoliubov-de Gennes (BdG) self-consistent equations [10]. They have been widely used to solve the problems related to impurity scattering, elementary excitations around vortex lines, surface states, and the Andreev reflection.

The BdG self-consistent equations are equivalent to the Green's function theory of superconductivity at the mean-field level. In a spatially inhomogeneous system, the Green's function $G(r, r')$ depends on both r and r', not just on their difference $r - r'$. In this case, it is usually more convenient to solve the BdG equation than the Green's function because the BdG wavefunction $(u_n(r), v_n(r))$ depends only on one coordinate r.

1.7 Charge and Probability Current Density Operators of Superconducting Quasiparticles

As mentioned previously, the Bogoliubov quasiparticles determined by Eq. (1.25) are not eigenstates of the electron number operator, and the total electron number is not conserved. This can be more clearly understood from the Bogoliubov transformation of quasiparticle operators given in Eq. (1.24). The breaking of the electron number conservation implies that the probability of quasiparticles, $\rho_P(r)$, is not proportional to the density of electrons, $\rho_Q(r)$. Correspondingly, the current density of quasiparticles, $J_P(r)$, is also not proportional to the electric current density, $J_Q(r)$. This is markedly different from the situation in a normal metal where the electron probability (current) density and the corresponding charge (current) density are essentially equivalent and satisfy the simple equations, $\rho_P(r) = e\rho_Q(r)$ and $J_Q(r) = eJ_P(r)$. The difference results from the Cooper pair condensation in the superconducting state. A thorough understanding to it is important to the understanding of the gauge invariance and the scattering problem of electrons in superconductors.

The definitions of the Bogoliubov quasiparticle current density and the electric current density depend on the symmetry of the gap function $\Delta(r, r')$. In an isotropic s-wave superconductor, the pairing interaction is entirely local, these quantities are simple to define and the Bogoliubov wavefunctions u and v are governed by the equation

$$i\hbar \frac{\partial}{\partial t} \begin{pmatrix} u(r) \\ v(r) \end{pmatrix} = \begin{pmatrix} H_0(r) & \Delta(r) \\ \Delta^*(r) & -H_0(r) \end{pmatrix} \begin{pmatrix} u(r) \\ v(r) \end{pmatrix}, \tag{1.27}$$

where $\Delta(r) = \Delta(r, r')\delta(r - r')$ is the superconducting order parameter. In a *d*-wave or other unconventional superconductor, the gap function becomes non-local, and a few off-diagonal non-local terms needs to be added to the definitions of these quantities.

In the isotropic *s*-wave superconductor, the gap function is independent of momentum, i.e. $\Delta_k = \Delta$, the quasiparticle density contains the contribution from both particles (u) and holes (v),

$$\rho_P(r) = |u(r)|^2 + |v(r)|^2. \tag{1.28}$$

Its time-derivative, $\partial\rho_P/\partial t$, can be obtained using Eq. (1.27). The conservation law of probability is described by the equation

$$\frac{\partial}{\partial t}\rho_P + \nabla \cdot J_P = 0. \tag{1.29}$$

Based on this equation, we find that the quasiparticle probability current is defined as

$$J_P = \frac{\hbar}{m}\text{Im}\,(u^*\nabla u - v^*\nabla v). \tag{1.30}$$

As expected, particles and holes have opposite contributions to the probability current density.

Particles and holes carry opposite charges. The charge density of superconducting quasiparticles is therefore defined by

$$\rho_Q(r) = e\left(|u(r)|^2 - |v(r)|^2\right). \tag{1.31}$$

From the time-evolution equation of u and v, Eq. (1.27), we find that the charge density satisfies the equation

$$\frac{\partial}{\partial t}\rho_Q(r) + \nabla \cdot J_Q(r) = \frac{4e}{\hbar}\text{Im}\,(\Delta u^*(r)v(r))\,, \tag{1.32}$$

where

$$J_Q(r) = \frac{e\hbar}{m}\text{Im}\,(u^*\nabla u + v^*\nabla v) \tag{1.33}$$

is the electric charge current density of quasiparticles. In comparison with the probability conservation equation, the electric charge conservation equation contains an extra term, contributed by the superconducting paired electrons.

This term is proportional to the product of both particle and hole wavefunctions of Bogoliubov eigenstates. If we define a supercurrent density operator J_S by the equation

$$\nabla \cdot J_S(r) = -\frac{4e}{\hbar}\mathrm{Im}\left(\Delta u^*(r)v(r)\right),$$ (1.34)

then the charge conservation law can be expressed as

$$\frac{\partial}{\partial t}\rho_Q(r) + \nabla \cdot [J_Q(r) + J_S(r)] = 0.$$ (1.35)

In a translational invariant system, momentum k is a good quantum number. The Bogoliubov quasiparticle wavefunctions are determined by the BCS mean-field equations, Eq. (1.16) and (1.17). In real space, they are given by

$$u(r) = \frac{1}{\sqrt{V}}e^{ik \cdot r}\sqrt{\frac{1}{2} + \frac{\xi_k}{2E_k}},$$ (1.36)

$$v(r) = -\frac{1}{\sqrt{V}}e^{ik \cdot r}\sqrt{\frac{1}{2} - \frac{\xi_k}{2E_k}}.$$ (1.37)

When k is real, the probability current density is equal to

$$J_P = \frac{\hbar \xi_k k}{mVE_k}.$$ (1.38)

In contrast, the normal charge current J_Q and the supercurrent J_S are given by

$$J_Q = \frac{e\hbar k}{mV},$$ (1.39)

$$J_S = 0.$$ (1.40)

It indicates that a Bogoliubov quasiparticle with a real momentum k will not decay to generate a supercurrent by forming a Cooper pair with another quasiparticle. Thus the supercurrent vanishes, $J_S = 0$. Both J_P and J_Q are proportional to the momentum $\hbar k$, but J_P contains the factor of ξ_k/E_k. The charge current density is in the same direction as k. For a particle-like quasiparticle with $\xi_k > 0$, its probability current density is also parallel to k. But for a hole-like quasiparticle with $\xi_k < 0$, its probability current density is anti-parallel to k.

On the other hand, if k contains a small imaginary part, say $k = k_0 + i\eta\hat{x}$ with $\eta > 0$, the wavefunction of quasiparticle decays exponentially along the x-direction. In this case, the charge current of quasiparticles becomes

$$J_Q = \frac{e\hbar k_0}{mV} e^{-2\eta x}. \tag{1.41}$$

It also decays along the x-direction. The supercurrent is still zero along the y- and z-directions, $J_S^y = J_S^z = 0$. However, it is finite along the x-direction

$$J_S^x = -\frac{e\Delta^2}{\eta\hbar V} \left(1 - e^{-2\eta x}\right) \text{Im} \frac{1}{E_k}. \tag{1.42}$$

In the limit $\eta \ll |k_0|$, J_S^x is approximately given by

$$J_S^x = \frac{e\Delta^2 \hbar \xi_{k_0} k_{0,x}}{mV E_{k_0}^3} \left(1 - e^{-2\eta x}\right). \tag{1.43}$$

It indicates that the charge current of quasiparticles is transformed into the supercurrent of Cooper pairs. The inverse of the imaginary part of the quasiparticle momentum η^{-1} is a characteristic length scale of quasiparticles to form condensed superconducting Cooper pairs.

1.8 Off-Diagonal Long Range Order

The superconducting transition is a continuous transition from a high temperature normal conducting phase to a low temperature macroscopic long-range ordered phase. In 1962, C. N. Yang pointed out [11] that the long-range order of superconductivity is an *off-diagonal long range order* (ODLRO), which is fundamentally different from a diagonal long-range order, such as the crystalline order of crystals. This kind of order is induced purely by quantum effects and there is no correspondence in classical systems.

The concept of ODLRO provides a mathematical foundation for the microscopic theory of superconductivity as well as the corresponding theory of macroscopic quantum phase transition. The variational wavefunction proposed by Bardeen-Cooper-Schrieffer, Eq. (1.20), actually possesses the ODLRO, which might not be very clear when this wavefunction was proposed. But the BCS

theory achieves great success just because it captures correctly the most important feature of superconductivity, namely the ODLRO. The ODLRO plays a
similar role as the diagonal long-range crystalline order in the study of physical
properties of crystals. It is impossible to establish the theory of superconductivity correctly if the superconducting ODLRO is not properly included in the
wavefunction.

The ODLRO exists only in quantum fluids, including quantum gases and
liquids, such as the Fermi liquid state of conducting electrons in metals. In
insulators, charge fluctuations are short-ranged and the ODLRO is suppressed.
But an ODLRO can exist with a diagonal long-range order. For example, a
superfluid long-range order, which is an ODLRO, can coexist with the diagonal
density-wave order in a supersolid.

In order to understand the ODLRO, let us consider the following two-
particle reduced density matrix

$$\rho_2(i\sigma_i, j\sigma_j; k\sigma_k, l\sigma_l) = \langle c_{i\sigma_i} c_{j\sigma_j} c_{k\sigma_k}^\dagger c_{l\sigma_l}^\dagger \rangle \equiv \mathrm{Tr}\left(c_{i\sigma_i} c_{j\sigma_j} \rho c_{k\sigma_k}^\dagger c_{l\sigma_l}^\dagger\right), \quad (1.44)$$

where

$$\rho = \frac{e^{-\beta H}}{\mathrm{Tr} e^{-\beta H}} \tag{1.45}$$

is the density matrix. It is simple to show that ρ and ρ_2 are semi-positive
definite, namely all their eigenvalues are always larger than or equal to zero.

In a normal metallic state of N electrons, the eigenvalues of ρ_2 are typically
of order 1, much smaller than N. Hence there is not a state which can be
occupied by macroscopical many pairs of electrons with the same quantum
numbers. In this case, an infinitesimal energy, in comparison with the total
energy which is proportional to N, is able to change the microscopic distribution
of ρ_2, and thus the system is dissipative. On the contrary, if ρ_2 has an eigenvalue of order N (assuming it to be αN with α a number of order 1), it
is no longer easy to change the behavior of electrons in this eigenstate by
applying a macroscopically small perturbation, implying that the system is
macroscopically coherent and dissipationless. This is just the most prominent
feature of superconductivity arising from the macroscopic pair condensation.

That eigenstate is just a superconducting condensed state. In this case, one can separate that eigenstate from ρ_2 and rewrite ρ_2 as

$$\rho_2(i\sigma_i, j\sigma_j; k\sigma_k, l\sigma_l) = \alpha N\phi(i\sigma_i, j\sigma_j)\phi^*(k\sigma_k, l\sigma_l) + \rho_2'(i\sigma_i, j\sigma_j; k\sigma_k, l\sigma_l),$$
$$(1.46)$$

where $\phi(i\sigma_i, j\sigma_j)$ is the normalized eigenfunction corresponding to the eigenvalue of αN. The normalization requires $\phi(i\sigma_i, j\sigma_j)$ to be inversely proportional to the system volume V. Thus $N\phi(i\sigma_i, j\sigma_j)\phi^*(k\sigma_k, l\sigma_l)$ is proportional to the electron density. ρ_2' is a regular reduced density matrix whose eigenvalues are all macroscopically small compared to N.

Eq. (1.46) suggests that electron pairs are long range correlated. This is because no matter how far a pair of electrons at sites i and j is from another pair of electrons at sites k and l, their correlation function $\langle c_{i\sigma_i} c_{j\sigma_j} c_{k\sigma_k}^\dagger c_{l\sigma_l}^\dagger \rangle$ generally remains finite. A superconducting state possesses the ODLRO because it has a finite probability to annihilate a local pair of electrons and simultaneously create another local pair of electrons separated in an arbitrary long distance. On the other hand, if a system possesses the ODLRO, it can be also shown that its two-particle reduced density matrix has at least one eigenvalue of order N. A superconducting transition emerges when the maximal eigenvalue ρ_2 changes from a number of order 1 to a number of order N. The coefficient α can be defined as the superconducting order parameter. It is finite in the superconducting state and becomes 0 at and above the superconducting transition temperature T_c.

In Eq. (1.46), if we take an approximation by neglecting the regular ρ_2' term, then ρ_2 becomes

$$\rho_2(i\sigma_i, j\sigma_j; k\sigma_k, l\sigma_l) \approx \alpha N\phi(i\sigma_i, j\sigma_j)\phi^*(k\sigma_k, l\sigma_l). \qquad (1.47)$$

In this case, the electron pair correlation function is factorized. This is just the basic assumption made in the BCS mean-field theory.

In textbooks and literatures, the quasiparticle excitation gap is generally defined as the superconducting order parameter. Rigorously speaking, this definition is not that accurate. A system is superconducting as long as it possesses the ODLRO, no matter whether it has an energy gap or not. In fact, there are gapless superconductors. For example, there is no gap in the quasiparticle excitation spectra in a superconductor with magnetic impurities. Conceptually, the superconducting energy gap and the superconducting order parameter are different. But in most superconductors, it is not necessary to distinguish these two concepts because the superconducting energy gap is proportional to the superconducting order parameter at least in the mean-field approximation.

In both one and two dimensions, it was proven by Hohenberg (see Appendix B) that there is no ODLRO at any nonzero temperatures if the f-sum rule is valid [12]. But for the BCS reduced Hamiltonian, defined by Eq. (1.7), the pairing potential is long-ranged and the f-sum rule is violated. Thus the ODLRO with the corresponding superconducting phase transition is allowed in this system even in one or two dimensions[13].

1.9 Ginzburg-Landau Free Energy

The BCS theory provides a microscopic framework to describe superconducting properties. But it is not convenient to be used in the study of the dynamics of magnetic fluxes, and in the quantitative characterization of superconducting phase transition as well as many other macroscopic phenomena of superconductors. In this case, it is technically simpler and conceptually more transparent to describe superconducting properties by adopting the phenomenological theory first proposed by Ginzburg and Landau (GL)[14].

The phenomenological GL theory was introduced before the establishment of the BCS microscopic theory. It relies on the assumption that the superconducting state is a macroscopic quantum state that can be described by an order parameter. Furthermore, it was assumed that the macroscopic properties of superconductors are completely governed by the free energy, independent on their microscopic details. Thus if the spatial variation of the order parameter is slow in comparison with the scale of coherence length, the free energy can be expanded as a functional of the order parameter ψ and its spatial gradient

$\nabla\psi$ as

$$f = f_n + \frac{1}{2m^*}\left|(-i\hbar\nabla - e^*A)\,\psi\right|^2 + \alpha\left|\psi\right|^2 + \frac{\beta}{2}\left|\psi\right|^4 + \frac{H^2}{2\mu_0}, \tag{1.48}$$

where f_n is the free energy in the normal state, $H = \nabla \times A$ is the external magnetic field and A is the associated vector potential. In obtaining this expression, the variance of the order parameter with time is assumed small and negligible.

In a homogeneous system without external magnetic fields, the GL free energy becomes

$$f = f_n + \alpha\left|\psi\right|^2 + \frac{\beta}{2}\left|\psi\right|^4. \tag{1.49}$$

In the normal state, $\alpha > 0$, and the order parameter is zero, $\psi = 0$, and $f = f_n$ is the free energy. In the superconducting state, α becomes negative ($\alpha < 0$) and the system is in an ordered state and the value of the order parameter ψ_0 is determined by the minimum of the free energy and given by

$$\left|\psi_0\right|^2 = -\frac{\alpha}{\beta}. \tag{1.50}$$

Substituting it into Eq. (1.49), we find the difference in the free energy between the superconducting and normal phases to be

$$f_s - f_n \equiv -\frac{H_c^2}{2\mu_0} = -\frac{\alpha^2}{\beta}, \tag{1.51}$$

where H_c is the thermodynamic critical field of the superconducting state.

In the GL theory, parameters α and β are unknown but can be determined from the BCS theory or alternatively from the measurement values of the critical field H_c and the magnetic penetration length λ using the formula

$$\alpha(T) = -\frac{2e^{*2}}{m^*}H_c^2(T)\lambda^2(T), \tag{1.52}$$

$$\beta(T) = \frac{4\mu_0 e^{*4}}{m^{*2}}H_c^2(T)\lambda^4(T). \tag{1.53}$$

In 1959, Gor'kov [15] showed that the GL free energy can be derived from the BCS theory around the transition temperature T_c under the condition that both ψ and A do not vary too fast over the coherence length scale ξ. He found that, as expected, the order parameter ψ is proportional to the quasiparticle energy gap Δ. Furthermore, he showed that the effective charge that ψ carries is $e^* = 2e$, and the corresponding effective mass $m^* \approx 2m$ under the free electron approximation, as a clear indication of pairing nature of ψ. Substituting $e^* = 2e$ into Eq. (1.48), the GL free energy becomes

$$f = f_n + \frac{1}{2m^*}\left|(-i\hbar\nabla - 2eA)\,\psi\right|^2 + \alpha\,|\psi|^2 + \frac{\beta}{2}\,|\psi|^4 + \frac{H^2}{2\mu_0}. \tag{1.54}$$

Gor'kov's work established a microscopic foundation for the GL theory. It clarifies the condition of validity and the limitation of the GL theory, and provides a clear guidance to the application of the theory in real materials.

In the equilibrium, the free energy is minimized. By taking the variance of the free energy f with respect to ψ^*, we obtain the first GL equation

$$\frac{1}{4m}\left(-i\hbar\nabla - 2eA\right)^2\psi + (\alpha + \beta\psi^*\psi)\,\psi = 0. \tag{1.55}$$

Furthermore, by taking the variation with respect to A and using the Ampere's law

$$J_s = \frac{1}{\mu_0}\nabla \times H, \tag{1.56}$$

we obtain the expression of the supercurrent,

$$J_s = \frac{ie\hbar}{2m}\left(\psi^*\nabla\psi - \psi\nabla\psi^*\right) + \frac{2e^2}{m}A\psi^*\psi, \tag{1.57}$$

which is also called the second GL equation.

1.10 Spontaneous Symmetry Breaking and Meissner Effect

Superconductors are not only ideal conductors but also perfect diamagnets. This is due to the formation of the ODLRO with spontaneous breaking of U(1) electromagnetic gauge symmetry. It is simple to show that the GL free energy is invariant under the following U(1) gauge transformation,

$$\psi \to \psi' = e^{i\varphi}\psi, \tag{1.58}$$

$$A \to A' = A + \frac{\hbar}{2e}\nabla\varphi, \tag{1.59}$$

where φ is an arbitrary single-valued scalar function. This U(1) gauge invariance of the GL free energy is a consequence of electric charge conservation. It is valid independent of the detailed formulism of the free energy. The free energy is invariant under the above gauge transformation because

$$(-i\hbar\nabla - 2eA')\,\psi' = e^{i\varphi}\,(-i\hbar\nabla - 2eA)\,\psi. \tag{1.60}$$

The gauge invariance implies that the electromagnetic field and the phase field of superconducting order parameter are interchangeable. They can be transformed into each other by the gauge transformation. The phase ϕ of the order parameter $\psi = |\phi|\exp(i\phi)$ is a Goldstone boson field. If we take the gauge in which φ in Eq. (1.58) equals the phase field ϕ (dubbed as the unitary gauge in literatures), then the free energy defined by Eq. (1.54) becomes

$$f = f_n + \frac{\hbar^2}{2m^*}\,(\nabla|\psi|)^2 + \frac{m_A^2}{2}A^2 + \alpha\,|\psi|^2 + \frac{\beta}{2}\,|\psi|^4 + \frac{H^2}{2\mu_0}, \tag{1.61}$$

where

$$m_A^2 = \frac{4e^2}{m^*}|\psi|^2.$$

Under this gauge, the phase field ϕ of the order parameter is completely absorbed by the gauge field A and does not appear explicitly in the expression of the GL free energy. But the gauge field A now acquires a mass m_A due to the spontaneous symmetry breaking, i.e. $|\psi| \neq 0$, in the superconducting phase. Hence, the onset of superconductivity generates a mass for the vector potential, so that the electromagnetic field become massive. This is just the celebrated Anderson-Higgs mechanism associated with the spontaneous breaking of the U(1) gauge symmetry in the context of superconductivity[13].

From the derivative of the free energy with respect to A, we find the equation that the gauge field satisfies

$$\nabla \times (\nabla \times A) + 2\mu_0 m_A^2 A = 0. \tag{1.62}$$

By further using the identity of vector algebra,

$$\nabla \times (\nabla \times H) = -\nabla^2 H,$$

we then obtain the equation that the magnetic field satisfies

$$\nabla^2 H = 2\mu_0 m_A^2 H. \qquad (1.63)$$

This is nothing but the London equation, i.e. Eq. (1.2), previously introduced in the description of the Meissner effect. The penetration depth is therefore given by

$$\lambda^{-2} = 2\mu_0 m_A^2. \qquad (1.64)$$

By comparison with Eq. (1.4), we find that the superfluid density is given by

$$n_s = 4|\psi|^2. \qquad (1.65)$$

Thus the square of the order parameter in the GL theory is proportional to the superfluid density.

The above results show that under the unitary gauge, the phase field (or the Goldstone boson field) ψ is completely absorbed by the gauge field and has no contribution to the free energy. It seems that the total degrees of freedom are reduced. This is in fact not the case. The massless U(1) gauge field, i.e. the electromagnetic field, is a transverse field with only two degrees of freedom. It has not the longitudinal component. After its acquiring of mass by absorbing the Goldstone boson, the longitudinal component of the gauge field emerges, which maintains the total degrees of freedom of the system.

The Meissner effect is therefore a consequence of spontaneous symmetry breaking. It is a manifestation of the Anderson-Higgs mechanism, resulting from the interplay between the phase field and the gauge field. Under an external magnetic field, the spatial variance of the phase field in ψ generates a persistent supercurrent to screen the applied field. Therefore, the applied field becomes massive and decays inside the superconductor in a length scale characterized by the penetration depth.

1.11 Two Characteristic Energy Scales

There are two important energy scales in superconductors. One is the quasi-particle excitation gap Δ, which is also the binding energy of Cooper pairs. The

other is the phase coherence energy, T_θ, which is determined by the phase fluctuation of Cooper pairs. Cooper pairs can develop global phase coherence only
when their phase fluctuation energy is lower than T_θ. Superconducting properties, in particular the superconducting transition temperature T_c, are strongly
influenced by the competition of these two energy scales. Both depairing (i.e.
breaking Cooper pairs) and dephasing (i.e. disrupting the phase coherence of
Cooper pairs) effects can suppress superconductivity.

In order to determine the energy scale of phase coherence, let us consider
a superconducting system in the absence of an external magnetic field. If we
ignore the amplitude fluctuation and keep the phase fluctuation of the order
parameter, the free energy, according to Eq. (1.54), is then given by

$$f = \frac{\hbar^2}{2m^*}|\psi|^2 \left(\nabla\phi\right)^2 . \tag{1.66}$$

The long-range correlation of Cooper pairs is completely suppressed if the phase
fluctuation over the coherence length ξ reaches the order of 2π. The energy scale
corresponding to this critical fluctuation is just the energy of phase coherence,
T_θ. Its value, which measures the phase stiffness of superconducting electrons,
is estimated to be

$$T_\theta \approx \frac{\hbar^2}{m^*}|\psi|^2\xi^3 \left(\frac{2\pi}{\xi}\right)^2 = \frac{4\pi^2\hbar^2\xi}{m^*}|\psi|^2 = \frac{\pi^2\hbar^2\xi}{2\mu_0 e^2\lambda^2} . \tag{1.67}$$

This expression agrees with the result given in Ref. [16] up to a constant factor
of order 1. Thus the phase coherence energy is proportional to the superfluid
density, $T_\theta \propto |\psi|^2$, which measures the capacity of paired electrons carrying
superconducting currents.

In a highly anisotropic system, for example, a high-T_c cuprate, λ in Eq.
(1.67) and the corresponding superfluid density n_s should be their values along
the c-axis. This is because the phase fluctuation is the strongest and hence n_s
is the smallest along this direction. Similarly, ξ is the correlation length along
the c-axis. If the correlation length is shorter than the interlayer distance d,
then ξ should be set to d.

If T_θ is much larger than Δ, electrons immediately become phase coherent
once they form Cooper pairs. In this case, the pair breaking is the main destruc-

tor of superconductivity, and T_c is entirely determined by the superconducting gap Δ. Thus the superconducting transition temperature is approximately proportional to the energy gap, $T_c \sim \Delta$. This is just the result of the BCS mean-field theory by neglecting phase fluctuations. In almost all conventional superconductors, made of metals or alloys, T_c is indeed found to scale approximately with the energy gap Δ. For the isotropic s-wave superconductor, the BCS mean-field theory predicts that

$$\Delta = 1.76T_c. \tag{1.68}$$

On the other hand, if $\Delta \gg T_\theta$, electrons form Cooper pairs well before they develop phase coherence, and the BCS mean-field approximation is no longer valid. In this case, superconductivity can be eliminated by destroying the phase coherence, but without breaking Cooper pairs. Hence, it is dephasing, rather than depairing, that becomes the main destructor of superconductivity. Consequently, the superconducting transition temperature T_c is controlled by the phase coherence energy, T_θ, instead of the pairing energy gap Δ. Hence T_c is roughly proportional to T_θ,

$$T_c \sim T_\theta. \tag{1.69}$$

This is a result predicted by the theory of preformed pairs, applicable to systems with strong phase fluctuations. It can be observed in the systems with small superfluid densities, where electrons can form pairs but without developing long-range phase coherence in relatively high temperatures. In underdoped high-T_c cuprates, it was found by experimental measurements that T_c is proportional to the superfluid density n_s, not the energy gap Δ [17]. This linear relationship between T_c and n_s is believed to be a consequence of strong phase fluctuations. It is a key experimental fact that should be seriously taken into account in the study of the phase diagram of high-T_c superconductors.

Table 1.1 shows the ratio T_θ/T_c estimated from experimental results for a number of superconductors. The smaller is T_θ/T_c, the stronger is the phase fluctuation. In conventional three-dimensional metal-based superconductors, T_θ/T_c is typically larger than 10^2, and T_c is hardly affected by phase fluctuations. On the contrary, in organic or underdoped high-T_c superconductors, T_θ/T_c is close to 1, and phase fluctuations are very strong.

Table 1.1   Phase fluctuation energy scale T_θ versus T_c for conventional metal-based, organic, and high-T_c superconductors (from Ref. [16])

materials	$T_c(K)$	T_θ/T_c
Pb	7	2×10^5
Nb_3Sn	18	2×10^3
UBe_{13}	0.9	3×10^2
$LaMO_6S_8$	5	2×10^2
$B_{0.6}K_{0.4}BiO_3$	20	50
K_3C_{60}	19	17
$(BEDT)_2Cu(NCS)_2$	8	1.7
$Nd_{2-x}Ce_xCu_2O_{4+\delta}$	21	16
$Tl_2Ba_2CuO_{6+\delta}$	80	2
	55	3.6
$Bi_2Sr_2CaCu_2O_8$	84	1.5
$Bi_2Pb_xSr_2Ca_2Cu_3O_{10}$	106	$0.8 \sim 1.4$
$La_{2-x}Sr_xCuO_{4+\delta}$	28	1
	38	2
$YBa_2Cu_3O_{7-\delta}$	92	1.4
$YBa_2Cu_4O_8$	80	0.7

Phase fluctuations can suppress the long-range phase coherence of superconducting order parameters. It can also induce particle number fluctuation to enhance the charge fluctuation, since the particle number is conjugate to the phase of Cooper pairs. This is purely a quantum effect, which is too weak to be observed in conventional metal-based superconductors. However, in a superconductor made of bad metal with poor Coulomb screening, the influence of this effect might become important.

1.12 Pairing Mechanism

As explained before, there are two steps for electrons to become superconducting: to form Cooper pairs and to develop phase coherence. Correspondingly, we need to address the following two questions in the study of pairing mechanism: (1) What is the main interaction that glues electrons to form Cooper pairs? (2) How do the Cooper pairs form phase coherence and condense? The first question has been thoroughly discussed in textbooks and literatures, although no consensus has been reached for high-T_c cuprates. Discussions on the second one are rather limited, and less quantitatively. In fact, our understanding to the dynamics of phase coherence is inadequate. This is not a serious problem in the study of metal- or alloy-based superconductors, because phase fluctuations in these materials are weak and electrons become condensed almost immediately after they form Cooper pairs. However, in underdoped high-T_c cuprates, phase fluctuations are very strong. A thorough understanding of phase coherence is indispensable for the understanding of microscopic mechanism of high-T_c superconductivity.

Investigation of pairing mechanism plays the central role in the establishment of microscopic theory of superconductivity. Once the pairing mechanism, especially the main interaction that drives electrons to superconduct, is determined, one can control and synthesize materials with certain targeted structures and chemical stoichiometries to enhance pairing interactions so that both the superconducting transition temperature and the critical current density can be improved.

In conventional superconductors, the pairing arises from the electron-phonon interaction. This has been verified by numerous experimental measurements. A frequently mentioned experimental evidence is the isotope effect. If one type of atoms is partially or completely replaced by its isotope in a superconductor, then the characteristic phonon frequency is changed due to the change of the atomic mass. Under the BCS mean-field approximation, it is predicted that the superconducting transition temperature T_c induced by electron-phonon interactions is inversely proportional to the square root of the atomic mass. This prediction was confirmed in a number of superconductors of simple metals. However, the transition temperature induced by electron-phonon interactions is generally not very high. It is estimated to be less than 40K according to the McMillan formula because the energy scale of the Debye frequency is just of the order of room temperature, and there is no much room to greatly enhance it in laboratory. The electron-phonon coupling cannot be significantly enhanced either. Otherwise, it may cause instability in crystal structures. However, if the pairing arises from the interaction of electrons with optical phonons, the superconducting transition temperature is not constrained by the Debye frequency. As the characteristic frequency of optical phonons can be significantly higher than the Debye frequency, T_c can in principle exceed 40K.

The mechanism of high-T_c cuprate superconductors remains one of the biggest challenge in contemporary condensed matter physics. As the transition temperature is much higher than that estimated from the electron-phonon interaction, it is unlikely that the pairing is due to the electron-phonon interaction. On the contrary, the electron-electron interaction, in particular the antiferromagnetic fluctuation, in high-T_c cuprates is very strong. It might be the driving force for high-T_c superconductivity.

It should be pointed out that once electrons form Cooper pairs with macroscopic phase coherence, their physical properties are universal, no matter whether the superconducting pairing is from the electron-phonon interaction or from the electron-electron interaction. As long as we know the characteristic energy scale of pairing interaction and the quasiparticle spectra function, we can accurately predict all dynamic and thermodynamic properties of superconducting states. This is the reason why we can still discuss and successfully predict physical properties of high-T_c superconductors without knowing clearly its microscopic pairing mechanism.

1.13 Classification of Pairing Symmetry

Superconductors can be classified according to the internal symmetry of Cooper pairs. The wavefunction of a Cooper pair depends on both the spatial coordinates and the spin configurations of two electrons. In the absence of spin-orbit coupling or other interactions that break the spin rotational symmetry, the total spin is conserved and the pairing wavefunction can be factorized as a product of the spatial and spin wavefunctions

$$\Psi(\sigma_1, r_1; \sigma_2, r_2) = \chi(\sigma_1, \sigma_2)\Delta(R, r), \tag{1.70}$$

where (σ_1, r_1) and (σ_2, r_2) are the spin and spatial coordinates of the first and second electrons, respectively. $R = (r_1 + r_2)/2$ is the coordinate of the center of mass and $r = r_1 - r_2$ is the relative coordinate of two electrons.

A Cooper pair can be either in a spin singlet or in a spin triplet state depending on whether the total spin is 0 or 1. The spin wavefunction is anti-symmetric, $\chi(\sigma_1, \sigma_2) = -\chi(\sigma_2, \sigma_1)$, for the spin singlet state, and symmetry, $\chi(\sigma_1, \sigma_2) = \chi(\sigma_2, \sigma_1)$, for the spin triplet state. Since the full pairing wave-function, $\Psi(\sigma_1, r_1; \sigma_2, r_2)$, is always anti-symmetric under the exchange of two electrons, the spatial wavefunction corresponding to the spin singlet and triplet pairing states should be symmetric and antisymmetric, respectively.

Under the exchange of two electrons, the coordinate of the center of mass R is invariant, but the relative coordinate r changes sign. The pairing symmetry is classified by the symmetry of the spatial wavefunction under the change of the relative coordinate r. If the Hamiltonian is rotationally invariant, the orbital angular momentum is conserved and $\Delta(R, r)$ is an eigenfunction of the orbital angular momentum $L = -i\hbar r \times \nabla_r$. Thus the spatial wavefunction can be classified according to the eigenvalues of L^2:

$$L^2\Delta(R, r) = l(l + 1)\hbar^2\Delta(R, r), \tag{1.71}$$

where l is an integer. The eigenstate of the orbital angular momentum is sym-metric if l is even or anti-symmetric if l is odd. Thus the orbital angular mo-mentum even for the spin singlet pairing state, and odd for the spin triplet pairing state. In a translation invariant system, $\Delta(R, r)$ can be further factor-ized as a product of the wavefunction for the coordinate of the center of mass, $\Delta_0(R)$, and that for the relative coordinate, $\phi(r)$,

$$\Delta(R, r) = \Delta_0(R)\phi(r), \tag{1.72}$$

where $\phi(r)$ is the eigenfunction of orbital angular momentum. The pairing symmetry is determined by $\phi(r)$.

Cuprate superconductors have *d*-wave symmetry whose orbital angular momentum equals 2, i.e. $l = 2$ by adopting the convention of atomic physics. Superconductors with pairing orbital angular momenta $l = 0, 1, 2, 3, 4$ are called *s*, *p*, *d*, *f* and *g*-wave superconductors, respectively. Among them, the *s*, *d*, and *g*-wave superconductors have spin singlet pairing, and the *p* and *f*-wave superconductors have spin triplet pairing. The $l = 0$ state of the orbital angular momentum is isotropic and non-degenerate. The corresponding *s*-wave pairing state is also non-degenerate and spatially isotropic. The $l = 2$ states are 5-fold degenerate, and the corresponding *d*-wave superconductors possess five different representations or pairing symmetries. They are generally denoted as d_{xy}, $d_{x^2-y^2}$, d_{xz}, d_{yz}, and $d_{3z^2-r^2}$ according to the eigenvalue of the third component of the angular momentum, respectively.

Physical properties of superconductors with different pairing symmetries are markedly different. The gap functions of *d*-wave superconductors (or any other superconductors with $l \neq 0$) can have gap nodes at which $\Delta(R, r) = 0$. In contrast, the gap function of *s*-wave superconductors is nodeless, namely $\Delta(R, r)$ is always finite. This is the major difference between the *s*-wave and *d*-wave superconductors, which can significantly affect their low energy properties. In *s*-wave superconductors, the density of states of Bogoliubov quasiparticles vanishes inside the gap, and thermodynamic quantities decay exponentially with temperature and energy. However, in *d*-wave superconductors with gap nodes, the low energy density of states is linear, and thermodynamic quantities exhibit power-law behaviors in low temperatures, qualitatively different from *s*-wave superconductors.

In solids, the continuous rotational symmetry is broken into the discrete lattice point group symmetries. The definition of orbital angular momentum must be consistent with the point group symmetries. If the superconducting pairing is local in real space and the size of Cooper pairs is comparable to the lattice constant, such as in a high-T_c superconductor, the lattice symmetry needs to be considered in order to determine the pairing symmetry. The pairing symmetry should be classified according to the eigenstates of the point group.

For quasi-two-dimensional materials with tetragonal symmetry, the gap function $\phi(r)$ is invariant under the rotation around the c-axis in an s-wave superconductor. However, in a p- or d-wave superconductor, the gap function changes sign under the rotation of 180° or 90° around the c-axis. The p-wave pairing have two degenerate representations, p_x or p_y. A p-wave superconductor can be in either one of these state, or in a combined $p_x \pm i p_y$ pairing state with spontaneous breaking of time-reversal symmetry. There are also two possible representations for a d-wave superconductor, namely d_{xy} and $d_{x^2-y^2}$, but they are generally not degenerate even if the lattice is tetragonal.

In momentum space, the gap function is defined by Eq. (1.13), i.e. $\Delta_k = \Delta_0 \phi_k$, where ϕ_k is the Fourier transformation of the gap function, $\phi(r)$. ϕ_k is also the form factor of the pairing interaction presented in Eq. (1.7). For the $d_{x^2-y^2}$ superconductor,

$$\phi_k = c_1(\cos k_y - \cos k_x),\qquad(1.73)$$

where c_1 is a normalization factor, which is defined as the inverse of the maximal value of $(\cos k_x - \cos k_y)$ on the Fermi surface. For the d_{xy}-wave superconductor, ϕ_k is defined by

$$\phi_k = c_2 \sin k_x \sin k_y,\qquad(1.74)$$

where c_2 is the inverse of the maximal value of $\sin k_x \sin k_y$ on the Fermi surface. The maximal value of ϕ_k is 1, and the minimum is -1.

The nodal points of the $d_{x^2-y^2}$ and d_{xy}-wave superconductors are different in the Brillouin zone. The gap nodes lie along the diagonal lines of the Brillouin zone, i.e. $k_x = \pm k_y$, in the former case, and along the two axes of the Brillouin zone, i.e. $k_x = 0$ or $k_y = 0$, in the latter case. Except this, physical properties of these two kinds of d-wave pairing states are similar. The conclusion drawn from a $d_{x^2-y^2}$-wave superconductor can be applied to a d_{xy}-wave supercon- ductor simply by rotating the axes by $\pi/4$, and vice versa. But it should be emphasized that the microscopic origins leading to these two kinds of states could be different.

In the study of low energy physics, only the quasiparticle excitations around the Fermi surface are physically important. In this case, the pairing function can be simplified. For the $d_{x^2-y^2}$-wave superconductor, ϕ_k can be approximately represented using the azimuthal angle of the wavevector, φ, and the gap function becomes

$$\Delta_\varphi = \Delta_0 \cos(2\varphi), \tag{1.75}$$

where $\varphi = \tan^{-1} k_y/k_x$. This simplified expression is convenient to use in analytic calculations. Around the nodal points, Δ_k can be also written as

$$\Delta_k = \Delta_0 \frac{k_x^2 - k_y^2}{k_F^2}. \tag{1.76}$$

For the $d_{x^2-y^2}$-wave superconductors, the above three expressions of ϕ_k are physically equivalent. One can use the most convenient one in dealing with a concrete problem.

The relative wavefunction $\phi(r)$, which discloses the internal structure of Cooper pairs, is determined by pairing interaction. Usually the gap energy decreases with the increase of the orbital angular momentum of paired electrons. Thus for all spin singlet superconductors, the s-wave pairing is generally more favored in energy and has the highest probability to be observed. Indeed most of superconductors discovered up to present are s-wave ones.

However, in strongly correlated electronic systems, a non-s-wave pairing might be energetically more favorable. This is because in strongly correlated systems, the local Coulomb repulsion is generally strong, which tends to reduce the probability of two electrons approaching each other. In a non-s-wave superconductor, the gap function vanishes at $r = 0$, i.e. $\phi(0) = 0$. This releases the energy raised by the Coulomb repulsion between two electrons. On the contrary, in an s-wave pairing state, the gap function is finite at $r = 0$, which is not favored by strong Coulomb repulsion.

Energetically, singlet Cooper pairings in the g-wave and even high angular momentum channel are unlikely to occur, but not completely impossible. Evidence supporting the g-wave pairing has been found in heavy fermion superconductors. However, in order to confirm this more convincingly, more experimental and theoretical investigations are desired.

Triplet pairing breaks the time reversal symmetry and is rarer to discover.

This kind of pairing is energetically favored in materials with strong ferromagnetic fluctuations, such as Sr_2RuO_4. Sr_2RuO_4 has a similar lattice structure as the Mott insulator La_2CuO_4. But unlike La_2CuO_4, Sr_2RuO_4 is a metal with strong ferromagnetic correlation. Sr_2RuO_4 is a spin triplet pairing superconductor has been confirmed by several experiments. But there is no consensus on the pairing symmetry of this material [18].

Pairing symmetry is determined not just by pairing interactions, but also by the lattice symmetry. It can be classified according to the value of orbital angular momentum as the s, d, or other pairing state only if the system possesses perfect tetragonal or other lattice symmetries in two dimensions. Otherwise, different pairing channels are mixed. The level of mixing is determined by the lattice anisotropy between the two principal axes. The mixing may result from an explicit breaking of lattice symmetry induced, for example, by an uniaxial stress. It may also arise from spontaneous breaking of lattice symmetry generated, for example, by some non-linear interactions in the GL free energy.

1.14 Pairing Symmetry of High-T_c Superconductors

To identify and verify the pairing symmetry of high-T_c Cooper pairs has been a great challenge and also one of the major achievements in the study of high-T_c superconductivity. It is an indispensable and key step towards the understanding of fundamental pairing mechanism and the establishment of microscopic theory of high-T_c superconductivity.

Different from conventional superconductors whose normal states are Landau Fermi liquids, the normal states of high-T_c superconductors are much more complicated and believed to be non-Fermi liquid. However, in the superconducting phase, the difference between these two kinds of superconductors in the superconducting state is small, except phase fluctuations are weaker and coherence lengths are longer in conventional superconductors. It is generally believed that the BCS theory of superconductivity is applicable to high-T_c superconductors, no matter whether the normal state is a Landau Fermi liquid or not. This is a basic assumption made in the analysis of experimental data of high-T_c cuprates. It implies that we can identify the pairing symmetry of high-T_c superconductivity by comparing experimental measurements with theoretical predictions from the BCS theory, without knowing its pairing mechanism.

Both the pairing mechanism and the symmetry of Cooper pairs are determined by low-energy electronic structures and electron-electron interactions. In conventional superconductors of metals, the pairing results from electron-phonon interactions, and Cooper pairs have isotropic energy gaps with s-wave symmetry. Low-energy physics of high-T_c cuprates is determined by the conducting electrons in the two-dimensional CuO-planes on which Cooper pairing is expected to arise. Without doping, high-T_c cuprates are antiferromagnetic insulators with strong antiferromagnetic exchange interactions. The pairing in high-T_c cuprates is likely to arise from antiferromagnetic fluctuations, rather than from electron-phonon interactions. Based on the scenario of antiferromagnetic fluctuations, it was predicted that high-T_c superconductivity has the $d_{x^2-y^2}$-wave pairing symmetry [19, 20, 21, 22]. As already mentioned, once pairing symmetry is fixed, physical properties of high-T_c superconductors can be understood without knowing the detailed microscopic mechanism of Cooper pairing.

To apply the BCS theory, one needs to first verify experimentally whether there exists Cooper pairs in high-T_c superconductors and whether the superconducting phase transitions therein are due to pair condensations. On the condition that Cooper pairing does exist, the next step is to determine the spin structure and paring symmetry of Cooper pairs.

To determine whether electrons are paired and condensed in low temperatures, one needs to examine their characteristic effects and compare with theoretical predictions. The main physical phenomena or effects that have been utilized to judge the existence of Cooper pairs in high-T_c cuprates include:

(1) The direct-current (DC) and alternating-current (AC) Josephson effects. In addition to the single electron tunneling, there is also the Josephson pair tunneling in a junction between two superconductors. The response of pair tunneling to an applied electric or magnetic field behaves differently from that of normal single electron tunneling. It exhibits a number of characteristic co-

herent effects which can be used to determine the pairing state and its phase coherence.

(2) Andreev reflection. When a beam of electrons is incident onto the surface of a metal, part of the beam will be reflected. However, when a beam of electrons is incident onto the surface of a superconductor, in addition to the reflection of normal electrons, there is also the reflection of holes due to the pair condensation, which enhances the reflection current. At zero bias, the reflection current can be twice of the incident electric current. Thus we can determine the pairing and phase coherence through the measurement of the Andreev reflection current.

(3) The Little-Parks magnetic flux quantization. The magnetic flux enclosed by a superconducting ring is quantized due to the phase coherence of superconducting order parameter, since the phase variable is gauge equivalent to a vector potential. The minimal quantized value of flux is $h/2e$ instead of h/e, determined by the total charge of a Cooper pair, $2e$, instead of the charge of a single electron. This experiment can be used to test if there is a flux quantization and if the minimal quantized flux is $h/2e$.

(4) The electron-hole mixing. In the superconducting states, the number of electrons is not conserved due to pair condensation. Electron and hole are mixed and manifested as Bogoliubov quasiparticle excitations. This mixing is also a strong evidence of superconducting pairing. It can be probed by angular resolved photo-emission spectroscopy (ARPES).

For high-T_c superconductors, there were a great deal of experimental investigations on the above four effects. All the experimental measurements on the Josephson effect [23, 24, 25], the Andreev reflection [26, 27], the flux quantization [28, 29, 30], and the electron-hole mixing [31] agree with the predictions of BCS theory. In addition, a large amount of measurements on thermodynamic and dynamic properties are also qualitatively consistent with the Cooper pairing picture. They all convincingly show that high-T_c superconducting transitions are still due to condensations of electron pairs, same as in conventional superconductors.

The spin structure of Cooper pairs can be determined from the Knight shift of nuclear magnetic resonance (NMR). The Knight shift measures the electron magnetic susceptibility. In a spin singlet superconductor, the spin excitation is gapped and the Knight shift is suppressed in low temperatures, exhibiting a thermally activated exponential behavior. On the contrary, in a triplet superconductor, the spin excitation is gapless, and the spin susceptibility in the superconducting states is comparable to or the same as in the normal state, hence the Knight shift is nearly temperature independent across the transition temperature. The Knight shift experiments in high-T_c superconductors are consistent with the prediction based on the spin singlet pairing picture[32, 33]. It shows that high-T_c pairing happens in the spin singlet channel and the gap function is spatially symmetric under the exchange of two electrons.

A variety of experimental techniques have been used to measure the orbital angular momentum or to probe the pairing symmetry of high-T_c superconductors. It also generates many interesting problems for theoretical studies. Useful information on the pairing symmetry, properties of quasiparticle excitations and their interactions can be drawn from nearly all kinds of thermodynamic and dynamical measurements. This is important not just for identifying the pairing symmetry, but also for exploring the origin of many anomalous behaviors of high-T_c cuprates in the normal state. The experimental results depend strongly on the quality of samples measured. If the sample quality is not that good so that measurement errors are large, experimental results might not reflect the intrinsic properties of superconductors, making the judgement on the pairing symmetry difficult or even wrong in some cases. For example, in the early years of high-T_c study, in particular before 1993, most experimental measurements on both thermodynamic and dynamic properties suggested that the high-T_c pairing has the *s*-wave symmetry, similar as in conventional phonon-mediated superconductors. However, the conclusion was completely changed after high quality single crystals became available. In the meanwhile, theoretical studies also achieved great progress, providing important guidance toward a thorough understanding of experimental results. Now more and more experimental and theoretical studies have overwhelmingly shown that the high-T_c pairing has *d*-wave rather than *s*-wave symmetry. To learn more about the early history in this respect, please refer to Ref. [34].

A d-wave superconductor differs from an s-wave one in two respects. First, the d-wave gap function changes sign under the rotation of $90°$ around the c-axis. In contrast, the s-wave gap function does not change sign under rotation. Second, there are nodes in the d-wave gap function and the low-energy density of states of superconducting quasiparticles scales linearly with energy. Consequently, all thermodynamic quantities of d-wave superconductors exhibit power-law behaviors as functions of energy or temperature in low temperatures. In contrast, the isotropic s-wave gap function is fully gapped over the entire Fermi surface, and all thermodynamic quantities exhibit activated behaviors in low temperatures. These qualitative differences set up criteria for identifying pairing symmetry in high-T_c cuprates. Correspondingly, experimental measurements can be divided into two categories:

The first category contains all the experiments that are sensitive to the phase of gap function, by detecting the phase variation over the Fermi surface through the measurement of quantum interference effects induced in various Josephson junctions. This kind of experiments is not sensitive to the gap amplitude, but can be used to detect the positions of gap nodes and the sign change of the phase variable. It provides an indisputable way to differentiate a $d_{x^2-y^2}$-wave from a strongly anisotropic s-wave pairing state.

The second category includes experimental measurements of ARPES, magnetic penetration depth, NMR, optical conductivity, thermal conductivity, and specific heat, etc. This category does not contain any experiment that is phase sensitive. It intends to identify the pairing symmetry by directly detecting the nodal positions and the gap anisotropy through the measurements of response functions of low energy excitations to various applied perturbations, like heat, light, electromagnetic fields, etc. In particular, ARPES can directly measure the momentum dependence of the gap function on the Fermi surface, from which the pairing symmetry can be inferred. The Raman scattering can selectively probe the gap function along different directions on the Fermi surface by changing the directions of incident and scattered lights. From the temperature dependence of the penetration depth, NMR, or the specific heat, one can determine the low-energy density of states of quasiparticles. Measurement results of magnetic penetration depth and NMR are relatively simple to interpret because these probes measure directly physical properties of superconducting electrons, without considering the contribution of phonons and other effects. In low temperatures, physical properties of *d*-wave superconductors are governed by low energy quasiparticle excitations around the gap nodes, not by the size and the shape of the Fermi surface. But disorder effects induced by sample inhomogeneities, impurities, and dislocations can strongly affect low-energy behaviors of *d*-wave superconductors. These extrinsic effects should be considered in the analysis of measurement data.

Many physical properties of superconductors in the vicinity of T_c are also sensitive to pairing symmetry. But experimental results are difficult to be analysed because both superconducting phase fluctuations and antiferromagnetic fluctuations become strong around T_c. A collective resonance may emerge in the neutron scattering spectroscopy when the momentum transfer equals the momentum difference between two gap nodes. This can be also used to identify the locations of gap nodes. However, as the momentum difference between the two gap nodes is close to the characteristic wavevector of antiferromagnetic fluctuations in high-T_c cuprate, it is not that simple to distinguish a neutron resonance peak from a peak induced purely by antiferromagnetic fluctuations.

It should be emphasized that different experimental techniques have their own limitations. It is impossible to draw a decisive conclusion simply based on a single experimental measurement. Instead, a unified and unbiased explanations to all experimental results is important in the analysis of high-T_c superconductivity.

Generally speaking, we have already achieved significant progress in the study of pairing symmetry in high-T_c superconductors. In hole-doped cuprate superconductors, most experimental and theoretical studies suggested that the gap function is strongly anisotropic and possesses the $d_{x^2-y^2}$-symmetry. However, there are still debates on whether the pairing symmetry has the same symmetry in electron-doped materials, which needs to be resolved in the future. In this book, we give a general introduction to the theory of *d*-wave superconductors, by taking high-T_c cuprates as a prototype system. We hope it can deepen our understanding on this class of novel quantum phenomena, and provide useful guidance for further exploration and analysis of novel superconductors.

Chapter 2
Microscopic Models for High Temperature Superconductors

2.1 Phase Diagram of Cuprate Superconductors

Perovskite cooper oxides, or cuprates, are the first class of high-T_c superconductors that have been discovered [35]. The parent compounds of these superconductors are antiferromagnetic Mott insulators, exhibiting an antiferromagnetic long-range order in low temperatures. Chemical doping, by element substitutions or by changing oxygen or atomic contents, introduces conducting electrons to the parent compounds. This suppresses antiferromagnetic fluctuations and drives copper oxides into the superconducting phase. For example, La_2CuO_4 and $YBa_2Cu_3O_6$ are two typical parent compounds. They are antiferromagnetic Mott insulators and become high-T_c superconductors upon hole-doping. In fact, they are the first two families of high-T_c superconductors discovered.

Cuprate superconductors can be obtained from the parent compounds by either hole or electron doping. The resultants are called hole- and electron-doped high-T_c superconductors, respectively. Most of high-T_c superconductors are hole-doped.

Up to now more than 10 families of high-T_c superconductors with different lattice and chemical structures have been discovered. They all have a layered structure. The layers are composed of CuO_2 planes (see Fig. 2.1), whose crystalline axes are denoted as a and b, respectively. The c-axis is perpendicular to the ab-plane. Strong anisotropy exists between the ab-plane and the c-axis. In the presence of free charge carriers, the conductivity in the CuO_2 plane is usually 2 to 4 orders of magnitude higher than that along the c-axis. As verified by both band structure calculations and numerous experimental measurements, transport properties and low energy thermal excitations are governed by electrons in the CuO_2 planes. This is a basic property of cuprates that should be considered in the analysis of high-T_c superconductivity.

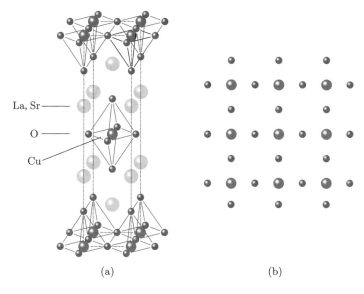

La, Sr

O

Cu

(a)   (b)

Figure 2.1   (a) Crystalline structure of high-T_c superconductor $La_{2-x}Sr_xCuO_4$. (b) Lattice structure of a CuO_2 plane.

As mentioned, cuprate superconductors are quasi-two dimensional materials, which possess two characteristic features: (1) quantum and thermal fluctuations are very strong; (2) the Coulomb screening is poor and hence electron-electron interactions are strong. These features are responsible for various strongly correlated and anomalous behaviors observed in high-T_c cuprates. High-T_c cuprates are prototype systems of strongly correlated electrons. Investigation into physical properties of cuprates is important not only for the understanding of microscopic mechanism of high-T_c superconductivity, but also for a comprehensive understanding of general low-dimensional strongly correlated systems.

The microscopic models introduced in this chapter serves as a starting point for understanding the effective low energy physics of high-T_c cuprates. At the current stage, it remains unclear whether the high-T_c problem can be solved simply based on these models. It is even unknown whether these models possess the superconducting off-diagonal long range order. Each of these models has its own limitations and regimes of validity. Before clarifying all these subtleties, we first give a brief overview on the physical properties, especially the phase diagram, of high-T_c superconductivity.

Physical properties of high-T_c cuprates are complicated, depending crucially on temperature as well as doping level. Applying pressure and strong electromagnetic fields can also strongly affect physical properties of high-T_c cuprates. Fig. 2.2 shows a typical phase diagram of high-T_c cuprates. The high-T_c cuprates are antiferromagnetic insulators at low doping levels. Superconductivity emerges when doping exceeds a critical level. The superconducting transition temperature T_c increases with the doping at the beginning, then drops after passing a maximal value. The doping level at which the transition temperature T_c reaches the maximum is called the optimal doping. The dop-

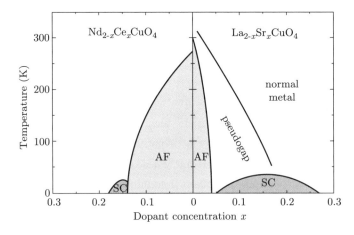

Figure 2.2   Phase diagram of La$_{2-x}$Sr$_x$CuO$_4$ and Nd$_{2-x}$Ce$_x$CuO$_4$ superconductors. The left- and right-hand sides represent electron- and hole-doped cases, respectively. AF represents the antiferromagnetic long-range ordered phase, and SC represents the superconducting phase. Reprinted with permission from [36]. Copyright (2003) by the American Physical Society.

ing above and below the optimal doping is called overdoping and underdoping, respectively.

The phase diagram of high-T_c superconductivity is asymmetric with respect to electron- and hole-dopings. In the hole-doped case, the antiferromagnetic insulating state disappears above 3%, and superconductivity emerges when the doping level exceeds 5%. The optimal doping takes place around 15%. On the other hand, in the electron-doped case, the antiferromagnetic insulating state disappears at the doping level higher than 13%. The superconductivity appears in a much narrower range than in the hole-doped case.

Cuprate superconductors behave very differently from conventional metal-based ones. Some of the phenomena discovered in cuprates can be explained satisfactorily, but many of them are lack of a unified and comprehensive explanation. This includes the spin-charge separation [37, 38], the pseudogap phenomenon observed in the underdoped cuprates [39], the intrinsic charge inhomogeneity [40], etc. Understanding these anomalous properties is important not only to the understanding of measurement data, but also crucial to the construction of high-T_c theory.

Among various anomalous properties, the pseudogap is one of the most important effects observed in the normal state. The pseudogap is a manifestation of the suppression in the density of states of low-lying electronic excitations of underdoped high-T_c cuprates. It shares many similarities with the superconducting gap of quasiparticle excitations in the superconducting state. For example, the pseudogap suppresses various physical quantities at low temperatures such as the specific heat, the magnetic susceptibility, the optical conductivity, and the spectra weight of electrons. It is also conceivable that the pseudogap has the same symmetry as the superconducting energy gap. However, the pseudogap is not a superconducting order parameter. There is no phase transition associate with a pseudogap. The transition from the normal metallic phase to the pseudogap phase is continuous without any singularities in the specific heat and other thermodynamic quantities. Therefore, it is very difficult to accurately determine the boundary temperature of the pseudogap phase. In the strongly underdoped regime, the onset temperature of the pseudogap is about one order of magnitude higher than the superconducting transition temperature T_c. But it drops with increasing doping.

The physical origin of the pseudogap remains unclear. One possibility is that it results from "preformed" Cooper pairs, but without developing global phase coherence. This scenario is consistent with the facts that the superfluid density is low and the phase fluctuation is strong in underdoped high-T_c superconductors, and supported by the experimental measurement on the Nernst effect of transverse thermal conductivity [41]. However, we are still lack of a quantitative understanding of phase fluctuations. It is difficult to make a conclusive judgment on the validity of this scenario. In addition, the pseudogap appears in the vicinity of the antiferromagnetic phase, where strong antiferromagnetic fluctuations exist which complicate the study on this puzzling phenomenon.

The stripe phase, or the intrinsic charge inhomogeneity, is another important effect observed in the underdoped high-T_c cuprates [42]. The key experimental evidence comes from the incommensurate peaks of spin structure factors measured by neutron scattering spectroscopy. These peaks appear near the characteristic wave vector (π, π) of antiferromagnetic fluctuations. The stripe phase is not observed in all underdoped cuprate superconductors. In most of high-T_c cuprates, the static stripe phase is not observed, and there is also no direct or strong evidence for the existence of dynamic stripes. Similar to the pseudogap, theoretical study on the stripe phase is immature, and a quantitative description to it is still not available.

In the overdoped regime, the pseudogap effect and antiferromagnetic fluctuations are weakened. The temperature and energy dependencies of various thermodynamic quantities and transport coefficients behave similarly as in a conventional metal, as predicted by the Landau-Fermi liquid theory. It seems that overdoped high-T_c cuprates are just "conventional" superconductors.

It is unclear if there is a quantum phase transition between the underdoped and overdoped high-T_c materials. This is an important question that needs to be resolved by experiments in the future. Experimentally, it was found that there might exist a critical regime that separate the underdoped pseudogap phase and the overdoped doped Fermi liquid phase, which implies the existence of a quantum critical point at zero temperature and the phases on the two sides of this critical point are different [43]. This quantum critical point lies in the slightly overdoped regime [44]. But the scaling behavior in the vicinity of this putative critical point and the associated discontinuity of thermodynamic quantities were not observed. It remains an open question whether this quantum critical point really exists.

2.2 Antiferromagnetic Insulating States

In the undoped insulating parent compounds, the copper and oxygen ions in the CuO_2 plane are in the Cu^{2+} and O^{2-} valence states, respectively. The outer shell electron configuration of O^{2-} is $2p^6$, whose three $2p$ orbitals are fully filled. The outer shell electron configuration of Cu^{2+} is $3d^9$. Among the five $3d$-orbitals of Cu^{2+}, four of them are fully filled, and the one with the highest energy, $3d_{x^2-y^2}$, is singly occupied, namely in the half-filled state (Fig. 2.3). In such a configuration, Cu^{2+} carries a spin of $S = 1/2$.

Based on the standard Bloch band theory, solid state materials with half-filled bands are metallic in the absence of Peierls-type lattice structure transitions. However, experimentally La_2CuO_4 and other parent compounds of cuprate superconductors are actually antiferromagnetic insulators in low temperatures, indicating that the $3d_{x^2-y^2}$ electrons of Cu^{2+} are localized around the copper sites and have no contribution to the charge current. This class of insulators are called Mott insulators, which are fundamentally different from the band insulators with either empty or fully filled bands. The Mott insulator results from the Coulomb interaction and is an effect of many-body strong correlations. In comparison, band insulators are purely a consequence of Pauli's exclusion principle of Fermi statistics.

The strongest interaction in the CuO_2 plane is the Coulomb interaction between two electrons at the outmost $3d$ orbital in a Cu^{2+} cation. The Coulomb repulsion between different Cu^{2+} cations are relatively weaker. Removing one electron from a Cu^{2+} site to one of its neighboring sites creates a doubly occupied site and an empty site which is energetically unfavored. This effective

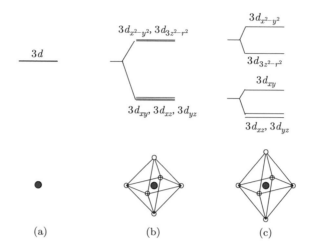

(a)   (b)   (c)

Figure 2.3   Crystal field splitting of the five 3d orbitals of Cu^{2+}. (a) In an isolated Cu^{2+} (blue solid circle), the five 3d-orbitals are degenerate. (b) In a CuO_6 octahedron with cubic symmetry, the five d-orbitals split into three t_{2g}-orbitals, $3d_{xy}$, $3d_{xz}$, and $3d_{yz}$, and two e_g-orbitals, $3d_{x^2-y^2}$ and $3d_{3z^2-r^2}$. The energy level of e_g-electrons is higher because the charge clouds of e_g-orbitals point towards the oxygen anions at the vertexes with stronger Coulomb repulsion. (c) Energy level splitting in a CuO_6 octahedron elongated along the c-axis. The energy level of $d_{3z^2-r^2}$ becomes lower because the wavefunction overlap between this orbital and the 2p orbitals of the two apical oxygens becomes smaller. The three t_{2g}-orbitals also become non-degenerate. d_{xy} has higher energy because the Coulomb repulsions between the other two t_{2g}-orbitals and the two apical oxygens are reduced. The energy levels of d_{xz} and d_{yz} remain degenerate if the octahedral is $\pi/2$ rotational symmetry along the c-axis, but can be split by the Jahn-Teller effect if the occupation numbers in these orbitals are different.

Coulomb interaction is modeled by the Hubbard interaction whose Hamiltonian is defined by $H_I = U \sum_i n_{i\uparrow} n_{i\downarrow}$, with $n_{i\uparrow}$ and $n_{i\downarrow}$ the up- and down-spin electron number operators in the $3d_{x^2-y^2}$-orbital at site i, respectively. In cuprate superconductors, the effective Coulomb repulsion energy U is about a few electron volt, larger than the band width of conducting electrons.

At half-filling, if the Hubbard interaction is strong enough, electrons are localized on lattice sites and do not conduct. In the meanwhile, the antiferromagnetic exchange interaction between two spins on the neighboring sites is unscreened and becomes the most important interaction that governs low energy excitations. It leads to the antiferromagnetic long-range-order in low temperatures.

The Mott insulators are intimately connected with the antiferromagnetic orders. In fact, the antiferromagnetic orders are discovered in nearly all the Mott insulating materials. The antiferromagnetic order is absent in the one dimensional Hubbard model at half filling, at which the antiferromagnetic correlations exhibit an algebraic decay. In two or three dimensions, the Mott insulating states without long-range antiferromagnetic ordering, namely the spin-liquid states, have not been found without doubt experimentally.

The antiferromagnetic Heisenberg model is the fundamental model describing the low energy antiferromagnetic exchange interactions. The corresponding Hamiltonian reads

$$H = J \sum_{\langle ij \rangle} S_i \cdot S_j, \tag{2.1}$$

where S_i is the spin operator of the Cu^{2+} site, $\langle ij \rangle$ represents the summation over the nearest neighboring sites i and j. According to the Wagner-Mermin theorem, there is not any long-range magnetic order at any finite temperature for this SU(2) invariant spin model in two dimensions. Nevertheless, high-T_c cuprates are not exactly two-dimensional materials. They exhibit a quasi-two-dimensional layered structure with weak couplings along the c-axis. The antiferromagnetic long-range order may appear in low temperatures.

The spin operator S_i can be expressed in terms of electron operators as

$$S_i = d_i^\dagger \frac{\sigma}{2} d_i, \tag{2.2}$$

where σ are Pauli matrices; $d_i = (d_{i\uparrow}, d_{i\downarrow})$ are the annihilation operators of Cu $3d_{x^2-y^2}$ electrons. At half-filling, every site is singly occupied, and d_i satisfies the constraint

$$d_i^\dagger d_i = 1. \tag{2.3}$$

In this case, the spin operator S_i is invariant under the following local SU(2) transformation

$$\begin{pmatrix} d_{i\uparrow} & d_{i\downarrow} \\ d_{i\downarrow}^\dagger & -d_{i\uparrow}^\dagger \end{pmatrix} \rightarrow g_i \begin{pmatrix} d_{i\uparrow} & d_{i\downarrow} \\ d_{i\downarrow}^\dagger & -d_{i\uparrow}^\dagger \end{pmatrix}, \tag{2.4}$$

where g_i is a local SU(2) transformation matrix. This local SU(2) symmetry is equivalent to the particle-hole symmetry, and is a consequence of the particle-hole invariance of the Hubbard model at half-filling. It plays an important role in the mean-field study of high-T_c superconductors.

2.3 The Three-Band Model

The electronic states of high-T_c cuprates are significantly changed after doping. In the hole-doped cuprates, holes are mainly doped onto the oxygen sites, and the valence configuration of the doped oxygen site changes from O^{2-} to O^-. Due to the hybridization between the oxygen p-orbitals and the copper d-orbitals, part of holes occupy the copper sites, which changes the valence configuration of Cu from Cu^{2+} to Cu^{3+}. In contrast, in the electron-doped case, most electrons are doped onto the copper sites, which changes the valence configuration of the copper cations from Cu^{2+} to Cu^+.

Doping opens new conducting channels which allow electrons to move without encountering the penalty of on-site Coulomb repulsion. In the hole-doped cuprates, a hole can hop from one site to another without creating double occupancy. Similarly, in electron-doped materials, the hopping of electrons between a doubly occupied site and a singly occupied one does not cost extra Coulomb repulsion. Therefore, doping, regardless of hole-doping or electron-doping, can always destabilize the Mott insulating state and enhance conductivity. The high-T_c superconductivity emerges when the doping reaches a critical level.

The high-T_c physics is determined by the orbitals in the CuO_2-plane, particularly the Cu $3d_{x^2-y^2}$-orbitals with their hybridized O $2p$-orbitals. However, for a given oxygen site, which orbital can couple to the $3d_{x^2-y^2}$-orbitals on the neighboring copper site is determined by the p-orbital orientation relative to that Cu site. Based on the symmetry analysis, only the $2p_{x(y)}$-orbital can form a σ-bond along the $x(y)$-direction with a Cu $3d_{x^2-y^2}$-orbital (Fig. 2.4). Other oxygen $2p$-orbitals do not couple to the copper $3d_{x^2-y^2}$-orbitals because the overlap integrals between these orbitals vanish.

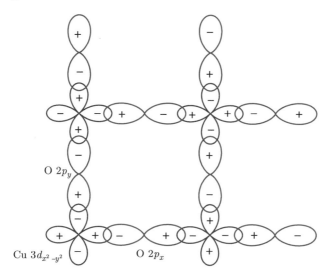

Figure 2.4   Three most important orbitals that govern the low-energy physics of high temperature superconductors: copper $3d_{x^2-y^2}$, oxygen $2p_x$ and $2p_y$ orbitals. The relative phases of these Wannier orbitals are not uniquely defined. The convention used here are determined by requiring that the overlap integrals between Cu $3d_{x^2-y^2}$ and O $2p$ orbitals are always negative.

Low energy excitations of high-T_c cuprates are governed by the electrons in the copper $3d_{x^2-y^2}$-orbitals, the oxygen $2p_x$-orbitals along the a-axis, and the oxygen $2p_y$-orbitals along the b-axis. In the hole-doped materials, the interactions among these orbitals are described by the Hamiltonian [45]:

$$H = -\sum_{\langle il \rangle} t_{pd} \left(p_l^\dagger d_i + d_i^\dagger p_i \right) + \sum_l \varepsilon_p p_l^\dagger p_l + \sum_i \varepsilon_d d_i^\dagger d_i$$
$$+ \sum_l U_p p_{l\uparrow}^\dagger p_{l\uparrow} p_{l\downarrow}^\dagger p_{l\downarrow} + \sum_i U_d d_{i\uparrow}^\dagger d_{i\uparrow} d_{i\downarrow}^\dagger d_{i\downarrow}, \qquad (2.5)$$

where i and l represent the coordinates of copper and oxygen sites, respectively. The summation $\langle il \rangle$ runs over the nearest neighboring copper cation and oxygen anion sites. $p_l = (p_{l\uparrow}, p_{l\downarrow})$ is the annihilation operator of holes in the oxygen 2p-orbital. The first term describes the hybridization between a copper $3d_{x^2-y^2}$-orbital and an oxygen 2p-orbital. The second and third terms are the on-site Coulomb potentials of holes in the oxygen 2p- and copper $3d_{x^2-y^2}$-orbitals, respectively. U_p and U_d are the Coulomb repulsions on the oxygen and copper sites, respectively. A convenient phase convention for the Wannier wavefunctions of theses orbitals is shown in Fig. (2.3).

The above Hamiltonian Eq. (2.5) is called the *three-band model* of high-T_c superconductors. It offers a starting point for the study of high-T_c mechanism. However, this model includes too many degrees of freedom and parameters, and is difficult to handle.

2.4 The *dp*-Model of Interacting Spins and Holes

As discussed previously, in the hole-doped cuprates, the $3d_{x^2-y^2}$-orbitals on the copper sites are singly occupied, and doped holes are predominantly located on the oxygen sites. This is valid if the following condition is satisfied,

$$U_d \gg \varepsilon_p - \varepsilon_d \gg |t_{pd}|. \tag{2.6}$$

Under this condition, one can take the first term in Eq. (2.5) as a perturbation and the other terms as the zeroth-order Hamiltonian, and use the degenerate perturbation theory introduced in Appendix 13.4 to simplify the three-band model as an effective low energy Hamiltonian. For this purpose, we define the zeroth-order Hamiltonian H_0 and the perturbation H_1 as

$$H_0 = \varepsilon_d \sum_i d_i^\dagger d_i + \varepsilon_p \sum_l p_l^\dagger p_l + U_d \sum_i d_{i,\uparrow}^\dagger d_{i,\uparrow} d_{i,\downarrow}^\dagger d_{i,\downarrow}, \tag{2.7}$$

$$H_1 = -t_{pd} \sum_{\langle il \rangle} \left(p_l^\dagger d_i + d_i^\dagger p_l \right). \tag{2.8}$$

In high-T_c cuprates, the hole density is low and the chance for two holes to occupy the same oxygen site is very low. Thus we can neglect the oxygen Coulomb repulsion term, i.e. the U_p-term, in Eq. (2.5).

In the CuO$_2$-plane, every unit cell contains one copper atom and two oxygen atoms. For convenience, we treat these two oxygen atoms separately. If we use

$p_{x,k}$ and $p_{y,k}$ to represent respectively these O $2p$ orbitals in the momentum space, the Fourier transform of the oxygen hole operators p_l is then defined as

$$p_{i+\hat{x}/2,\sigma} = \frac{1}{\sqrt{N}} \sum_k p_{x,k,\sigma} \exp\left[ik \cdot \left(R_i + \frac{\hat{x}}{2} \right) \right], \tag{2.9}$$

$$p_{i+\hat{y}/2,\sigma} = \frac{1}{\sqrt{N}} \sum_k p_{y,k,\sigma} \exp\left[ik \cdot \left(R_i + \frac{\hat{y}}{2} \right) \right]. \tag{2.10}$$

Substituting them into Eqs. (2.7) and (2.8) and after simplification, we obtain

$$H_0 = \varepsilon_d \sum_i d_i^\dagger d_i + \varepsilon_p \sum_i \left(a_i^\dagger a_i + b_i^\dagger b_i \right) + U_d \sum_i d_{i,\uparrow}^\dagger d_{i,\uparrow} d_{i,\downarrow}^\dagger d_{i,\downarrow}, \tag{2.11}$$

$$H_1 = -t_{pd} \sum_{ij} u\left(i - j\right) \left(a_j^\dagger d_i + d_i^\dagger a_j^\dagger \right), \tag{2.12}$$

where

$$a_{i\sigma} = \frac{1}{\sqrt{N}} \sum_k \frac{c_x p_{x,k,\sigma} + c_y p_{y,k,\sigma}}{\sqrt{c_x^2 + c_y^2}} e^{ikR_i}, \tag{2.13}$$

$$b_{i\sigma} = \frac{1}{\sqrt{N}} \sum_k \frac{c_y p_{x,k,\sigma} - c_x p_{y,k,\sigma}}{\sqrt{c_x^2 + c_y^2}} e^{ikR_i}, \tag{2.14}$$

$$u\left(r\right) = \frac{2}{N} \sum_k \sqrt{c_x^2 + c_y^2} e^{ikr}, \tag{2.15}$$

and $c_x = \cos(k_x/2)$, $c_y = \cos(k_y/2)$. a_i and b_i are independent fermion operators, $\{a_i, b_i\} = \{a_i, b_i^\dagger\} = 0$. $u(r)$ satisfies the equation

$$\sum_i u\left(i - j\right) u\left(i - j'\right) = 4\delta_{j,j'} + \delta_{\langle j,j' \rangle}.$$

$|u(r)|$ is a fast-decay function of r. When $r \gg 1$, $u(r)$ approximately decays as $1/r^3$. The first three largest values of $u(r)$ are $u(0,0) = 1.91618$, $u(1,0) = 0.280186$, and $u(1,1) = -0.0470135$.

The above equations show that the interactions only exist between a- and d-electrons, and there is no interaction between b- and d-electrons. Thus b_i is a non-bonding orbital and a_i represents a bonding orbital. The energy of b-electron lies above the Fermi energy, which has no contribution to dynamics and can be neglected. This leads to an effective two-band model which contains only *a*- and *d*-electrons. This equivalence between the two-band model and the three-band one is based on the assumption that the Coulomb interaction on the oxygen orbitals is negligible. If this term is included, *a*- and *b*-electrons are mixed. In this case, the three-band model cannot be reduced to a two-band model.

The ground states of H_0 are highly degenerate. All the states in which the copper $3d$ orbitals are singly occupied are the ground states of H_0. Below we use the degenerate perturbation theory to project the Hamiltonian into this degenerate ground state subspace to derive the low energy effective model. We use P to denote the projection operator for the ground state of H_0. Its effect is to project the Hamiltonian into the physical subspace in which all copper $3d$ orbitals are singly occupied, i.e. $d_i^\dagger d_i = 1$.

The hopping terms in H_1 changes the occupation number of copper $3d$ orbitals. Thus the first order correction of H_1 to the ground state is 0, i.e.

$$H_{eff}^{(1)} = PH_1P = 0. \tag{2.16}$$

Similarly, it can be shown that all odd perturbation terms of H_1 vanish.

The second order perturbation contribution from H_1 is given by

$$H_{eff}^{(2)} = PH_1(1-P)\frac{1}{E_0 - H_0}(1-P)H_1P.$$

After neglecting an irrelevant constant term, we find that

$$H_{eff}^{(2)} = -t_P \sum_{\langle ij \rangle} Pa_i^\dagger a_j P + J_P \sum_i Pd_{i,\sigma}^\dagger d_{i,\sigma'} \tilde{a}_{i\sigma'}^\dagger \tilde{a}_{i\sigma} P, \tag{2.17}$$

where

$$t_P = \frac{t_{pd}^2}{\varepsilon_p - \varepsilon_d}, \tag{2.18}$$

$$J_P = \frac{t_{pd}^2}{\varepsilon_p - \varepsilon_d} + \frac{t_{pd}^2}{U_d - \varepsilon_p + \varepsilon_d}, \tag{2.19}$$

$$\tilde{a}_i = \sum_j u(i-j)a_j. \tag{2.20}$$

$H_{eff}^{(2)}$ contains both the hopping and interaction terms of oxygen holes. In the undoped system, $H_{eff}^{(2)} = 0$. In order to study the interaction between Cu spins in the low doping limit, we need to calculate the contribution from the 4th order perturbation in H_1.

$H_{eff}^{(4)}$ contains more terms than $H_{eff}^{(2)}$. Some of them are just to renormalize the coupling constants in $H_{eff}^{(2)}$. These terms can be absorbed into $H_{eff}^{(2)}$ just by modifying the coefficients. The terms, which are new and important at low doping, include the Heisenberg exchange interactions among copper spins, and the hopping terms of oxygen a-electrons between the next-nearest neighboring and the next-next-nearest neighboring sites:

$$H_{eff}^{(4)} = J \sum_{\langle ij \rangle} P S_i \cdot S_j P + t_P' \sum_{\langle ij \rangle'} P(a_i^\dagger a_j + h.c.)P + t_P'' \sum_{\langle ij \rangle''} P(a_i^\dagger a_j + h.c.)P,$$

(2.21)

where $\langle \rangle'$ and $\langle \rangle''$ represents summation over the next-nearest and next-next-nearest neighbor sites, respectively. The Heisenberg exchange constant is given by

$$J = \frac{t_{pd}^4}{\varepsilon_p - \varepsilon_d} \left(\frac{1}{U_d} + \frac{1}{\varepsilon_p - \varepsilon_d} \right).$$

(2.22)

The sum of $H_{eff}^{(2)}$ and $H_{eff}^{(4)}$ gives the low energy effective Hamiltonian for cuprates. It describes the interaction among electrons in the copper $3d_{x^2-y^2}$ and oxygen $2p$ orbitals. It is correct up to the 4th order of H_1, represented as

$$H_{dp} = H_{eff}^{(2)} + H_{eff}^{(4)}.$$

(2.23)

2.5   The Zhang-Rice Singlet

In H_{dp}, J_P-term is a relatively large energy scale. It describes the interaction between the local spins in the copper $3d$ orbitals and the holes in the oxygen $2p$ orbitals. Because the off-site interaction between a d-electron and an a-hole is much smaller than the on-site interaction, we can take the approximation $u(r) \approx u(0)\delta_{r,0}$. In this case, the J_P-term becomes

$$H_{J_P} = J_P u^2(0) \sum_i P \left(a_i^\dagger a_i - 2e_i^\dagger e_i \right) P,$$

(2.24)

where e_i is a spin singlet operator formed by a- and d-electrons:

$$e_i = \frac{1}{\sqrt{2}} \left(d_{i\uparrow} a_{i\downarrow} - d_{i\downarrow} a_{i\uparrow} \right). \qquad (2.25)$$

In the limit J_P is much larger than t_P and other parameters in H_{dp}, the above equation shows that it is energetically favored for a copper $3d$ electron (d_i) and an oxygen $2p$ hole (a_i) to form a spin singlet bound state. It has an energy lower than both a unbounded state and a spin triplet one. In the low energy limit, e_i should be treated as a composite operator, and the two-band model can be further simplified as a single-band model. Based on this observation, Zhang and Rice derived an effective single-band model for high-T_c superconductors in 1988 [46, 47]. We call a localized spin singlet formed by the copper $3d$ spin and the oxygen hole a Zhang-Rice singlet. The energy difference between a Zhang-Rice singlet and the corresponding triplet is given by

$$E_{ZR} = 2J_P u^2(0). \qquad (2.26)$$

The single-band model is obtained by projecting the dp-model H_{dp} onto the subspace spanned by the ground states of H_{J_P}. In this subspace, each lattice site is either in a state with singly occupied d-orbitals, or in a Zhang-Rice singlet, namely it is limited by the constraint,

$$e_i^\dagger e_i + d_i^\dagger d_i = 1. \qquad (2.27)$$

If P_{ZR} is the corresponding projection operator, the effective single-band model is then determined by

$$H = P_{ZR} H_{dp} P_{ZR}. \qquad (2.28)$$

The rule of projection is simple. A Zhang-Rice singlet exists at site i if and only if there is an oxygen hole at that site. d_i is invariant after the projection, i.e. $P_{ZR} d_i P_{ZR} = d_i$. The hole operator a_i, after projection, becomes

$$P_{ZR} a_{i\sigma} P_{ZR} = -\frac{1}{\sqrt{2}} \sigma d_{i\bar{\sigma}}^\dagger e_i.$$

It simply means that annihilating an oxygen hole with spin σ is equivalent to annihilating a Zhang-Rice singlet and at the same time creating an electron with opposite spin. The coefficient $1/\sqrt{2}$ is due to the fact that in the Zhang-Rice singlet state, the spin of the oxygen hole a_i only has half probability in the state of σ. Applying these results to H_{dp}, and after neglecting some dynamically irrelevant constant terms, we find the effective single-band Hamiltonian to be

$$H = -\sum_{ij} t_{ij} d_i^\dagger d_j e_j^\dagger e_i + J \sum_{\langle ij \rangle} \left(S_i \cdot S_j - \frac{1}{4} n_i n_j \right), \qquad (2.29)$$

where

$$t_{ij} = t\delta_{\langle ij \rangle} + t'\delta_{\langle ij \rangle'} + t''\delta_{\langle ij \rangle''},$$

and $t = t_P/2$, $t' = t'_P/2$, $t'' = t''_P/2$.

Equations (2.29) and (2.27) are just the t-J model in the slave-boson representation, in which e_i plays the role of slave-boson.

2.6 The Hubbard Model

The one-band Hubbard model is a fundamental model of interacting electrons. It has been widely used to investigate magnetism and metal-insulator transitions. Soon after the discovery of high-T_c superconductors, P. W. Anderson first proposed to use the Hubbard model to study the mechanism of high-T_c superconductivity[46]. The Hubbard model is an effective low energy model. For sufficiently strong Coulomb repulsion, its ground state is a Mott insulator at half-filling, exhibiting an antiferromagnetic long-range order with strong antiferromagnetic fluctuations, similar as in the undoped high-T_c cuprates. The one-band Hubbard model is much simpler to analyze than the three-band one. In the strong coupling limit, the one-band Hubbard model is equivalent to the t-J model at low doping.

The Hubbard model is defined by the Hamiltonian

$$H = -t \sum_{\langle ij \rangle} \left(d_i^\dagger d_j + d_j^\dagger d_i \right) + U \sum_i d_{i\uparrow}^\dagger d_{i\uparrow} d_{i\downarrow}^\dagger d_{i\downarrow}, \qquad (2.30)$$

where the t-term describes the hopping of electrons between two neighboring sites and the U-term describes the on-site Coulomb repulsion. In high-T_c

cuprates, $U \gg t$, the Coulomb repulsive energy is much higher than the kinetic energy.

In spite of simple, the Hubbard model cannot be exactly solved in more than one dimension. In one dimension, it can be solved by employing the Bethe-ansatz [48]. In the limit $U \gg t$, we can treat the U-term as the zeroth order Hamiltonian and the hopping term as the perturbation. By applying the degenerate perturbation theory, this model can be simplified by projecting out all high energy states with doubly occupied sites. Up to the second order in t/U, the low energy effective Hamiltonian is just the t-J model defined by Eq. (2.29).

In the t-J model, the Hilbert space of each site contains three states and the double occupation state is excluded. Thus the Hilbert space of the t-J model is constrained. This is an advantage in the numerical study of the t-J model. But it is difficult to treat this constraint analytically.

In the constrained Hilbert space, the electron operator $\tilde{d}_{i\sigma}$ does not satisfy the usual anti-commutation relation of fermions, and the standard method of quantum field theory does not apply. Of course, we can force $\tilde{d}_{i\sigma} = d_{i\sigma}$ to satisfy the fermion anti-commutation relation. But in this case, the constraint becomes an inequality

$$d_i^\dagger d_i \leqslant 1.$$

It is difficult to implement this constraint analytically. Usually, it is to introduce the slave-boson or slave-fermion representation to change this constraint to an equality. But this needs to pay the price of introducing unphysical degrees of freedom. Once this constraint is not implemented rigorously, non-physical states will be included to artificially enlarge the Hilbert space.

Thus the analysis of the t-J model may not be easier than the Hubbard model. Does this mean we should abandon this model to study directly the Hubbard model? The answer is no because the t-J model only contains the low energy degrees of freedom that are most relevant to the high-T_c physics. It is easier to catch the key physical properties of cuprate superconductors by taking some approximations for the t-J model than for the Hubbard model. Moreover, as shown previously, the t-J model is also a low energy effective model of the three-band Hubbard model.

2.7 Electronic Structure along the c-Axis

Dynamics of electrons along the c-axis is dramatically different from that along the ab-plane in high-T_c cuprates. The difference is not only quantitative, but also qualitative in many aspects. For example, in underdoped cuprates, the in-plane resistivity is metal-like, while the c-axis resistivity is semiconductor-like. Various theories were proposed to explain the difference between the in-plane and c-axis charge dynamics. It was conjectured that the interlayer hopping of electrons is incoherent, i.e. electron momentum is not conserved. It was also proposed that in analogy to the quark confinement, electrons could be dynamically confined to the CuO_2-plane. These phenomenological hypotheses are simplified interpretations to experimental results. It is not a genuinely microscopic description of electron motion along the c-axis. With the progress of high-T_c study, it has been gradually realized that, in order to correctly describe electron dynamics along the c-axis, a comprehensive understanding of the microscopic picture of electron hopping along the c-axis is desired.

As mentioned, there are three key orbitals that are responsible for the low energy physics in high-T_c cuprates: the copper $3d_{x^2-y^2}$ orbital, the oxygen $2p_x$ and $2p_y$ orbitals. These orbitals couple to each other and determine the low energy physics of each CuO_2-plane. However, these orbitals have strong two-dimensional characters. Their charge clouds extend mainly along the CuO_2-plane. The characteristic length scale of these orbitals along the c-axis is less than 1Å. The overlap between these orbitals on different CuO_2-planes are almost zero. Thus electrons can hardly hop along the c-axis. This is the reason why the c-axis conductivity is so small in the layered cuprates. However, in real high-T_c materials, the hopping along the c-axis is not exactly zero. Electrons in the copper $3d_{x^2-y^2}$ and oxygen $2p_x, 2p_y$ orbitals can hop between CuO_2 layers via other orbitals. Among them, the most important one is the Cu $4s$ orbital, which is rotationally symmetric with respect to the c-axis.

On the same site, the Wannier wave functions of Cu $3d_{x^2-y^2}$ and $4s$ orbitals are orthogonal to each other. Therefore, the copper $4s$ orbital cannot assist electrons in the copper $3d_{x^2-y^2}$ orbital to hop along the c-axis. Rather, it can facilitate the interlayer hopping of electrons between two oxygen $2p$ orbitals. In fact, this is the main channel through which electrons hop along the c-axis. The microscopic hopping process [49, 50] is

$$(O\,2p)_1 \rightarrow (Cu\,4s)_1 \rightarrow (*)_{12} \rightarrow (Cu\,4s)_2 \rightarrow (O\,2p)_2,$$

where the subscripts denote the indices of CuO_2 planes, $(*)_{12}$ represents the orbitals assisting electrons hopping between two neighboring CuO_2 planes. This is a virtual hopping process because the energy of Cu $4s$-orbital is above the Fermi energy.

The effective interlayer hopping integral, t_c, between the oxygen $2p$-orbitals of the first and the second layers, is proportional to the product of the matrix elements of all virtual hopping steps, that is

$$t_c \sim \langle (O\,2p)_2 | (Cu\,4s)_2 \rangle \langle (Cu\,4s)_2 | (*)_{12} \rangle \langle (*)_{12} | (Cu\,4s)_1 \rangle \langle (Cu\,4s)_1 | (O\,2p)_1 \rangle,$$
(2.31)

where $\langle a | b \rangle$ represents the hopping integral between the Wannier orbitals $|a\rangle$ and $|b\rangle$. The values of these integrals depend on the crystal and electronic structures. But these overlaps possess certain symmetry, which holds generally, independent on detailed properties of materials. In particular, the overlap between the copper $4s$-orbital and the oxygen $2p$ orbitals within the same CuO_2-plane, i.e. $\langle O\,2p | Cu\,4s \rangle$, possesses the $d_{x^2-y^2}$ symmetry under the rotation around the c-axis. This symmetry can be identified from the phase structure of the overlap between Cu $4s$ and O $2p_x$ or $2p_y$ orbitals shown in Fig. 2.5: the overlap between Cu $4s$ and O $2p_x$ orbitals is positive, while that between Cu $4s$ and O $2p_y$ orbitals is negative. This wave function overlap has precisely the $d_{x^2-y^2}$ symmetry. In momentum space, it implies that the corresponding overlap can be represented as

$$\langle (Cu\,4s)_n | (O\,2p)_n \rangle \propto \cos k_a - \cos k_b,$$

where $n = 1$ or 2. The right-hand side of the equation is just the wave function of $d_{x^2-y^2}$-orbital in momentum space.

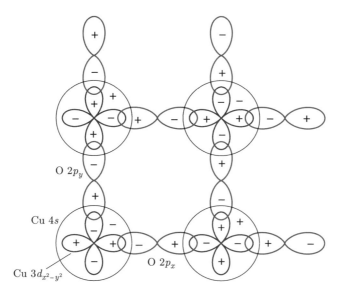

Figure 2.5　Wave functions of Cu $3d_{x^2-y^2}$, Cu $4s$, and the bonding O $2p_x$ and $2p_y$ orbitals. The overlap integrals between Cu $4s$ and O $2p$ orbitals possess the $d_{x^2-y^2}$ symmetry.

The other two overlap integrals, $\langle(Cu\,4s)_2|(*)_{12}\rangle$ and $\langle(*)_{12}|(Cu\,4s)_1\rangle$, are related to the crystal and chemical structures between two neighboring CuO_2 planes. Generally, they do not possess specific symmetry. Here we treat them as constants. Therefore, we have

$$t_c \sim t_\perp \left(\cos k_a - \cos k_b\right)^2. \tag{2.32}$$

It shows that the interlayer hopping of electrons strongly depends on the momentum direction in the CuO_2-plane. For the in-plane momentum along the diagonal lines, i.e. $|k_a| = |k_b|$, the c-axis hopping integral equals zero. In other words, when $|k_a| = |k_b|$, electrons are dispersionless along the c-axis, which is a peculiar and important property of high-T_c cuprates. The coincidence of the zeros of t_c and the nodal line of the $d_{x^2-y^2}$-wave pairing gives leads to many anomalous effects observed in experiments. It is still unclear whether this coincidence is also related to the pairing symmetry in high-T_c cuprates.

Eq. (2.32) is a general property of high-T_c cuprates, independent of specific crystalline structures and chemical ingredients. It is valid for all the monolayer, bilayer, trilayer, and even infinite-layer compounds, because the $d_{x^2-y^2}$ symmetry of the overlap integral between Cu 4s- and O 2p-orbitals results simply from a symmetry property of wave functions within each CuO_2-plane, independent of the interlayer coupling. For $Bi_2Sr_2CaCu_2O_8$, or other high-T_c cuprates whose unit cell contains two CuO_2 planes, the coupling between two CuO_2 plane leads to a bilayer splitting of the energy bands with a splitting energy scale of the order of $2t_c$. This splitting, as confirmed by the angle-resolved photoemission spectroscopy(ARPES) experimental observation, is highly anisotropic. It vanishes along the nodal line of the $d_{x^2-y^2}$-wave pairing gap, but takes a maximal value along the anti-nodal direction. The value of t_c, measured by experiments agrees quantitatively with Eq. (2.32) within experimental errors [51].

For $La_{2-x}Sr_xCuO_4$ or other cuprates with the body-centered lattice symmetry, the coefficient of $(\cos k_a - \cos k_b)^2$, i.e. $t_\perp$, also depends on k_a and k_b. t_c generally has the form

$$t_c \propto \cos \frac{k_a}{2} \cos \frac{k_b}{2} (\cos k_a - \cos k_b)^2 . \tag{2.33}$$

It also vanishes when $k_a = \pi$ or $k_b = \pi$. This is a general property of cuprates with the body-centered lattice symmetry in the tight-binding approximation. It has been verified experimentally[52].

2.8 Systems Doped with Zn- or Ni-Impurities

Doping magnetic or non-magnetic impurities is an important approach to perturb and probe high-T_c superconductors, in addition to measuring the response of a system to a perturbation generated by an external electric, magnetic, or thermal field. Both theoretical and experimental studies on the impurity effects have greatly deepened our understanding on the mechanism of high-T_c superconductivity.

There are various ways to dope impurities into high-T_c materials. The most common one is the element substitution. Depending on different types of dopants and elements substituted, the responses of superconductors to impurities are different. The impurity that affects strongly physical properties of high-T_c superconductors include both zinc and nickel elements. They are also the impurity elements that have been systematically studied.

The zinc and nickel substitutions affect strongly physical properties of high-T_c cuprates because they replace mainly the copper elements in the CuO_2-plane. The zinc impurity is a strong scattering center, and is known the strongest pair-breaker. Experimentally it was found that the scattering phase shift induced by the zinc impurity potential approaches the limit of resonant scattering $\pi/2$. For $YBa_2Cu_3O_{7-\delta}$ superconductors, around 7% zinc impurity concentration can completely suppresses the superconducting long range order and reduces the transition temperature T_c to zero. In contrast, the nickel impurity has a weaker influence on the high-T_c properties. It suppresses T_c three times weaker than Zn, indicating that the nickel impurity is a weak scattering center.

Same as for Cu, both Zn and Ni are divalent elements. Substituting Cu^{2+} by Zn^{2+} or Ni^{2+} neither increases nor reduces the carrier number in the system. Fig. 2.6 shows the $3d$ electron configurations of these cations. The $3d$ shell of Zn^{2+} is fully occupied, hence Zn^{2+} is non-magnetic. Ni^{2+} is different, both the $3d_{x^2-y^2}$ and $3d_{3z^2-r^2}$ orbitals are singly occupied. According to the Hund's rule, the spin states in these two orbitals are parallelized. Thus Ni^{2+} carries a magnetic moment and serves as a magnetic impurity.

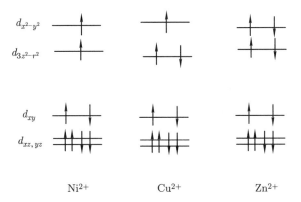

Figure 2.6   Configurations of the $3d$ electrons in Ni^{2+}, Cu^{2+}, and Zn^{2+} cations in an octahedral crystal field.

For conventional *s*-wave superconductors, the pair-breaking effect from a magnetic impurity is much stronger than that from a non-magnetic one. However, in high-T_c superconductors, the non-magnetic zinc impurity suppresses superconductivity much stronger than the magnetic nickel impurity. This implies that the impurity scattering potential of zinc is much stronger than that of nickel. The underlying physics can be understood only by correctly constructing the microscopic model for these impurities.

In a background with strong antiferromagnetic fluctuations, non-magnetic impurities can induce magnetic moments around it. Therefore, they exhibit many features of magnetic impurities. Based on this reasoning, some theoretical and experimental works tend to attribute the strong scattering effect of zinc to the induced magnetic moments. This sounds to be a correct picture. However, it could not explain why the scattering from zinc is stronger than nickel. First, the magnetic moment induced by a zinc impurity is a secondary effect in comparison with the intrinsic magnetic moment of nickel. It is unlikely that a zinc impurity can exhibit stronger pair-breaking effect than a nickel impurity. Second, in the overdoped high-T_c superconductors, the antiferromagnetic correlations are significantly weakened. The argument of induced magnetic moments by the zinc impurity should not work in this regime. A weak pair breaking effect from the zinc impurity is expected, but it is not consistent with experimental observations. Therefore, the pair-breaking effect from the zinc impurity comes predominantly from the non-magnetic potential scattering. In fact, the difference in the suppression of the transition temperature T_c by both zinc and nickel impurities result mainly from the potential scattering effect.

As already mentioned, the scattering potential of the zinc impurity is in the limit of the resonant scattering and the scattering phase shift δ_0 approaches $\pi/2$, and the substitution of Cu^{2+} by Zn^{2+} does not change the total carrier number. These two seemingly unrelated facts are actually inconsistent with each other. They violate the Friedel sum rule [9]:

$$\Delta Z = \frac{2}{\pi} \sum_l (2l + 1)\delta_l. \tag{2.34}$$

On the left-hand side of Eq. (2.34), $\Delta Z = 0$ because the zinc substitution does not change the number of charge carriers. The right-hand side equals 1 because $\delta_l \approx \pi/2$ if $l = 0$ and negligibly small if $l \geqslant 1$. As will be discussed, this discrepancy results from the correlated effect of high-T_c superconductors. It can be resolved by considering the correction to the Zn scattering potential by the correlation effect of electrons on the CuO_2 planes.

Below we derive the effective one-band model for a zinc or nickel impurity, starting from the corresponding three-band model [53]. The idea guiding to the derivation holds generally, and can be extended to apply to other impurities similar to zinc or nickel.

2.8.1   The Zn Impurity

Physically it is interesting to just consider a system with low Zn concentration so that the interplay among these impurities is small and negligible. In this case, we just need to solve a single impurity problem. The result can be readily generalized to a many-impurity system.

As the five $3d$ electron orbitals of Zn^{2+} are fully occupied, the total spin of Zn^{2+} is zero. This cation is very stable against valence fluctuations. It is difficult to change Zn^{2+} to Zn^{3+} by removing one electron, or to Zn^+ by adding one more electron. Therefore, Zn^{2+} is an inert non-magnetic impurity and has no charge transfer with the surrounding O^{2-} anions and Cu^{2+} cations.

The three-band Hamiltonian, corresponding to Eqs. (2.7) and (2.8), for the system including a zinc impurity is defined by

$$H^{Zn} = H_0^{Zn} + H_1^{Zn}, \tag{2.35}$$

$$H_0^{Zn} = \varepsilon_p \sum_l p_l^\dagger p_l + \sum_{i \neq i_0} \left(\varepsilon_d d_i^\dagger d_i + U_d d_{i\uparrow}^\dagger d_{i\uparrow} d_{i\downarrow}^\dagger d_{i\downarrow} \right), \tag{2.36}$$

$$H_1^{Zn} = - \sum_{\langle il \rangle i \neq i_0} t_{pd} \left(p_l^\dagger d_i + d_i^\dagger p_l \right), \tag{2.37}$$

where i_0 is the position of the zinc impurity. Using the bonding and non-bonding operators of the oxygen holes, the above equations can be expressed as

$$H_0^{Zn} = \varepsilon_p \sum_i \left(a_i^\dagger a_i + b_i^\dagger b_i \right) + \sum_{i \neq i_0} \left(\varepsilon_d d_i^\dagger d_i + U_d d_{i\uparrow}^\dagger d_{i\uparrow} d_{i\downarrow}^\dagger d_{i\downarrow} \right), \quad (2.38)$$

$$H_1^{Zn} = -t_{pd} \sum_{i \neq i_0, j} u\,(i - j) \left(a_j^\dagger d_i + d_i^\dagger a_j \right). \quad (2.39)$$

In the limit of $U_d \gg \varepsilon_p - \varepsilon_d \gg t_{pd}$, we can take H_1^{Zn} as a perturbation to project the above Hamiltonians onto the ground state subspace spanned by the Zhang-Rice singlets and the unpaired copper spins using the method introduced in Sections 2.4 and 2.5. Following the derivation steps previously introduced, the effective low energy one-band Hamiltonian is found to be

$$H_{Zn} = \sum_i V_{Zn}(i) d_i^\dagger d_i - \sum_{i \neq j} t_{ij}^{Zn} d_j^\dagger d_i + \sum_{\langle ij \rangle \neq i_0} J S_i \cdot S_j. \quad (2.40)$$

Similar as in the standard *t-J* model, the *d*-electrons at site $i \neq i_0$ satisfy the constraint

$$d_i^\dagger d_i \leqslant 1. \quad (2.41)$$

At the impurity site, the operator d_{i_0} is not the annihilation operator of $3d$ electrons. Instead it is defined by the bonding operator a_{i_0} of the oxygen hole at that site:

$$d_{i_0 \sigma} = -\sigma a_{i_0 \bar{\sigma}}^\dagger.$$

Unlike the *d*-electron operators at the other lattice sites, there is no constraint for d_{i_0}. Therefore, two electrons with opposite spins can occupy the impurity site. Formally, this is consistent with the fact that there is only one electron at the $3d_{x^2-y^2}$ orbital in Cu^{2+} and two electrons at that orbital in Zn^{2+}. However, it should be emphasized that there is no unoccupied electrons at the zinc site in the original three-band Hamiltonian Eq. (2.35).

The electron hopping integral is

$$t_{ij}^{Zn} = \frac{\tilde{t}_{ij}}{2} \delta_{i \neq i_0, j \neq i_0} + \frac{\tilde{t}_{i_0 j}}{\sqrt{2}} \delta_{i, i_0} + \frac{\tilde{t}_{i_0 i}}{\sqrt{2}} \delta_{j, i_0} - t' \delta_{\langle ij \rangle' \neq i_0} - t'' \delta_{\langle ij \rangle'' \neq i_0}, \quad (2.42)$$

where

$$\tilde{t}_{ij} = t_P \delta_{\langle i,j \rangle} - t_P u(i_0 - i) u(i_0 - j).$$

The first term is the hopping integral in the absence of the impurity. The second term is the correction introduced by the zinc impurity, which is small but non-local.

The effective impurity potential is defined by

$$V_{Zn}(i) = -t_P u^2 (i_0 - i)\delta_{i \neq i_0} - (t_P + J_P)\, u^2(0)\, \delta_{i,i_0}, \qquad (2.43)$$

which is an attractive interaction for electrons. This is not an on-site potential. It decays approximately as $|V_{Zn}(i)| \propto |i - i_0|^{-6}$ in the large $|i - i_0|$ limit. This potential at the impurity site, $|V_{Zn}(i_0)| \approx (t_P + J_P)\, u^2(0)$, is about two orders of magnitude larger than that on the nearest neighboring sites, which is about $0.0785 t_P$. It is also more than one order of magnitude larger than the effective hopping integral $t_P/2$. Thus the zinc impurity is a strong scattering center. The effective attractive potential arises from the strong repulsion of the Zn^{2+} cation to the bonding oxygen holes on the impurity site, for the following two reasons. First, Zn^{2+} is spinless and cannot form a Zhang-Rice singlet with an oxygen hole. This leads to a relative increase of the oxygen hole energy, $J_P u^2(0)$, at the impurity site. Second, electrons cannot hop between O $2p$ and Zn $3d$-orbitals, resulting in a lose of kinetic energy, $t_P u^2(0)$. Therefore, the total energy lost is $(t_P + J_P)u^2(0)$, which is equivalent to having a repulsive potential for oxygen holes, or an attractive potential for electrons, at the impurity site.

The above discussion reveals an intimate connection between the scattering potential of the zinc impurity and the Zhang-Rice singlet. It shows that the bonding energy of the Zhang-Rice singlet can be determined by measuring the Zn impurity scattering potential.

Assuming the total number of doped holes is N_h, then

$$\sum_i a_i^\dagger a_i = N_h. \qquad (2.44)$$

Because the impurity potential is strongly repulsive to holes, no oxygen hole in low energy can exist on this site, thus we have

$$\sum_i d_i^\dagger d_i = (N - N_h) + 1, \qquad (2.45)$$

where N is the total number of lattice sites. This expression shows that, in the effective low energy one-band model, each Zn impurity contributes an extra electron to the system. It also shows that although both Cu^{2+} and Zn^{2+} ions are divalent, Zn^{2+} should be treated as having one more electron than Cu^{2+} in the effective single-band model.

Therefore, in the strongly correlated CuO_2-plane, the phase space of electrons is enlarged by the substitution of Zn impurities. Effectively, each Zn introduces an extra electron to the system so that $\Delta Z = 1$, rather than $\Delta Z = 0$, which modifies the Friedel sum rule Eq. (2.34). This is a consequence of strongly correlated effect. It shows that the resonant scattering phase shift induced by the zinc impurity scattering, $\delta_0 = \pi/2$, is consistent with the Friedel sum rule, resolving the aforementioned puzzle about the Friedel sum rule.

2.8.2 The Ni Impurity

A Ni^{2+} cation has eight electrons in its $3d$ shell. Due to the strong Hund's coupling between $3d_{3z^2-r^2}$ and $3d_{x^2-y^2}$ electrons, the total spin is 1. Similarly to Zn^{2+}, Ni^{2+} is very stable. However, as the $3d_{3z^2-r^2}$ and $3d_{x^2-y^2}$ orbitals are not fully filled, the hybridizations between these two orbitals and the surrounding oxygen $(2p_x, 2p_y)$ orbitals are strong.

Similar as for the zinc impurity, the three-band model for the system with one nickel impurity is defined by the Hamiltonian

$$H^{Ni} = H^{Zn} + \sum_\alpha \varepsilon_\alpha^{Ni} c_\alpha^\dagger c_\alpha - \sum_{\langle l i_0 \rangle \alpha} t_\alpha^{Ni} \left(p_l^\dagger c_\alpha + h.c. \right)$$
$$- J_H c_1^\dagger \frac{\sigma}{2} c_1 \cdot c_2^\dagger \frac{\sigma}{2} c_2 + \sum_\alpha U_\alpha^{Ni} c_{\alpha\uparrow}^\dagger c_{\alpha\uparrow} c_{\alpha\downarrow}^\dagger c_{\alpha\downarrow}, \qquad (2.46)$$

where $\alpha = 1$ and 2 represent the $3d_{x^2-y^2}$ and $3d_{3z^2-r^2}$ orbital of Ni^{2+}, respectively. $c_\alpha = (c_{\alpha\uparrow}, c_{\alpha\downarrow})$ are the electron annihilation operators of these orbitals. The corresponding onsite energy and Coulomb repulsion are denoted as ε_α^{Ni} and U_α^{Ni}, respectively. t_α^{Ni} is the hybridization between the Ni $3d$ and the surrounding oxygen $2p$ electrons. J_H is the Hund's coupling constant between $3d_{x^2-y^2}$ and $3d_{3z^2-r^2}$ spins.

Compared to the Hund's coupling, the hopping term t_{α}^{Ni} is relatively a small quantity and can be treated as perturbation. The effective one-band model of the nickel impurity can be derived in a similar way as for the zinc impurity. However, as the total spin of Ni^{2+} is one, the Ni spin cannot form a Zhang-Rice singlet with an oxygen hole. Instead, they will form a Zhang-Rice-like spin doublet, which reduces strongly the scattering potential so that Ni^{2+} behaves like a weak scattering center. This is a subtle difference between Zn and Ni impurities.

Using the degenerate perturbation theory, the effective one-band Hamiltonian for describing a system with one nickel impurity is found to be

$$H_{Ni} = \sum_{i \neq j \neq i_0} t_{ij}^{Ni} d_j^{\dagger} d_i + \sum_i V_{Ni}(i) d_i^{\dagger} d_i + \sum_{\langle ij \rangle,} J_{ij} S_i \cdot S_j, \qquad (2.47)$$

where $d_{i_0} = (d_{i_0\uparrow}, d_{i_0\downarrow})$ are the annihilation operators of the spin doublet formed by the Ni^{2+} spin and the oxygen hole at the impurity site. Similarly, if $i \neq i_0$, the d-electrons satisfy the constraint $d_i^{\dagger} d_i \leqslant 1$. However, on the impurity site, the nickel spin and the oxygen hole spin form a spin doublet. In this effective single-band model, the Ni^{2+} spin is partially screened, and can be effectively identified as a spin-$1/2$ magnetic impurity.

The hopping integrals in the first term of H_{Ni} are given by

$$t_{ij}^{Ni} = t_{ij}^{Zn} + t_P' u(i_0 - i) u(i_0 - j), \qquad (2.48)$$

$$t_P' = \sum_{\alpha} \frac{\left(t_{\alpha}^{Ni}\right)^2}{U_{\alpha}^{Ni} - \varepsilon_p + \varepsilon_{\alpha}^{Ni} + J_H/4}. \qquad (2.49)$$

The exchange energy $J_{ij} = J$ when neither i nor j equals i_0. When either i or j equals i_0, $J_{ij} \neq J$ but remains at the same order.

The scattering potential of the nickel impurity is given by

$$V_{Ni}(i) = \left(t_P + J_P - \frac{1}{2}J_P' - \frac{3}{2}t_P'\right) u^2(0)\delta_{i,i_0} - (t_P - t_P')u^2(i_0 - i)\delta_{i \neq i_0}, \qquad (2.50)$$

$$J_P' = \sum_{\alpha} \frac{\left(t_{\alpha}^{Ni}\right)^2}{\varepsilon_p - \varepsilon_{\alpha}^{Ni} + J_H/4}.$$

The J'_P term arises from the hybridization between the nickel $3d$-orbitals and the oxygen $2p$ orbitals. The t'_P term is the binding energy of the local spin doublet formed by the Ni^{2+} spin and the oxygen hole. These terms suppress the scattering potential generated by the t_P and J_P terms. On the impurity site, $V_{Ni}(i) = \left(t_P + J_P - \frac{1}{2}J'_P - \frac{3}{2}t'_P\right)u^2(0)$, which is much smaller than the corresponding Zn scattering potential. Thus the nickel impurity is a weaker scattering center compared in comparison with the zinc impurity, consistent with the experimental result.

The above discussions demonstrate the complexity of strong correlated electronic systems. The difference between the zinc and nickel impurities does not arise from the distribution of electrons in their $3d$ orbitals, but from their correlation effects with the surrounding oxygen holes. It indicates that the influence of zinc and nickel impurities to high-T_c superconductors results predominantly from the potential scattering, rather than from the magnetic interaction generated by the effective spin of the nickel impurity or the induced magnetic moment around the zinc impurity.

Chapter 3

Fundamental Properties of d-Wave Superconductor

3.1 Gap Function

The temperature dependence of the gap function ϕ_k is determined by the equation

$$\frac{g}{V}\sum_k \frac{\phi_k^2}{2\sqrt{\xi_k^2 + \Delta_k^2}} \tanh \frac{\beta\sqrt{\xi_k^2 + \Delta_k^2}}{2} = 1. \qquad (3.1)$$

If $\phi_k = \gamma_\varphi$ depends only on the azimuthal angle φ of the wave vector k and the energy dispersion ξ_k is isotropic, independent on φ, the above equation then becomes

$$g\int\frac{\mathrm{d}\varphi}{2\pi}\int_0^{\omega_0}\mathrm{d}\xi\rho_0(\xi)\frac{\gamma_\varphi^2}{\sqrt{\xi^2 + \Delta^2\gamma_\varphi^2}} \tanh \frac{\beta\sqrt{\xi^2 + \Delta^2\gamma_\varphi^2}}{2} = 1, \qquad (3.2)$$

where $\rho_0(\xi)$ is the normal state density of states of electrons. γ_φ equals $\cos 2\varphi$ and 1 for the d- and s-wave superconductors, respectively. ω_0 is the characteristic energy scale of the pairing interaction, which is generally much larger than the superconducting transition temperature T_c. In a metal-based or other conventional superconductor induced by the electron-phonon interaction, ω_0 approximately equals the Debye frequency of phonons. For high-T_c superconductors, it is unclear what the energy scale ω_0 represents.

The normal state density of states $\rho_0(\xi)$ is determined by the band width which is usually much larger than ω_0. In this case, $\rho_0(\xi)$ can be approximated by its value at the Fermi energy, i.e. $\rho_0(\xi) \approx \rho_0(\xi_F) = N_F$, and Eq. (3.2) can be further expressed as

$$gN_F \int \frac{d\varphi}{2\pi} \int_0^{\omega_0} d\xi \frac{\gamma_\varphi^2}{\sqrt{\xi^2 + \Delta^2\gamma_\varphi^2}} \tanh \frac{\beta\sqrt{\xi^2 + \Delta^2\gamma_\varphi^2}}{2} = 1. \qquad (3.3)$$

It is difficult to solve this integral equation analytically. However, at both zero temperature and the superconducting transition point, the integral in Eq. (3.3) can be simplified, allowing us to derive explicitly the expressions for the maximal gap value Δ_0 at zero temperature and the critical temperature T_c.

At the critical transition temperature, $\Delta = 0$, Eq. (3.3) can be simplified as

$$\int_0^{\omega_0/2k_BT_c} dx \frac{\tanh x}{x} = \alpha, \qquad (3.4)$$

where α is a constant, which is equal to $1/(gN_F)$ and $2/(gN_F)$ for the s- and d-wave superconductor, respectively. After integration by parts, the equation becomes

$$\ln \frac{\omega_0}{2k_BT_c} \tanh \frac{\omega_0}{2k_BT_c} - \int_0^{\omega_0/2k_BT_c} dx \ln x \operatorname{sech}^2 x = \alpha. \qquad (3.5)$$

The integral on the left-hand side of the equation converges as $x \to \infty$. Thus the upper limit of the integral can be safely set to $+\infty$ in the limit $\omega_0 \gg T_c$. The transition temperature T_c is then found to be

$$k_BT_c \approx \begin{cases} c_0\omega_0 \exp(-1/gN_F), & s\text{-wave}, \\ c_0\omega_0 \exp(-2/gN_F), & d\text{-wave}, \end{cases} \qquad (3.6)$$

where

$$c_0 = \frac{1}{2} \exp\left(-\int_0^\infty dx \ln x \operatorname{sech}^2 x\right) \approx 1.134.$$

At zero temperature, after integrating out ξ in Eq. (3.3), it is simple to show that the zero-temperature gap $\Delta_0 = \Delta(T = 0)$ is determined by the equation

$$\frac{gN_F}{\pi} \int_0^\pi d\varphi \gamma_\varphi^2 \ln \frac{\omega_0 + \sqrt{\omega_0^2 + \Delta_0^2\gamma_\varphi^2}}{\Delta_0|\gamma_\varphi|} = 1. \qquad (3.7)$$

For the isotropic s-wave superconductor, $\gamma_\varphi = 1$, the solution is

$$\Delta_0 = \frac{2\omega_0 \exp\left(-\dfrac{1}{gN_F}\right)}{1 - \exp\left(-\dfrac{2}{gN_F}\right)}. \tag{3.8}$$

For the d-wave superconductor, it is difficult to obtain an analytical expression for the integral in Eq. (3.7). But in the limit $\omega_0 \gg \Delta_0$, this equation can be approximately written as

$$\frac{gN_F}{\pi} \int_0^\pi d\varphi \gamma_\varphi^2 \ln \frac{2\omega_0}{\Delta_0|\gamma_\varphi|} \approx 1. \tag{3.9}$$

The solution of this equation is given by

$$\Delta_0 = c_1 \omega_0 \exp(-2/gN_F), \tag{3.10}$$

where

$$c_1 = 2\exp\left(-\frac{4}{\pi} \int_0^{\pi/2} d\varphi \cos^2 \varphi \ln \cos \varphi\right) = 4e^{-0.5} \approx 2.426.$$

The superconducting transition temperature T_c and the zero temperature gap Δ_0 are the two fundamental parameters of superconductors. Δ_0^2 is proportional to the condensation energy of superconducting state. Given g and ω_0, Δ_0 in the s-wave superconductor is larger than that in the d-wave one. Thus the s-wave pairing is energetically favored. However, if the corrections from the Coulomb repulsion among electrons are included, the d-wave pairing may win over the s-wave one. The d-wave pairing can reduce the on-site Coulomb repulsion energy because the probability for two paired electrons occupying the same lattice site in a d-wave superconductor vanishes.

The above results indicate that the ratio between Δ_0 and T_c depends on the pairing symmetry but not on the detailed band structures in the limit $\omega_0 \gg k_B T_c$. In a d-wave superconductor,

$$\frac{2\Delta_0}{k_B T_c} \approx 4.28, \tag{3.11}$$

whereas in an *s*-wave superconductor,

$$\frac{2\Delta_0}{k_B T_c} \approx 3.53. \tag{3.12}$$

The ratio for the *s*-wave superconductor is smaller than that for the *d*-wave one.

In a conventional metal-based superconductor, the deviation of $2\Delta_0/k_B T_c$ from 3.53 is a characteristic quantity to determine the coupling strength of Cooper pairs. In high-T_c cuprates, however, the values of $2\Delta_0/k_B T_c$ extracted from different experimental measurements are different. They exhibit strong sample and doping dependence. In the optimal or overdoped regime, this ratio is around 4.28, close to the value predicted by the BCS theory. But in the underdoped regime, $2\Delta_0/k_B T_c$ increases rapidly with decreasing doping and is generally much larger than the BCS value. The suppression of T_c is likely to be induced by strong phase fluctuations as well as the gap anisotropy.

Rigorously speaking, $\Delta_k = \cos 2\varphi$ is a good approximation only in the vicinity of the *d*-wave gap nodes. The value of $2\Delta_0/k_B T_c$ obtained from this approximation is valid in the low energy limit. However, the value of $2\Delta_0/k_B T_c$ obtained by experiments is generally an average of $2\Delta(\varphi)/k_B T_c$ over the entire Fermi surface. It could also be a value of $2\Delta(\varphi)/k_B T_c$ along a particular momentum direction. This is the reason why the experimental values of $2\Delta_0/k_B T_c$ exhibit large fluctuations. Strong correlation effects, such as antiferromagnetic fluctuations, may also change significantly the value of this ratio. It is not reliable to determine the pairing symmetry simply based on the value of $2\Delta_0/k_B T_c$.

In order to determine the temperature dependence of Δ, we take the difference between Eq. (3.3) and the corresponding equation in the limit of $\beta \to \infty$. By utilizing Eq. (3.7) in the limit $\omega_0 \gg \Delta_0$, we find that the gap function is determined by the equation

$$\langle \gamma_\varphi^2 \rangle_{FS} \ln \frac{\Delta_0}{\Delta(T)} = \int_0^{\omega_0} d\xi \int_{-\pi}^{\pi} \frac{d\varphi}{2\pi} \frac{\gamma_\varphi^2}{\sqrt{\xi^2 + \Delta^2 \gamma_\varphi^2}} \left(1 - \tanh \frac{\beta\sqrt{\xi^2 + \Delta^2 \gamma_\varphi^2}}{2} \right),$$

$$\tag{3.13}$$

where $\langle\gamma_\varphi^2\rangle_{FS}$ is the average of γ_φ^2 on the Fermi surface. It is equal to 1 and $1/2$ for the s- and d-wave superconductors, respectively.

In an s-wave superconductor, there are very few low-lying excitations and $\Delta(T)$ approaches $\Delta(0)$ exponentially in low temperatures:

$$\Delta(T) = \Delta_0 - \sqrt{2\pi k_B T \Delta_0} \exp\left(-\frac{\Delta_0}{k_B T}\right) \qquad (T \ll T_c). \tag{3.14}$$

In a d-wave superconductor, Eq. (3.13) can be rewritten as

$$\pi \ln\frac{\Delta_0}{\Delta(T)} = \int_0^{\omega_0} d\xi \int_0^1 \frac{dx}{\sqrt{1-x^2}} \frac{4x^2}{\sqrt{\xi^2 + \Delta^2 x^2}} \left(1 - \tanh\frac{\beta\sqrt{\xi^2 + \Delta^2 x^2}}{2}\right). \tag{3.15}$$

In low temperatures, the integral contributes mainly from the domain of $x < k_B T/\Delta$ where $\sqrt{1-x^2}$ in the denominator is approximately equal to 1, and Eq (3.15) becomes

$$\pi \ln\frac{\Delta_0}{\Delta(T)} = \frac{1}{\beta^3 \Delta^3} \int_0^{\beta\omega_0} dy \int_0^{\beta\Delta} dx \frac{4x^2}{\sqrt{x^2 + y^2}} \left(1 - \tanh\frac{\sqrt{x^2 + y^2}}{2}\right). \tag{3.16}$$

In the limit $\omega_0 \gg k_B T$ and $\Delta \gg k_B T$, the upper limits of the above two integrals can be set to infinity, we have

$$\pi \ln\frac{\Delta_0}{\Delta(T)} = \frac{1}{\beta^3 \Delta^3} \int_0^\infty dy \int_0^\infty dx \frac{4x^2}{\sqrt{x^2 + y^2}} \left(1 - \tanh\frac{\sqrt{x^2 + y^2}}{2}\right). \tag{3.17}$$

Solving this equation yields

$$\Delta(T) = \Delta_0 \exp\left[-\alpha_0 \frac{k_B^3 T^3}{\Delta^3(T)}\right] \approx \Delta_0 \left(1 - \alpha_0 \frac{k_B^3 T^3}{\Delta_0^3}\right), \tag{3.18}$$

where

$$\alpha_0 = \int_0^\infty r^2 \left(1 - \tanh\frac{r}{2}\right) \approx 3.606. \tag{3.19}$$

Thus $\Delta(T)$ approaches $\Delta(0)$ cubically with temperature, different from the s-wave case.

In the vicinity of the superconducting transition temperature, the gap value $\Delta(T)$ is small and we can expand Eq (3.3) in terms of $\Delta(T)$. Take the approximation up to the second order term in $\Delta(T)$, we have

$$gN_F\left[\langle\gamma_\varphi^2\rangle_{FS} \int_{-\omega_0}^{\omega_0} d\xi \frac{\tanh(\beta\xi/2)}{\xi} + \langle\gamma_\varphi^4\rangle_{FS} \Delta^2 p(T)\right] \approx 1, \tag{3.20}$$

where

$$p(T) = \int_0^{\omega_0} d\xi \left[\frac{\beta \text{sech}^2(\beta\xi/2)}{4\xi^2} - \frac{\tanh(\beta\xi/2)}{2\xi^3} \right]. \tag{3.21}$$

Using Eq. (3.4), Eq. (3.20) can be further simplified as

$$\langle \gamma_\varphi^2 \rangle_{FS} \ln \frac{T_c}{T} + \langle \gamma_\varphi^4 \rangle_{FS} \Delta^2 p(T) \approx 0. \tag{3.22}$$

Thus the solution of $\Delta(T)$ is given by

$$\Delta(T) = \left[\frac{\langle \gamma_\varphi^2 \rangle_{FS}}{\langle \gamma_\varphi^4 \rangle_{FS} p(T)} \ln \frac{T}{T_c} \right]^{1/2} = \left[-\frac{\langle \gamma_\varphi^2 \rangle_{FS} k_B^2 T^2}{\langle \gamma_\varphi^4 \rangle_{FS} g_0} \ln \frac{T}{T_c} \right]^{1/2}, \tag{3.23}$$

where

$$g_0 = -\int_0^\infty dx \left[\frac{\text{sech}^2(x/2)}{4x^2} - \frac{\tanh(x/2)}{2x^3} \right] \approx 0.107.$$

In the limit $T_c - T \ll T_c$,

$$\ln \frac{T}{T_c} \approx -\left(1 - \frac{T}{T_c} \right).$$

$\Delta(T)$ is approximately given by

$$\Delta(T) \approx c_2 k_B T_c \sqrt{1 - \frac{T}{T_c}}, \tag{3.24}$$

where

$$c_2 = \left[\frac{\langle \gamma_\varphi^2 \rangle_{FS}}{\langle \gamma_\varphi^4 \rangle_{FS} g_0} \right]^{1/2}. \tag{3.25}$$

c_2 equals 3.063 and 3.537 for the s- and d-wave superconductors, respectively.

The gap equation in the whole superconducting phase can be solved numerically. Fig. 3.1 shows the temperature dependence of $\Delta(T)$ for both s- and d-wave superconductors. The difference between these two kinds of superconductors is small. The main difference occurs in low temperatures, where $\Delta(T)$ scales as T^3 in the d-wave state but varies exponentially with temperature in the s-wave state.

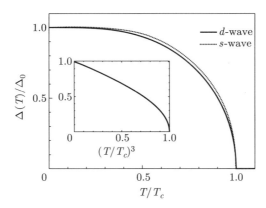

Figure 3.1   Temperature dependence of the gap parameter $\Delta(T)/\Delta_0$. The inset shows $\Delta(T)/\Delta_0$ as a function of $(T/T_c)^3$ for the d-wave superconductor.

3.2 Density of States

The density of states $\rho(\omega)$ of superconducting quasiparticle excitations is an important quantity of superconductors. Given the quasiparticle energy dispersion, $E_k = \sqrt{\xi_k^2 + \Delta_k^2}$, the density of states is defined as

$$\rho(\omega) = \frac{1}{V}\sum_k \delta(\omega - E_k) = \int_0^{2\pi}\frac{d\varphi}{2\pi}\int d\xi\, \rho_0(\xi)\delta(\omega - \sqrt{\xi^2 + \Delta^2\gamma_\varphi^2}). \quad (3.26)$$

In the low energy limit, $\rho_0(\xi) \approx N_f$, the above expression can be written as

$$\rho(\omega) = N_F \int_0^{2\pi}\frac{d\varphi}{2\pi}\mathrm{Re}\frac{\omega}{\sqrt{\omega^2 - \Delta^2\gamma_\varphi^2}}. \quad (3.27)$$

In the isotropic s-wave superconductor, $\gamma_\varphi = 1$. Eq. (3.27) shows that the density of states is finite only when $\omega > \Delta$ as a result of the opening of an isotropic energy gap on the whole Fermi surface:

$$\rho(\omega) = \frac{N_F\omega\theta(\omega - \Delta)}{\sqrt{\omega^2 - \Delta^2}}. \quad (3.28)$$

As expected, $\rho(\omega)$ approaches the normal state density of states N_F in the limit $\omega \gg \Delta$. However, right at the gap edge, $\omega = \Delta$, $\rho(\omega)$ diverges as $(\omega - \Delta)^{-1/2}$.

This divergence affects strongly physical properties of superconductors. For example, it yields a coherence peak in the spin-lattice relaxation rate of nuclear magnetic resonances (NMR) at the critical transition temperature, and a divergence in the optical conductivity at $\omega = 2\Delta$. The divergence happens at $\omega = 2\Delta$ not at $\omega = \Delta$ because in the optical conductivity measurement it is always a pair of quasiparticles are excited due to the momentum conservation.

In the *d*-wave superconductor, the density of states is determined by

$$\rho(\omega) = N_F \int \frac{d\varphi}{2\pi} \mathrm{Re} \frac{\omega}{\sqrt{\omega^2 - \Delta^2 \cos^2(2\varphi)}}. \tag{3.29}$$

When $\omega > \Delta$, $\rho(\omega)$ can be further expressed as

$$\rho(\omega) = \frac{2N_F}{\pi} \int_0^1 dx \frac{1}{\sqrt{1 - x^2}\sqrt{1 - (\Delta/\omega)^2 x^2}}, \tag{3.30}$$

the right-hand side is an elliptical integral. On the other hand, when $\omega < \Delta$, $\rho(\omega)$ is given by

$$\rho(\omega) = \frac{2N_F\omega}{\pi\Delta} \int_0^1 dx \frac{1}{\sqrt{1 - x^2}\sqrt{1 - (\omega/\Delta)^2 x^2}}. \tag{3.31}$$

As the quasiparticle dispersion is linear around the gap nodes in a *d*-wave superconductor, it is simple to show from this equation that the low energy density of states varies linearly with ω linear:

$$\rho(\omega) \approx \frac{N_F\omega}{\Delta}, \qquad \text{at} \quad \omega \ll \Delta. \tag{3.32}$$

This is a characteristic property of the *d*-wave or other two-dimensional superconductors with gap nodes. It has strong impact on low temperature properties of superconductors.

As ω approaches Δ from high frequency, the elliptical integral in Eq. (3.30) can be approximately integrated out, we obtain

$$\rho(\omega \to \Delta^+) \approx \frac{N_F}{\pi} \ln \frac{8}{1 - \Delta/\omega}. \tag{3.33}$$

As ω approaches Δ from low frequency, the density of states is given by

$$\rho(\omega \to \Delta^-) \approx \frac{N_F\omega}{\pi\Delta} \ln \frac{8}{1 - \omega/\Delta}. \tag{3.34}$$

Thus from either direction, $\rho(\omega)$ diverges logarithmically at $\omega = \Delta$, weaker than the square root divergence in the s-wave superconductor(see Fig. 3.2). This divergence of $\rho(\omega)$ can also induce a coherence peak in the NMR spin-lattice relaxation rate at the superconducting transition temperature and a divergent optical conductivity at $\omega = 2\Delta$. However, in real d-wave superconductors, this divergence is often smeared out by strong coupling effects or by impurity scattering, and is difficult to be observed experimentally.

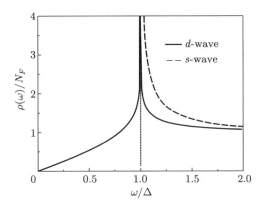

Figure 3.2   The quasiparticle density of states of the s- and d-wave superconductors.

3.3 Entropy

The entropy is an important quantity characterizing thermodynamic fluctuations. It can be determined from the free energy F and the internal energy U of quasiparticles. In the weak-coupling BCS theory, the free-energy is given by

$$F = -\frac{2}{V\beta} \sum_{k} \ln\left(1 + e^{-\beta E_k}\right).$$ (3.35)

The prefactor 2 comes from the spin degeneracy of quasiparticles. Using the quasiparticle density of states, the free energy can be also expressed as

$$F = -\frac{2}{\beta} \int d\omega \rho(\omega) \ln\left(1 + e^{-\beta\omega}\right).$$ (3.36)

Clearly, the temperature dependence of the free energy is purely determined by the quasiparticle density of states. From the free energy, one can in principle determine all other thermodynamic quantities.

The internal energy U can be also calculated directly from the quasiparticle density of states. From the definition, we have

$$U = \frac{2}{V} \sum_k E_k f(E_k) = 2 \int d\omega \rho(\omega) f(\omega). \tag{3.37}$$

The entropy is determined by the ratio between $U - F$ and temperature

$$S = \frac{U - F}{T}.$$

Substituting Eqs. (3.36) and (3.37) into the above equation, we obtain

$$S = \frac{2}{VT} \sum_k \left[E_k f(E_k) + k_B T \ln\left(1 + e^{-\beta E_k}\right) \right],$$

$$= 2k_B^2 T \int dx \rho(k_B T x) \left[\frac{x}{1 + e^x} + \ln\left(1 + e^{-x}\right) \right]. \tag{3.38}$$

The temperature dependence of the entropy can be evaluated numerically based on the above formulae. Fig. 3.3 compares the entropy as a function of temperature for the s- and d-wave superconductors. The difference lies mainly in the low temperature region where the entropy of the s-wave superconductor drops much faster than the d-wave one. As the low energy density of states is linear for the d-wave superconductor:

$$\rho(k_B T x) \approx \frac{N_F k_B T x}{\Delta(T)}, \tag{3.39}$$

the low temperature entropy varies quadratically with temperature

$$S \approx \frac{\alpha_1 N_F k_B^3 T^2}{\Delta(T)}, \qquad T \ll T_c. \tag{3.40}$$

This is different from the activated temperature dependence of the entropy in the s-wave superconductor. In Eq. (3.40),

$$\alpha_1 = 2 \int_0^\infty dx \left[\frac{x^2}{1 + e^x} + x \ln\left(1 + e^{-x}\right) \right] = \frac{9\zeta(3)}{2} \approx 5.41.$$

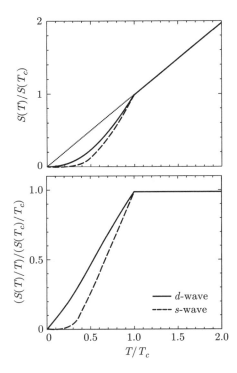

Figure 3.3   Temperature dependence of the normalized entropy for the d- and s-wave super-conductors.

Corresponding to this quadratic temperature dependence of the entropy, both the free energy F and the internal energy U scale as T^3 in low temperatures.

Below but close to the critical temperature T_c, Δ_k is small and the entropy S can be expanded in terms of Δ_k. To the second order approximation in Δ_k, the entropy of the d-wave superconductor is

$$S(T) \approx S_N(T) - \frac{1}{V k_B T^2} \sum_k \frac{e^{\beta \xi_k} \Delta_k^2}{\left(1 + e^{\beta \xi_k}\right)^2}$$
$$\approx S_N(T) - c_1^2 k_B^2 N_F \langle \gamma_\varphi^2 \rangle (T_c - T), \tag{3.41}$$

where $S_N(T)$ is the normal state entropy obtained by linear extrapolation from the entropy above T_c. The superconducting state is more ordered than the normal state. Thus its entropy is smaller than the extrapolated normal state value

$S_N(T)$. As shown in Fig. 3.3, $S(T)/T$ varies linearly with temperature in both *s*- and *d*-wave superconductors, but with different slopes, when T approaches T_c in the superconducting state.

Above the critical temperature, the superconductivity disappears, and $S(T)$ equals S_N. Assuming $\rho_0(\omega) \approx N_F$, independent on the energy, it is simple to show that $S_N(T)$ varies linearly with temperature:

$$S_N \approx \frac{2\pi^2}{3} N_F k_B^2 T, \qquad T > T_c. \tag{3.42}$$

It crosses the origin of coordinates if it is extrapolated to zero temperature.

In the superconducting state, the entropy is lowered due to the Cooper pair condensation. But the entropy should recover its normal state value at T_c. This is a consequence of the conservation of total degrees of freedom. If at T_c, the entropy does not reach the extrapolated value based on the high temperature data, it simply means that there is an entropy loss below T_c. This entropy loss does not exist in the conventional BCS superconductors. It must result from other physical effects, such as competing orders.

Fig. 3.4 shows the temperature dependence of the entropy for $YBa_2Cu_3O_{6+x}$ at several different doping levels [54]. The entropy is obtained from the temperature integration of the specific heat measured by experiments. In the overdoped regime, for example the curve of $x = 0.97$, the entropy behaves similarly as in an ideal BCS superconductor shown in Fig. 3.3. Above T_c, the entropy varies linearly with temperature and is extended to the origin if it is extrapolated to low temperatures.

In the underdoped regime, for example the curve of $x = 0.38$, the entropy behaves differently. First, there is not any cusp on the entropy curve at T_c. It is impossible to determine T_c based on the singularity in the entropy. Second, if we linearly extrapolate the entropy curve around $T \approx 300K$ to low temperatures, the extrapolated line will have a negative interception at zero temperature. This means that there is an entropy loss even above the superconducting transition temperature. This entropy loss is clearly not due to the superconducting condensation.

The suppression of the low energy density of states in the normal state in

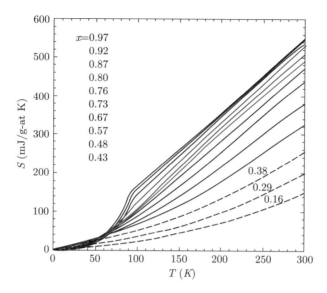

Figure 3.4   Temperature dependence of the entropy of $YBa_2Cu_3O_{6+x}$ at different doping levels. The larger is x, the higher is the doping level. (The experimental data are from Ref. [54])

high-T_c cuprates suggests that an energy gap similar to the superconducting gap exists in the normal state. It is this normal state gap that is responsible for the suppression of the low density of states of electrons. But unlike the superconducting gap, there is no coherent condensation associated with this normal state energy gap. In literature, this gap is called pseudogap. The pseudogap affects strongly physical properties of high-T_c cuprates. It yields various anomalous behaviors in the specific heat, magnetic susceptibility, resistance, and many other thermodynamic and transport coefficients. The entropy loss in the normal state is one of them.

3.4 Specific Heat

The specific heat is determined by the temperature derivative of the entropy. From Eq. (3.38), we have

$$C = T\frac{\partial S}{\partial T} = \frac{2}{V}\sum_k E_k \frac{\partial f(E_k)}{\partial T}. \tag{3.43}$$

It should be noted that the energy dispersion of the superconducting quasi-particle E_k is temperature dependent, i.e. $\partial E_k/\partial T \neq 0$, and $C \neq \partial U/\partial T$ since the particle number is not conserved in the superconducting state. The above expression can be also written as

$$C = \frac{2}{k_B V}\sum_k \frac{e^{\beta E_k}}{(1+e^{\beta E_k})^2}\left(\frac{E_k^2}{T^2} - \frac{E_k}{T}\frac{\partial E_k}{\partial T}\right). \tag{3.44}$$

In low temperatures, Δ_k depends weakly on temperature and the specific heat is simply proportional to the number of quasiparticles within the energy scale of $k_B T$. In this case, the specific heat is mainly contributed by the first term in Eq. (3.44). Since the low-energy density of states is linear in a d-wave superconductor, the low temperature specific heat varies quadratically with temperature

$$C \approx \frac{2\alpha_1 N_F k_B^3 T^2}{\Delta(T)}. \tag{3.45}$$

This can be compared with the low temperature specific heat of s-wave super-conductors. In an s-wave superconductor, since the low energy density is zero, the specific heat decays exponentially with temperature in low temperatures:

$$C \approx \frac{2\sqrt{2\pi}N_F \Delta_0^{5/2}}{k_B T^{3/2}}e^{-\Delta_0/k_B T}. \tag{3.46}$$

Around T_c, $\Delta(T)$ is given by Eq. (3.24). The derivative of E_k with respect to temperature is finite, and the specific heat has a finite jump at T_c. Above T_c, C is just the specific heat of electrons in the normal state:

$$C_N(T) = \frac{2}{k_B T^2 V}\sum_k \frac{\xi_k^2 e^{\beta \xi_k}}{(1+e^{\beta \xi_k})^2} \approx \frac{2\pi^2}{3}k_B^2 N_F T, \tag{3.47}$$

independent on the pairing symmetry. The jump of the specific heat at T_c,

$$\Delta C(T_c) = C(T_c^-) - C(T_c^+) = C(T_c^-) - C_N(T_c)$$

is determined by

$$\Delta C(T_c) = \frac{c_1^2 k_B}{V}\sum_k \frac{e^{\beta \xi_k}}{(1+e^{\beta \xi_k})^2}\gamma_\varphi^2 \approx c_1^2 k_B^2 N_F T_c \langle \gamma_\varphi^2 \rangle_{FS}. \tag{3.48}$$

Using the result of $C_N(T)$, we find the ratio between ΔC and C_N at T_c to be

$$\frac{\Delta C(T_c)}{C_N(T_c)} = \frac{3c_1^2 \langle \gamma_\varphi^2 \rangle_{FS}}{2\pi^2}. \tag{3.49}$$

This ratio equals 1.43 and 0.95 for the s- and d-wave superconductors, respectively. The specific heat jump of the d-wave superconductor is smaller than that of the s-wave one.

The above discussion indicates that there are two major differences in the specific heat between s- and d-wave superconductors. First, the specific heat of the s-wave superconductor decays exponentially in low temperatures. But it drops only quadratically with temperature in the d-wave superconductor. Second, the specific heat jump in the s-wave superconductor is larger than that in the d-wave one. These differences can be clearly seen from Fig. 3.5, in which the temperature dependence of the specific heat coefficient, $\gamma = C/T$, is depicted.

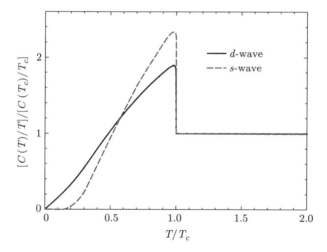

Figure 3.5   Temperature dependence of C/T for the d-wave and s-wave superconductors.

Fig. 3.6 shows the specific heat coefficient γ as a function of temperature for $YBa_2Cu_3O_{6+x}$ at different doping obtained from the differential heat capacity measurements [54]. Similar to the entropy, the specific heat behaves differently in the overdoped and underdoped regimes. In fact, the entropy curves shown in

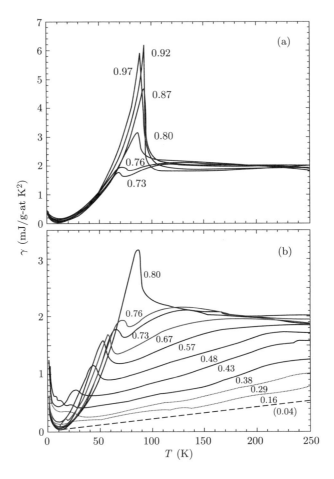

Figure 3.6   Temperature dependence of the specific heat coefficient $\gamma = C(T)/T$ for $YBa_2Cu_3O_{6+x}$ at several different doping levels. (From reference [54])

Fig. 3.4 were obtained by the temperature integration of the specific heat data shown in Fig. 3.6. In the overdoped regime, for example the case $x = 0.97$, the experimental result agrees qualitatively with the theoretical prediction for the *d*-wave superconductor shown in Fig. 3.5. However, there are two differences. First the specific heat jump at T_c is larger than the theoretical value for the *d*-wave superconductor, but closer to the value for the *s*-wave superconductor. It is unknown whether this relatively large jump is an intrinsic property of high-T_c

superconductivity or is simply due to measurement errors. Second, $\gamma(T)$ at low temperatures in the d-wave superconductor should be a linear function of T. But experimental data scale as T^2. This difference may result from the disorder effect. In the presence of impurity scattering, the temperature dependence of the low energy density of states changes from linear to quadratic, which can alter the low temperature specific heat coefficient γ changes from T to T^2. A detailed discussion on the correction of impurity scattering to the specific heat is given in Chapter 8.

In the underdoped regime, the specific heat jump at T_c is dramatically suppressed. Unlike in the overdoped samples, the specific heat coefficient begins to drop above T_c. Moreover, the temperature at which the specific heat coefficient begins to drop increases with decreasing doping. In the superconducting state, the decrease of γ is due to the opening of the quasiparticle energy gap. The drop of γ in the normal state implies a suppression in the normal state density of states, which is a manifestation of the pseudogap effect. As the microscopic origin of pseudogap is unknown, it is difficult to perform a quantitative analysis for the temperature dependence of the specific heat in the underdoped regime.

In low temperatures, the specific heat of $YBa_2Cu_3O_{6+x}$ shows an upturn. This is probably caused by magnetic impurities. In $La_{2-x}Sr_xCuO_{4+\delta}$, the contribution from magnetic impurities is significantly weakened, and the specific heat curve does not show any upturn. Instead, it follows the T^2-law in low temperatures [55] as predicted, in support of theory of d-wave superconductivity.

3.5 Gap Operators in the Continuum Limit

In a homogeneous system, the gap function of d-wave superconductors is diagonal in momentum space. However, if the translational symmetry is broken by, for example, vortex lines or disorders, it is more convenient to study the superconducting state using the Bogoliubov-de Gennes equation in real space. For high-T_c cuprates, the coherence length is short, the Bogoliubov-de Gennes equation can be discretized according to the lattice symmetry and solved numerically. However, the physical properties of low energy quasiparticles are not sensitive to the lattice structure because their de Broglie wavelengthes are very long. In this case, the Bogoliubov-de Gennes equations can be simplified by linearizing the Hamiltonian in the continuum limit. This is a commonly used approach in the study of low energy electromagnetic response functions and scaling behaviors of d-wave superconductors. To do this, we define the energy gap operator $\hat{\Delta}$ through its action on the wavefunction $\psi(r)$ as

$$\hat{\Delta}\psi(r_1) = \int dr_2 \Delta \left(\frac{r_1 + r_2}{2}, r_1 - r_2 \right) \psi(r_2). \tag{3.50}$$

The order parameter Δ is determined by the gap equation

$$\Delta(R, r) = -v(r)\langle \psi_\uparrow(r_1)\psi_\downarrow(r_2)\rangle, \tag{3.51}$$

where $R = (r_1 - r_2)/2$ and $r = r_1 - r_2$. $v(r)$ is the pairing interaction between electrons.

Using the gap operator, the Bogoliubov-de Gennes equation (1.25) can be expressed as

$$i\hbar\partial_t\psi = H\psi, \tag{3.52}$$

where $\psi = (u, v)^T$ and ε_F is the Fermi energy. H is the BCS mean-field Hamiltonian defined by

$$H = \begin{pmatrix} \hat{h} + U(r) - \varepsilon_F & \hat{\Delta} \\ \hat{\Delta}^\dagger & -\hat{h}^* - U(r) + \varepsilon_F \end{pmatrix}. \tag{3.53}$$

$U(r)$ is a scattering potential and $\hat{h}$ is the kinetic energy operator:

$$\hat{h} = \frac{1}{2m}\left(-i\hbar\nabla - \frac{e}{c}A\right)^2.$$

In order to determine the expression of the gap operator of d-wave superconductors, we take the Fourier transformation for the relative coordinate r and rewrite the gap parameter Δ as

$$\Delta(R, k) = \int dr\Delta(R, r)e^{ik\cdot r}.$$

The pairing symmetry is determined by the relative momentum dependence of $\Delta(R, k)$. For the $d_{x^2-y^2}$-wave superconductor, $\Delta(R, k)$ in the continuum limit is defined by

$$\Delta(R, k) = \Delta_0(R) \frac{k_x^2 - k_y^2}{k_F^2}, \tag{3.54}$$

where k_F is the Fermi momentum. $\Delta_0(R)$ measures the center of mass distribution of the order parameter, independent of pairing symmetry.

Taking an inverse transformation to convert $\Delta(R, k)$ back to the coordinate space, we find that the gap operator can be expressed as

$$k_F^2 \hat{\Delta}\psi(r_1)$$
$$= -\int dr_2 \Delta_0 \left(\frac{r_1 + r_2}{2} \right) \left(\partial_{xx}^2 - \partial_{yy}^2 \right) \delta(r_1 - r_2)\psi(r_2),$$
$$= -\partial_{xx}^2 \Delta_0(r_1)\psi(r_1) + \partial_x \frac{\partial \Delta_0(r_1)}{\partial x}\psi(r_1) - \frac{1}{4}\frac{\partial^2 \Delta_0(r_1)}{\partial x^2}\psi(r_1) - (\partial_x \to \partial_y).$$

Using the identity

$$\{\partial_x, f(r)\} = 2\frac{\partial}{\partial x}f(r) - \frac{\partial f(r)}{\partial x},$$

we can further express $\hat{\Delta}$ as

$$\hat{\Delta} = \frac{1}{4p_F^2}\{p_x, \{p_x, \Delta_0(r)\}\} - \frac{1}{4p_F^2}\{p_y, \{p_y, \Delta_0(r)\}\}, \tag{3.55}$$

where $p_F = \hbar k_F$. p_x and p_y are momentum operators. $\{a, b\} = ab + ba$ is the anti-commutator.

In the study of low energy properties, only low energy quasiparticles around the gap nodes are important. Therefore, we can expand the Hamiltonian in the nodal region, by just keeping the terms linear in momentum and neglecting all other high order terms. The linearization needs to be performed around each of the four nodes in the $d_{x^2-y^2}$-wave superconductor. Here, as an example, we consider the expansion around the node at $k_1 = (k_F/\sqrt{2}, k_F/\sqrt{2})$. It is straightforward to generalize the derivation to other three nodal points.

The linearization is to perform a Galilean transformation to change the origin of coordinates to the frame that moves with the wave vector k_1. The

Hamiltonian is projected onto this reference frame by expressing the wavefunction $\psi(r)$ as a product of the plane-wave of momentum k_1 and a wavefunction
$\tilde{\psi}(r)$ in the new reference frame

$$\psi(r) = e^{ik_F(x+y)/\sqrt{2}}\tilde{\psi}(r). \tag{3.56}$$

The linearization is to find the equation that $\tilde{\psi}(r)$ satisfies.

Substituting this expression into Eq. (3.52) and keeping the leading order
terms in momentum, we then obtain the following linearized Bogoliubov-de
Gennes equation

$$i\hbar\partial_t\tilde{\psi} \approx \hat{H}_0\tilde{\psi}, \tag{3.57}$$

where $\hat{H}_0$ is given by

$$\hat{H}_0 = \begin{pmatrix} \frac{v_F}{\sqrt{2}}(\hat{x}+\hat{y})\cdot\left(p-\frac{e}{c}A\right)+U & \frac{1}{\sqrt{2}p_F}\{p_x-p_y,\Delta_0(r)\} \\ \frac{1}{\sqrt{2}p_F}\{p_x-p_y,\Delta_0^*(r)\} & \frac{v_F}{\sqrt{2}}(\hat{x}+\hat{y})\cdot\left(p+\frac{e}{c}A\right)-U \end{pmatrix}. \tag{3.58}$$

The Hamiltonian that is neglected,

$$\hat{H}_1 = \begin{pmatrix} \hat{h} & \hat{\Delta} \\ \hat{\Delta}^\dagger & -\hat{h}^* \end{pmatrix}, \tag{3.59}$$

contains only the higher order terms in momentum.

Around the gap nodes, the energy dispersion of the quasiparticle excitations
is approximately given by the eigenvalues of $\hat{H}_0$:

$$\varepsilon_p = \sqrt{(v_F\hbar k_\perp)^2 + \left(\frac{\Delta_0 k_\parallel}{k_F}\right)^2}, \tag{3.60}$$

where $k_\perp$ and $k_\parallel$ are the momenta parallel and perpendicular to the tangent
direction of the Fermi surface at the nodal point, respectively. At a given temperature T, the thermal energy scales linearly with T:

$$\langle\hat{H}_0\rangle \approx \varepsilon_p \approx k_BT. \tag{3.61}$$

This implies that the corresponding wavevectors have the scaling properties

$$k_\perp \approx \frac{T}{\sqrt{2}v_F}, \tag{3.62}$$

$$k_\parallel \approx \frac{T\hbar k_F}{\sqrt{2}\Delta_0} \gg k_\perp. \tag{3.63}$$

The inequality holds under the condition $\varepsilon_F \gg \Delta_0$. The energy scale of $\hat{H}_1$ can be estimated using the above expressions. As $k_\parallel \gg k_\perp$, it can be shown that the leading term in $\hat{H}_1$ is given by

$$\langle \hat{H}_1 \rangle \approx \frac{1}{2} \left(\frac{k_B T}{\Delta_0} \right)^2 \varepsilon_F. \tag{3.64}$$

Thus the ratio between $\langle \hat{H}_1 \rangle$ and $\langle \hat{H}_0 \rangle$ scales linearly with temperature:

$$\frac{\langle \hat{H}_1 \rangle}{\langle \hat{H}_0 \rangle} \approx \frac{k_B T \varepsilon_F}{2\Delta_0^2}. \tag{3.65}$$

By requesting $\langle \hat{H}_1 \rangle \ll \langle \hat{H}_0 \rangle$, we then obtain the condition at which the linear approximation is valid:

$$T \ll \frac{2\Delta_0^2}{k_B \varepsilon_F}. \tag{3.66}$$

The above derivation can be readily extended to the d_{xy}-wave superconductor. In this case, $\Delta(R, k)$ is defined by

$$\Delta(R, k) = \Delta_0(R) \frac{k_x k_y}{k_F^2}, \tag{3.67}$$

and the corresponding gap operator is given by

$$\hat{\Delta} = -\frac{1}{4k_F^2} \left\{ \frac{\partial}{\partial y}, \left\{ \frac{\partial}{\partial x}, \Delta_0(r) \right\} \right\} = \frac{1}{4p_F^2} \left\{ p_y, \{ p_x, \Delta_0(r) \} \right\}. \tag{3.68}$$

This expression of the gap operator can be also obtained from Eq. (3.55) by taking 45°-rotation for the coordinates. The linearized Hamiltonian now becomes

$$\hat{H}_0 = \begin{pmatrix} v_F(p_x - \dfrac{e}{c} A_x) + U & \dfrac{1}{2p_F} \{ p_y, \Delta_0(r) \} \\[2ex] \dfrac{1}{2p_F} \{ p_y, \Delta_0^*(r) \} & -v_F(p_x + \dfrac{e}{c} A_x) - U \end{pmatrix}. \tag{3.69}$$

3.6 The Probability Current and Electric Current Operators

The probability density and charge density of superconducting quasiparticles for d-wave superconductors are similarly defined as for s-wave superconductors. From the probability or charge conservations, the continuity equation for the probability or charge conservation can be derived from the time-dependent Bogoliubov-de Gennes equation. From these equations, we can define the expressions of the probability current density J_P and the electric current density J_Q for d-wave superconductors. A major difference between s- and d-wave superconductors is that the gap function is non-local in the latter case, and this non-local gap function has also contribution to the probability current operator.

The derivation of J_P and J_Q for the d-wave superconductor is similar to the s-wave one. Here we skip the detail of derivation. For the $d_{x^2-y^2}$-wave superconductor, the probability current density J_P includes both the diagonal term from the kinetic energy and the off-diagonal term from the pairing energy, and is given by

$$J_P = J_P^{(1)} + J_P^{(2)}, \tag{3.70}$$

$$J_P^{(1)} = \frac{\hbar}{m}\text{Im}(u^*\nabla u - v^*\nabla v), \tag{3.71}$$

$$J_P^{(2)} = \frac{2}{\hbar k_F^2}\text{Im}\left[u^*(\hat{x}\partial_x - \hat{y}\partial_y)\Delta_0(r)v + v^*(\hat{x}\partial_x - \hat{y}\partial_y)\Delta_0^*(r)u\right], \tag{3.72}$$

where $J_P^{(1)}$ is independent on the pairing symmetry. $J_P^{(2)}$ is absent in the s-wave superconductor.

The charge current density of quasiparticles J_Q does not depend on the pairing symmetry. It has the same form as in the s-wave superconductor:

$$J_Q = \frac{e\hbar}{m}\text{Im}\left(u^*\nabla u + v^*\nabla v\right). \tag{3.73}$$

However, the supercurrent density J_S is determined by the equation

$$\nabla \cdot J_S = \frac{e}{2\hbar k_F^2}\text{Im}\left(u^*\{\partial_x, \{\partial_x, \Delta_0(r)\}\}v - v^*\{\partial_x, \{\partial_x, \Delta_0^*(r)\}\}u\right) - (\partial_x \to \partial_y). \tag{3.74}$$

For a translation invariant d-wave superconductor, $\Delta_0(r) = \Delta_0$, the quasiparticle wavefunctions $u(r)$ and $v(r)$ are given by

$$u(r) = \frac{1}{\sqrt{V}}e^{ik\cdot r}\sqrt{\frac{1}{2} + \frac{\xi_k}{2E_k}}, \tag{3.75}$$

$$v(r) = -\frac{\text{sgn}(k_x^2 - k_y^2)}{\sqrt{V}}e^{ik\cdot r}\sqrt{\frac{1}{2} - \frac{\xi_k}{2E_k}}. \tag{3.76}$$

If k is real, the wavefunctions do not decay, and the probability current and the electric current vectors become

$$J_P^{(1)} = \frac{\hbar \xi_k k}{m V E_k},$$ (3.77)

$$J_P^{(2)} = -\frac{2\Delta_0 \Delta_k (k_x \hat{x} - k_y \hat{y})}{\hbar k_F^2 V E_k},$$ (3.78)

$$J_Q = \frac{e\hbar k}{mV},$$ (3.79)

$$J_S = 0.$$ (3.80)

Compared with the results for s-wave superconductors, only the definition of the probability current vector is changed. It contains an extra term $J_P^{(2)}$. All other terms are unchanged.

Chapter 4
Quasiparticle Excitation Spectra

4.1 Single-Particle Spectral Function

The single-particle spectral function $A(k, \omega)$ is an important quantity characterizing physical properties of interacting electrons. It measures the weight of
a system after adding an electron with a given momentum k and energy ω, or
removing a hole with opposite momentum and energy. $A(k, \omega)$ in the superconducting state contains a great deal of information on the quasiparticle excitations. It can be used to extract the energy-momentum dispersion relation, the
momentum dependence of the energy gap, and the scattering lifetime of superconducting quasiparticles. It is also a basic parameter for describing interaction
effects and electromagnetic response functions. Experimentally, $A(k, \omega)$ can be
measured through the angle-resolved photoemission spectroscopy (ARPES).

In order to study the behavior of single-particle excitations in d-wave superconductors, let us define the Matsubara Green's function of superconducting
quasiparticles at finite temperatures as

$$G(k, \tau - \tau') = - \left\langle T_\tau \begin{pmatrix} c_{k\uparrow}(\tau) \\ c_{-k\downarrow}^\dagger(\tau) \end{pmatrix} \left(c_{k\uparrow}^\dagger(\tau') \; c_{-k\downarrow}(\tau') \right) \right\rangle. \tag{4.1}$$

$G(k, \tau)$ is a 2×2 matrix. Its diagonal terms are the normal propagators of
electrons. The off-diagonal terms of $G(k, \tau)$ are the anomalous propagators of
electrons which contain the information of superconducting pairing. Here the
Nambu spinor representation of electrons is invoked, which treats electron creation and annihilation operators on the equal footing. It provides a convenient
representation to study superconducting states in which the fermion number is
not conserved. The Green's function in the Nambu representation can be similarly treated as for the conventional Green's function in a system with electron

number conservation.

Taking the Fourier transformation with respect to τ, we obtain the Matsubara Green's function in the frequency space as

$$G(k, i\omega_n) = \int_0^\beta d\tau e^{i\omega_n \tau} G(k, \tau). \tag{4.2}$$

In the BCS weak coupling limit, the single-electron Green's function without considering the correction from disorder or other physical effects can be derived by diagonalizing the mean-field Hamiltonian defined by Eq. (1.11). It gives

$$G^{(0)}(k, i\omega_n) = \frac{1}{i\omega_n - \xi_k \sigma_3 - \Delta_k \sigma_1}, \tag{4.3}$$

where σ_1 and σ_3 are Pauli matrices.

This Green's function, as indicated by the superscript "0" in $G^{(0)}(k, \omega)$, can be used as the 0'th order or unperturbed Green's function to analyze electromagnetic, impurity, and thermodynamic response functions of superconducting electrons. Clearly, the mean-field Green's function $G^{(0)}(k, \omega)$ will be modified in the presence of electron-electron interaction, impurity, and phonon scatterings. The renormalized Green's function is determined by the Dyson equation

$$G^{-1}(k, i\omega_n) = \left[G^{(0)}(k, i\omega_n)\right]^{-1} - \Sigma(k, i\omega_n). \tag{4.4}$$

$\Sigma(k, i\omega_n)$ is the self-energy. It is an important quantity that describes the effect not included in the BCS mean-field approximation. But it is generally difficult to calculate it rigorously.

Physical measurement quantities are related to retarded Green's functions. They can be obtained from the Matsubara Green's function through analytical continuation by taking the Wick rotation

$$G^R(k, \omega) = G(k, i\omega_n \to \omega + i0^+). \tag{4.5}$$

The imaginary part of the diagonal component $G^R(k, \omega)$ is proportional to the spectral function of electrons,

$$A(k, \omega) = -\frac{1}{\pi} \text{Im} G_{11}^R(k, \omega). \tag{4.6}$$

The integral of $A(k,\omega)$ over the momentum k equals the electron density of states

$$\rho(\omega) = \frac{1}{V} \sum_k A(k,\omega). \tag{4.7}$$

On the other hand, the integral of $A(k,\omega)$ over the frequency gives rise to the following sum rules:

$$\int_{-\infty}^{\infty} d\omega A(k,\omega) = 1, \tag{4.8}$$

$$\int_{-\infty}^{\infty} d\omega f(\omega) A(k,\omega) = n(k) = \sum_\sigma \langle c_{k\sigma}^\dagger c_{k\omega} \rangle, \tag{4.9}$$

where $n(k)$ is the momentum distribution function of electrons.

For an ideal BCS superconductor, the self energy vanishes and the spectral function reads

$$A^{(0)}(k,\omega) = u_k^2 \delta(\omega - E_k) + v_k^2 \delta(\omega + E_k), \tag{4.10}$$

where u_k and v_k are defined by Eq. (1.16) and Eq. (1.17), respectively. The two δ-functions represent the contribution from electron and hole excitations with the corresponding spectra weight given by u_k^2 and v_k^2, respectively.

4.2 ARPES

When light shines on a solid, if its frequency exceeds a threshold determined by the work function of that material, electrons are emitted out from the surface. This is nothing but the photoelectric effect first discovered by Hertz in 1887. The measurement on the cross-section, or the electric current density of photoelectrons with specific momentum and energy, gives rise to the angle-resolved photoemission spectra. This measurement reveals the property of electron spectral functions. ARPES is an important method to analyze properties of superconducting electrons. It has played an important role in the study of high-T_c superconductivity.

In an ARPES measurement, the photoelectric current is proportional to the flux of incoming light, or, the square of the vector potential $|A(r,t)|^2$ of the light. The photoelectric spectra measure the non-linear response to an external electromagnetic field. We can treat the interaction between incident photons and electrons as a perturbation if the light intensity is not too strong, and evaluate the photoelectric current intensity using the perturbation theory. For this purpose, we divide the Hamiltonian into two parts:

$$H = H_0 + H_{int}, \tag{4.11}$$

where H_0 is the Hamiltonian of electrons, and H_{int} is the minimal electron-photon interaction defined by

$$H_{int} = \int dr A(r,t) \cdot J(r). \tag{4.12}$$

$A(r,t)$ is the vector potential of the external electromagnetic field. $J(r)$ is the electric current density operator. r is the coordinate of electron. We neglect the interaction among electrons already escaping the surface and their interactions with the external electromagnetic field.

We consider the perturbative contribution from H_{int} to the photoelectric current density $\langle J(R,t)\rangle$ at the location of detector R. At the zeroth order, the photoelectric current is proportional to the expectation value $\langle J(R,t)\rangle$ of the current operator under the ensemble average with respect to H_0. This contribution is zero since there is no current in the detector in the absence of applied electromagnetic field. The first order perturbation is proportional to $\langle J_\alpha(R,t)J_\beta(r',t')\rangle$, or $\langle J_\beta(r',t')J_\alpha(R,t)\rangle$. However, the matrix elements of $J_\alpha(R,t)$ is nonzero if and only if there is an electron at R. Thus, $\langle J_\alpha(R,t)J_\beta(r', t')\rangle = \langle J_\beta(r',t')J_\alpha(R,t)\rangle = 0$, and the first order perturbation of H_{int} has no contribution either. The finite contribution starts from the second order perturbation.

Using the theory of second order perturbation, it can be shown that the photoelectric current is determined by the correlation function of three current operators [56, 57]:

$$\langle J_\alpha(R,t)\rangle \propto \sum_{\mu,\nu} \int dr'dt'dr''dt'' A_\mu(r',t')A_\nu(r'',t'')$$
$$\langle J_\mu(r',t')J_\alpha(R,t)J_\nu(r'',t'')\rangle. \tag{4.13}$$

As there are no photoelectric currents at R in both the initial and final states, $J_\alpha(R,t)$ should be sandwiched between $J_\mu(r',t')$ and $J_\nu(r'',t'')$, otherwise the above current-current-current correlation function should be zero:

$$\langle J_\mu(r',t')J_\nu(r'',t'')J_\alpha(R,t)\rangle = \langle J_\alpha(R,t)J_\mu(r',t')J_\nu(r'',t'')\rangle = 0.$$

Eq. (4.13) is the basic formula for evaluating photoelectric current intensity. However, in order to accurately calculate the correlation function of three current operators, we need to know accurately the band structure of electrons. At the same time, we need also consider comprehensively the corrections to this correlation function from electron-electron, electron-phonon, electron-impurity, and electron-surface interactions. This type of calculation is formidable for strongly correlated electronic systems. Moreover, the physical picture governing this formula is not transparent and in some case even counter-intuitive.

The most commonly used model for the interpretation of photoemission spectra is the so-called three-step model [58, 59, 36]. It is a phenomenological approach, which has nonetheless proven to be quite successful. It breaks up the complicated photoemission process into three steps: (1) the excitation of electrons by the incident light, (2) its passage and relaxation from the solid to the surface, and (3) penetration through the surface into the vacuum, where it is detected. The intensity of the photoelectric current is determined by the product of the probabilities for these three processes, namely the probability of electron excited by the light, the scattering probability of the excited electrons during the propagation from the bulk to the surface, and the probability of tunneling through the surface barrier to the vacuum. Compared to the complicated perturbative result of Eq. (4.13), this simplified model neglects the interference effect between the bulk electrons and the excited surface electrons, as well as the quantum interference effect of electrons during the relaxation within the bulk. Nevertheless, the physical picture revealed by this simplified model is clear. It contains all essential points of photoelectric scattering and is a valuable empirical model for analyzing the ARPES experiments.

In the three-step model, the angle-resolved photoelectric spectra are obtained by calculating the scattering probability of photoelectrons. The scattering process of photoelectrons is determined by the energy and momentum of electrons in both the initial and final states, and the energy and direction of the incident photon. From the Fermi golden rule, the transition probability is approximately found to be

$$w_{fi} = \frac{2\pi}{\hbar} \left| \langle \Psi_f^N | H_{int} | \Psi_i^N \rangle \right|^2 \delta \left(E_f^N - E_i^N - h\nu \right), \tag{4.14}$$

where ν is the frequency of the incident photon. The energy of the final state is the sum of the kinetic energy of the outgoing electron E_{kin}, the surface work function ϕ, and the total energy after one electron escaping from the system:

$$E_f^N = E_{kin} + \phi + E_f^{N-1}. \tag{4.15}$$

According to the energy conservation, the kinetic energy of the photoelectron and the frequency of the incident photon satisfy

$$h\nu = E_{kin} + \phi + E_B, \tag{4.16}$$

where $E_B = -\varepsilon_k$ is the band energy of electron with respect to the Fermi level. Using these equations, we find that w_{fi} can be expressed as

$$w_{fi} = \frac{2\pi}{\hbar} \left| \langle \Psi_f^N | H_{int} | \Psi_i^N \rangle \right|^2 \delta \left(\varepsilon_k + E_f^{N-1} - E_i^N \right). \tag{4.17}$$

Both E_{kin} and ϕ can be determined experimentally. From the measurement of the frequency of incident photon and the energy of outgoing electron, we can determine the energy dispersion of electrons ε_k.

In order to further analyze the angle-resolved photoelectron spectra, we need to simplify Eq. (4.17) using the "sudden" approximation. This approximation assumes that the time scale for the photon excited electrons to escapes from the location of excitation to the vacuum is much shorter than the typical scattering time in the bulk, so that the excited electron is not scattered by other particles, including electrons, phonons and photons, in this process. It implies that the interaction between the excited electron and other electrons is negligible and the wave function of the final state is a direct product of the wave function of excited photoelectron and that of other electrons:

$$\Psi_f^N = \mathcal{A}\phi_f^k \Psi_f^{N-1}, \tag{4.18}$$

where $\mathcal{A}$ is the anti-symmetrization operator to ensure the wavefunction to be completely antisymmetric under the exchange of any two electrons. ϕ_f^k is the wave function of the photoelectron. Ψ_f^{N-1} is the wave function of the $N-1$ electrons not being excited by photons. It can be at any one of the eigenstates of electrons Ψ_m^{N-1}. To obtain the total transition probability, we need to sum over all possible eigenstates Ψ_m^{N-1} of these $N-1$ electrons.

The "sudden" approximation can be examined by comparing the escape-time with the average scattering time of electrons in the bulk. The kinetic energy of photoelectron is typically of the order of 20 eV and the corresponding velocity is about $v \approx 3 \times 10^8$cm/s. The escape depth of photoelectrons is typically of the order of 10Å. Thus the escape-time is approximately of the order $t_e \approx 3 \times 10^{-16}$sec. In solids, the characteristic scattering time induced by electron-electron interactions can be estimated from the plasma frequency ω_p. For high-T_c cuprate, $\omega_p \sim 1$eV, and the corresponding interaction time scale is approximately $t_s = 2\pi/\omega_p \sim 4 \times 10^{-15}$s. As t_e is about one order smaller than t_s, the "sudden" approximation is justified.

In the treatment of the initial state, the independent particle approximation is generally used. The initial wave function is a direct product of the wave function of the excited electron and that of other electrons:

$$\Psi_i^N = \mathcal{A}\phi_i^k \Psi_i^{N-1}, \tag{4.19}$$

and

$$\Psi_i^{N-1} = c_{k\sigma} \Psi_i^N.$$

Unlike Ψ_f^{N-1}, Ψ_i^{N-1} is generally not an eigenstate of electrons.

Under the above approximation, the scattering matrix element between the initial and final states becomes

$$\langle \Psi_f^N | H_{int} | \Psi_i^N \rangle = \langle \phi_f^k | H_{int} | \phi_i^k \rangle \langle \Psi_f^{N-1} | \Psi_i^{N-1} \rangle. \tag{4.20}$$

Substituting it into Eq. (4.17), we obtain the expression of the photoelectric current density

$$I(k, \varepsilon_k) \propto \frac{1}{Z} \sum_{i,f} e^{-\beta E_i^N} w_{f,i}$$

$$= \frac{I_0(k, \nu)}{Z} \sum_{i,f} e^{-\beta E_i^N} \left| \langle \Psi_f^{N-1} | c_{k\sigma} | \Psi_i^N \rangle \right|^2 \delta \left(\varepsilon_k + E_f^{N-1} - E_i^N \right),$$

$$(4.21)$$

where $I_0(k, \nu)$ is proportional to the single-electron dipole matrix element $\langle \phi_f^k | H_{int} | \phi_i^k \rangle$. To obtain the equality in Eq. (4.21), $I_0(k, \nu)$ is assumed to be independent on Ψ_i^{N-1} and Ψ_f^{N-1}.

The right-hand side of Eq. (4.21) is related to the electron spectral function. From the theory of Green's functions, it can be shown that the following equation holds:

$$\frac{1}{Z} \sum_{fi} e^{-\beta E_i^N} \left| \langle \Psi_f^{N-1} | c_{k\sigma} | \Psi_i^N \rangle \right|^2 \delta \left(\varepsilon_k + E_f^{N-1} - E_i^N \right) = f(\varepsilon_k) A(k, \varepsilon_k),$$

$$(4.22)$$

where $f(\varepsilon)$ is the Fermi distribution function. Substituting Eq. (4.22) into Eq.(4.21), we have

$$I(k, \omega) = I_0(k, \nu) f(\omega) A(k, \omega). \qquad (4.23)$$

This is just the formula that is used in the analysis of ARPES measurement data. $A(k, \omega)$ contains the information of electron-phonon and electron-electron interactions. It should bear in mind that $A(k, \omega)$ measured by ARPES contributes mainly from the surface states, which might be different from the bulk ones. But here we assume that they are the same for simplicity. Otherwise, the correction should be included. In the analysis of ARPRS data using Eq. (4.23), one should also consider the contribution of the background arising from the inelastic scattering of photoelectrons before they escape to the vacuum.

Eq. (4.23) shows that the photoelectron current density is proportional to the single-electron spectral function. Thus the single-electron spectra can be deduced from ARPES. This is an advantage of ARPES in comparison with other experimental measurements which probe physical quantities related to two- or multi-particle correlation functions. Theoretical analysis of measurement data in the latter case is clearly much more involved.

In Eq. (4.23), $I_0(k,\omega)$ depends on k, the polarization as well as the frequency of the incident photon. It is less sensitive to the photoelectron energy ω and temperature T. Thus the ω dependence of the ARPES lineshape is determined purely by the electron spectral function and the Fermi distribution function. In this case, it is convenient to analyze the momentum distribution function of electrons from the measurement data using the sum rule presented in Eq. (4.9).

The ARPES results are usually presented in terms of the energy distribution curve (EDC), which shows $I(k,\omega)$ as a function of ω for a given k. It is intuitive and convenient to use EDC to analyze properties of superconducting gaps and pseudogaps in high-T_c cuprates. But the EDC does not exhibit a Lorentzian lineshape. It is difficult to use it to analyze quantitatively experimental data unless we have a good understanding on the self-energy of electrons. There are two reasons that may cause the EDC lineshape non-Lorentzian. First, the Fermi distribution function $f(\omega)$ suppresses the spectral density at $\omega > 0$, which introduces an asymmetry in the lineshape. Second, the energy dependence of the spectral function $A(k,\omega)$ is generally complicated, and it may not be Lorentzian itself. Moreover, the contribution of the background to the spectra may also show a complicated ω-dependence, making a quantitative analysis of experimental data even more difficult.

In recent years, with the improvement in the momentum or angular resolution, it becomes feasible to analyze the ARPES data using the momentum distribution curve (MDC). The MDC is generally symmetric and has approximately a Lorentzian lineshape, which is more convenient to analyze than the EDC. In particular, it is more convenient to use the MDC to analyze the spectral functions of gapless excitations near the Fermi surface. Generally the spectral function of electron can be expressed as

$$A(k,\omega) = -\frac{1}{\pi}\frac{\mathrm{Im}\Sigma(k,\omega)}{[\omega - \varepsilon_k - \mathrm{Re}\Sigma(k,\omega)]^2 + [\mathrm{Im}\Sigma(k,\omega)]^2}. \qquad (4.24)$$

If we assume that $I_0(k,\omega)$ is momentum independent and $\Sigma(k,\omega)$ depends weakly on the momentum component $k_\perp$ that is perpendicular to the Fermi surface, then the MDC lineshape is a Lorentzian. The center of this Lorentzian peak is located at $k = k_F + [\omega - \mathrm{Re}\Sigma(\omega)]/v_F^0$ and the half-width is given by $\mathrm{Im}\Sigma(\omega)/v_F^0$. The peak position measures the renormalized energy-momentum dispersion of electrons. The half-width is proportional to the scattering rate or the inverse lifetime of electrons. Moreover, the contribution from the background scattering to the MDC depends weakly on momentum. Therefore, MDC provides a useful approach to extract information on the self-energy from ARPES.

The ARPES is an unique technique that can directly detects the single-particle spectral function with both momentum and energy resolutions. It has played an important role in determining the Fermi surface structures, the characteristic energy scale of low energy excitations, the momentum dependence of the superconducting gap and the pseudogap, and the pairing symmetry for cuprate superconductors. In recent years, the momentum and energy resolutions of ARPES have been significantly improved. The ARPES has become an indispensable experimental tool for studying various anomalous physical phenomena in high-T_c cuprates and other low dimensional properties of electrons. For the experimental progress of ARPES, please refer to the recent three review articles [36, 60, 61].

4.3 Fermi Surface and Luttinger Sum Rule

The Fermi wave vector and the Fermi surface are the two basic physical quantities characterizing interacting electron systems. As only the electrons around the Fermi level can be excited in low energy, the Fermi surface structure is important to the understanding of both electromagnetic and thermodynamic properties. In a normal metallic state, the volume enclosed by the Fermi surface is fixed regardless of the geometry of the Fermi surface. This volume is proportional to the electron density and is referred to as the Luttinger volume. The formula that reveals the relationship between the volume of the Fermi surface and the density of electrons is called the Luttinger sum rule, or the Luttinger theorem, which reads

$$\frac{N}{V} = 2 \int_{G(k,\omega=0)>0} \frac{\mathrm{d}^d k}{(2\pi)^d}, \tag{4.25}$$

where d is the spatial dimension. This equation shows that in a many-electron system, the momentum space volume in which the zero energy single-particle Green's function takes positive values, is proportional to the electron density N/V, independent of interaction. This formula is applicable not only to normal electrons but also to superconducting electrons. A proof of this theorem can be found in Ref. [62, 63].

The domains with $G(k, 0) > 0$ might be connected or disconnected. No matter in which case, the domains between $G(k, 0) > 0$ and $G(k, 0) < 0$ are separated by one or a few $(d-1)$-dimensional surfaces. There are two possibilities for this surface:

(1) $G(k, 0)$ diverges on this surface, approaching $+\infty$ and $-\infty$ from both sides of the surface. In a normal metal, this surface is just the Fermi surface ususally defined. On the Fermi surface, the lifetime of quasiparticles is infinite and the imaginary part of the Green's function vanishes. When the momentum k crosses through the Fermi surface, the divergence in the real part of the zero-energy Green's function changes sign. In this case, the right-hand side of Eq. (4.25) is just the phase space volume enclosed by the Fermi surface, which equals the density of electrons. This is the statement of the Luttinger theorem in the normal metallic state. When $G(k, 0)$ diverges, its residue is finite and proportional to the spectral weight of electrons. In this case, the zero-energy spectral function of electrons diverges and manifests as a sharp resonance peak in the ARPES data. The Fermi wave vector can be determined from the measurement of this sharp resonance peak.

(2) $G(k, 0) = 0$ on this surface. $G(k, 0)$ changes sign on the two sides of this surface. This kind of surface does not exist in conventional metals. Nevertheless, one can still regard it as a Fermi surface. It is actually a remnant of the usual Fermi surface in a gapped state. It can be found in the superconducting state. Under the BCS mean-field approximation, the diagonal component of the Green's function is given by

$$G_{11}(k, \omega) = \frac{\omega + \xi_k}{\omega^2 - \xi_k^2 - \Delta_k^2}. \tag{4.26}$$

On the Fermi surface, $\xi_k = 0$ but $\Delta_k \neq 0$. Hence, $G_{11}(k, 0) = 0$ and there is no divergence on the Fermi surface. Similar situations also exist in other gapped systems. In this case, the spectra weights of electrons are completely suppressed on the Fermi surface, and no low energy resonance peaks appear in the ARPES measurement. The divergence in G_{11} happens at $\omega = \pm\sqrt{\xi_k^2 + \Delta_k^2}$, but it is not located at the Fermi surface. Hence, the Fermi surface cannot be determined simply from low energy peaks of ARPES.

In the overdoped high-T_c cuprates, the ARPES measurement shows that there exists a closed Fermi surface, similar as in the conventional metallic state. The normalized area enclosed by the Fermi surface equals the electron density $1 - x$ (x is the doped hole concentration) within experimental errors. Thus the Luttinger theorem is obeyed. On the other hand, in the underdoped regime, the Fermi surface is not closed in low temperatures. Instead, it exhibits four disconnected arcs near the momenta $(\pm\pi/2, \pm\pi/2)$ [64], and the Luttinger theorem seems to be violated. Nevertheless, in high temperatures, these Fermi arcs

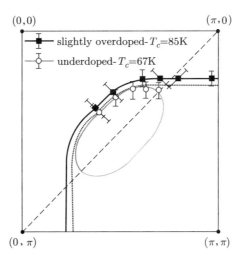

Figure 4.1   Fermi surfaces of slightly overdoped and underdoped high-T_c cuprates Bi$_2$Sr$_2$CaCu$_2$O$_{8+\delta}$ determined by the ARPES measurements. The Fermi surface of the underdoped cuprate is an open arc, which is markedly different from the Fermi surface in a normal metal. Reprinted with permission from [64]. Copyright (1996) by the American Physical Society.

become connected [65] and the Luttinger theorem seems to be restored. Fig. 4.1 shows the Fermi surface structure for the overdoped and underdoped high-T_c cuprates $Bi_2Sr_2CaCu_2O_{8+\delta}$. The anomalous behavior of the Fermi surface in the underdoped cuprate is apparently related to the pseudogap phenomenon. It is a consequence of strong correlation effect. It leads to a variety of anomalous phenomena. The microscopic origin of the Fermi arc and the pseudogap remains unclear and need to be further explored both experimentally and theoretically.

4.4 Particle-Hole Mixing and Superconducting Energy Gap

In the normal metallic state, the spectral function of electrons $A(k, \omega)$ exhibits approximately a Lorentzian peak centered at $\omega = \varepsilon_k$. When the momentum of electron passes through the Fermi surface from a point below the Fermi level to a point above, the peak energy changes from negative to zero, and then to positive. Hence from ARPES, we can see how the spectral peak moves towards the Fermi energy from an energy below. This corresponds to the shift of momentum k from a point below the Fermi level to the Fermi momentum. However, when k is above the Fermi surface, the spectral intensity is suppressed by the Fermi function $f(\omega)$ and the corresponding peak is difficult to be probed by ARPES.

The evolution of the ARPES lineshape with momentum k is markedly different in the superconducting state due to the particle-hole mixing. In an ideal BCS superconductor, the single-particle spectral function is given by Eq. (4.10). However, in real materials, the δ-function-like spectral peak in $A^{(0)}(k, \omega)$ is broadened or even completely suppressed by the self-energy correction induced by the scattering of electrons. In this case, in order to describe correctly the behavior of spectral function, an accurate evaluation of the Green's function of electrons is desired but difficult to fulfill. To take a quantitative analysis of experimental results, a commonly adopted approximation is to assume that the broadened peak has a Lorentzian lineshape and the spectral function in the superconducting state is approximately given by

$$A(k, \omega) \approx \frac{1}{\pi} \left[\frac{u_k^2 \Gamma}{(\omega - E_k)^2 + \Gamma^2} + \frac{v_k^2 \Gamma}{(\omega + E_k)^2 + \Gamma^2} \right], \qquad (4.27)$$

where u_k^2 and v_k^2 measure the spectral weights of the particle and hole excitations, respectively. Γ is the scattering rate. Eq. (4.27) shows that when the momentum k moves towards the Fermi surface from a point below the Fermi level, the peak energy grows with increasing energy. The peak stops to move upwards when the peak energy reaches a maximum at $|\omega| = |\Delta_k|$ and the momentum k touches the normal state Fermi surface. Unlike in the normal state, the spectral peak does not disappear immediately when the momentum is further increased. Instead, the peak moves towards a lower energy and disappears gradually. It suggests that in the superconducting phase, the ARPES measurement can detect not only the spectral peaks of normal electrons below the Fermi surface, but also those above.

Fig. 4.2 shows the evolution of the energy distribution curves with momentum in both the normal and superconducting states of $Bi_2Sr_2CaCu_2O_{8+\delta}$ [31]. The momentum dependence of the spectra in the superconducting phase is quite different from that in the normal state. In the normal state, the spectral peak is invisible above the Fermi surface. However, in the superconducting phase, the spectral peaks exist on both sides of the Fermi momentum. The experimental results agree qualitatively with the theoretical description based on the picture of BCS quasiparticles. It implies that the particle-hole mixing induced by the superconducting pairing exists in high-T_c superconductors, similar as in conventional metal-based superconductors.

The single-particle excitation spectra and the quasiparticle energy gap can be measured through ARPES. The pairing symmetry can be also determined by analyzing the momentum dependence of the superconducting gap on the Fermi surface. This is an advantage of ARPES. It plays an important role in the study of high-T_c cuprates. There are two approaches to determine the values of superconducting gaps from ARPES. The simplest and most intuitive one is to determine the gap directly from the relative shift in the leading edges of the ARPES spectra between the superconducting state and the normal state. This

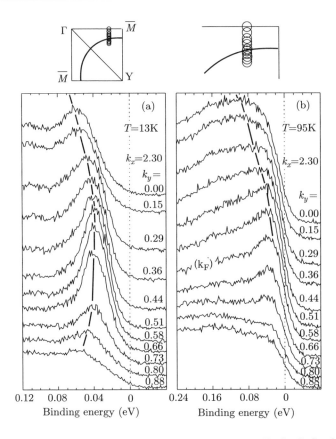

Figure 4.2   The ARPES spectral lines of the high-T_c cuprates $Bi_2Sr_2CaCu_2O_{8+\delta}$ at the momentum k marked by the circles in the figures above. (a) $T = 13K$ in the superconducting state, (b) $T = 95K$ in the normal state. Reprinted with permission from [31]. Copyright (1996) by the American Physical Society.

approach does not depend on the detailed formulism of quasiparticle spectra, and can be used even in the case that a comprehensive understanding of the superconducting quasiparticles is not available. It is based on this approach that Z. X. Shen *et. al.* discovered the anisotropy in the high-T_c gap function [66]. In particular, they found that the ARPES spectra do not change much in the middle point of the leading edge in both the normal and superconducting states for $Bi_2Sr_2CaCu_2O_{8+\delta}$, along the diagonal direction of the Brillouin zone, indicating that there is no gap along that direction. However, around $(0, \pi)$, the ARPES spectra in the normal and superconducting states behave very differently. In comparison with the normal state, the leading edge in the superconducting state clearly shifts down to lower energy, exhibiting a large energy gap [66]. Their results showed unambiguously that the gap function of high-T_c superconductors is strongly anisotropic, consistent with the $d_{x^2-y^2}$-wave symmetry.

A more quantitative approach is to directly extract the energy gaps by fitting the low-energy data using a spectral function which is obtained under proper assumptions and with a full consideration of the energy and momentum resolutions. Assuming the spectral function is phenomenologically determined by Eq. (4.27), Ding $et.$ $al.$ analyzed the variation of the gap function with the momentum on the Fermi surface. They found that the momentum dependence of the gap function agrees well with the $d_{x^2-y^2}$-wave formulae $\Delta_k = \Delta_0(\cos k_x - \cos k_y)$, as shown in Fig. 4.3 [67].

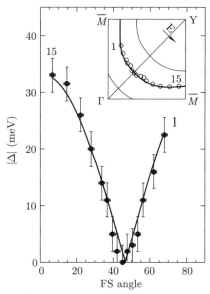

Figure 4.3   Angular dependence of superconducting energy gap on the Fermi surface of $Bi_2Sr_2CaCu_2O_{8+\delta}$ ($T_c = 87K$). The bold solid line in the inset is the Fermi surface. The open circles represent the momenta measured by ARPES. Reprinted with permission from [67]. Copyright (1996) by the American Physical Society.

4.5 Scattering between Quasiparticles

The interplay between superconducting quasiparticles can lower the lifetime
and change the behaviors of quasiparticles, which can affect thermodynamic
and transport properties of superconductors. The quasiparticle lifetime can
be probed by ARPES or through the measurements of electric conductivity,
thermal conductivity or other transport coefficients. In general, the quasiparticle lifetime probed by electric conductivity is different from that probed by
ARPES. The ARPES measures the single-electron excitations and the quasiparticle lifetime is determined by the scattering between electrons. However,
the electric conductivity, or resistivity, is determined by the current-current
correlation function of electrons, which is described by a two-electron Green's
function. As the total momentum of two electrons is conserved, the normal
momentum-conserving scattering process between electrons has no contribution to the resistivity. Only the Umklapp scattering between electrons, which
breaks the momentum conservation, contribute to the resistivity. The Umklapp
scattering happens when the total momenta of two electrons before and after
scattering differ by an integer multiple of the reciprocal lattice vector. In a *d*-
wave superconductor, this condition of Umklapp scattering is generally difficult
to fulfill and the Umklapp scattering is strongly suppressed. The quasiparticle
transport lifetime is dominated by the impurity scatterings and thus is different
from the single particle lifetime. It is important to understand this difference
in the analysis of different experimental results for *d*-wave superconductors.

The interaction among quasiparticles can be generally expressed as

$$H' = \frac{1}{2} \sum_{k_1 k_2 k_3 k_4} V_{k_1 k_2 k_3 k_4} \delta_{k_1 - k_2 + k_3 - k_4, G} C_{k_1}^\dagger \sigma_3 C_{k_2} C_{k_3}^\dagger \sigma_3 C_{k_4}, \qquad (4.28)$$

where C_k is the Nambu spinor fermion operator,

$$C_k = \begin{pmatrix} c_{k\uparrow} \\ c_{-k\downarrow}^\dagger \end{pmatrix},$$

and G is the reciprocal lattice vector. $V_{k_1 k_2 k_3 k_4} = V_{k_3 k_4 k_1 k_2}$ is the interaction
vertex function, which is assumed to be real for convenience.

The correction from the quasiparticle scattering to the Green's function is determined by the Dyson equation, Eq. (4.4). By neglecting the correction to the chemical potential under the second order perturbation, the self-energy function is given by

$$
\Sigma\left(k, \omega_n\right) = \frac{2}{\beta^2} \sum_{\omega_m \omega_t \omega_s} \sum_{k_2 k_3 k_4} V_{kk_2 k_3 k_4} \delta_{k-k_2+k_3-k_4, G} \delta\left(\omega_m + \omega_s - \omega_t - \omega_n\right)
$$

$$
\left[V_{k_2 k_3 k_4 k} \sigma_3 G^{(0)}\left(k_2, \omega_m\right) \sigma_3 G^{(0)}\left(k_3, \omega_t\right) \sigma_3 G^{(0)}\left(k_4, \omega_s\right) \sigma_3 \right.
$$

$$
\left. - V_{k_4 k_3 k_2 k} \sigma_3 G^{(0)}\left(k_2, \omega_m\right) \sigma_3 \mathrm{Tr} G^{(0)}\left(k_3, \omega_t\right) \sigma_3 G^{(0)}\left(k_4, \omega_s\right) \sigma_3 \right].
$$

$$(4.29)$$

Substituting the expression of $G^{(0)}$, Eq. (4.3), into Eq. (4.29), we arrive at

$$
\Sigma\left(k, \omega_n\right)
$$

$$
= \frac{2}{\beta^2} \sum_{\omega_t \omega_s} \sum_{k_2 k_3 k_4} \frac{V_{kk_2 k_3 k_4}\left(i\omega_n + i\omega_t - i\omega_s + \xi_{k_2}\sigma_3 - \Delta_{k_2}\sigma_1\right)}{\left[\left(i\omega_n + i\omega_t - i\omega_s\right)^2 - E_{k_2}^2\right]\left[\left(i\omega_t\right)^2 - E_{k_3}^2\right]\left[\left(i\omega_s\right)^2 - E_{k_4}^2\right]}
$$

$$
\left[V_{k_2 k_3 k_4 k}\left(i\omega_t + \xi_{k_3}\sigma_3 + \Delta_{k_3}\sigma_1\right)\left(i\omega_s + \xi_{k_4}\sigma_3 - \Delta_{k_4}\sigma_1\right) \right.
$$

$$
\left. - 2V_{k_4 k_3 k_2 k}\left(-\omega_t \omega_s + \xi_{k_3}\xi_{k_4} - \Delta_{k_3}\Delta_{k_4}\right)\right] \delta_{k-k_2+k_3-k_4, G}, \qquad (4.30)
$$

where the summation over ω_s and ω_t can be obtained using the standard method. The result is generally very complicated. But at zero temperature, it can be simplified and the imaginary part, which is proportional to the electron scattering rate τ^{-1}, reads

$$
\mathrm{Im}\Sigma\left(k, \omega > 0\right)
$$

$$
= \sum_{k_2 k_3 k_4} \frac{\pi V_{kk_2 k_3 k_4}}{4E_{k_2}E_{k_3}E_{k_4}} \delta_{k-k_2+k_3-k_4, G}\delta\left(\omega - E_{k_2} - E_{k_3} - E_{k_4}\right)
$$

$$
\left(E_{k_2} + \xi_{k_2}\sigma_3 - \Delta_{k_2}\sigma_1\right)\left[2V_{k_4 k_3 k_2 k}\left(-E_{k_3}E_{k_4} + \xi_{k_3}\xi_{k_4} - \Delta_{k_3}\Delta_{k_4}\right)\right.
$$

$$
\left. - V_{k_2 k_3 k_4 k}\left(-E_{k_3} + \xi_{k_3}\sigma_3 + \Delta_{k_3}\sigma_1\right)\left(E_{k_4} + \xi_{k_4}\sigma_3 - \Delta_{k_4}\sigma_1\right)\right]. \qquad (4.31)
$$

In an isotropic s-wave superconductor, there is a finite gap in the quasiparticle excitation spectrum, $E_k > \Delta$. Eq. (4.31) shows that the quasiparticle scattering has finite contribution to $\mathrm{Im}\Sigma$ only when $\omega > 3\Delta$. Hence, there is a threshold in the excitation energy of quasiparticles being scattered, $\omega = 3\Delta$, due to the opening of the pairing gap. This is because at zero temperature, the quasiparticle scattering needs to break a Cooper pair, and the quasiparticles after scattering can survive only when their energies are above the gap.

In a *d*-wave superconductor, the energy dependence of the scattering rate τ^{-1} behaves differently. In the low energy limit, the momentum summation in Eq. (4.31) needs to be done just around the nodal region, in which the momentum dependence of the quasiparticle energy E_k is linear. In the absence of the Umklapp scattering, i.e. $G = 0$, one can do a linear transformation to separate ω from the summation. If the interaction matrix elements $V_{k_1 k_2 k_3 k_4}$ are not strongly momentum dependent, one can show from dimensional analysis that the scattering rate scales cubically with ω [68]

$$\tau^{-1} = D_\tau \omega^3. \tag{4.32}$$

The ω-independent coefficient D_τ can be obtained by integrating over the momenta in Eq. (4.31). Compared with the energy dependence of τ^{-1} in the normal metal, $\tau^{-1} \sim \omega^2$, the scattering rate in the *d*-wave superconductors has one more power in ω. This extra ω comes from the linear density of states of quasiparticles.

In finite temperatures, the quasiparticle scattering rate can be obtained from Eq. (4.30). In the limit $\omega \to 0$, T (or more precisely $k_B T$) is the only parameter that has the dimension of energy. In this case, it is simple to show by dimensional analysis that, similar to Eq. (4.32), the quasiparticle lifetime scales cubically with temperature in the low temperature limit [69]:

$$\tau^{-1} \sim T^3. \tag{4.33}$$

Thus the quasiparticle scattering rate τ^{-1} scales as ω^3 at zero temperature and as T^3 in the zero frequency limit. This is consistent with the temperature and frequency dependencies of the scattering rate obtained from the ARPES and thermal conductivity measurements [70, 71, 72]. It shows that the low energy scattering between quasiparticles is an important channel of scatterings and should be considered seriously in the analysis of low energy behaviors of quasiparticles.

However, the quasiparticle lifetime probed by the microwave conductivity does not equal that determined by the usual electron-electron scattering. The formulae $\tau^{-1} \sim T^3$ cannot be used to explain the microwave conductance measurements. Instead, the Umklapp scattering with $G \neq 0$ should be considered to account for the contribution of electron-electron scattering to the electric conductance.

In a d-wave superconductor, quasiparticles near the gap nodes generally do not satisfy the condition of Umklapp scattering. The Umklapp scattering emerges only when the total energy of two quasiparticles becomes larger than a threshold Δ_U. Hence the Umklapp scattering is thermally activated, and the quasiparticle scattering rate should scale exponentially with temperature [69]

$$\tau_t^{-1} \sim e^{-\Delta_U/k_B T}. \tag{4.34}$$

This exponential behavior of τ_t^{-1} agrees qualitatively with the measurement data of microwave conductivity for high-T_c superconductors [73, 69].

Chapter 5
Tunneling Effect

5.1 Electron Scattering on the Surface of a Superconductor

Tunneling effect is an important quantum phenomenon. It plays an important role in the study of microscopic mechanism of superconductivity. A superconducting junction is formed by a superconductor separated by a thin insulating layer or vacuum from a normal metal or another superconductor. Electric current flows across the junction under a small bias voltage. By measuring the electric current response to the applied bias, or the differential conductance, one can extract information on the density of states of superconducting quasiparticles as well as their interactions.

The contact between a superconductor and a metal or an insulator layer or film in a tunneling junction can be either in a face-to-face or in a point-to-face configuration. For example, the contact between the tip of a scanning tunneling microscope (STM) and the surface of a superconductor is of the point-to-face type. For a face-to-face contact, the tunneling current is distributed throughout the whole interface, and the differential conductance measures the average density of states of superconducting quasiparticles on the interface. For a point-to-face contact, the tunneling current concentrates around the probe. It measures the local density of states of superconducting quasiparticles around the contact. By scanning the tip across the entire sample, one can obtain the spatial distribution of local density of states.

The tunneling effect of normal electrons relies strongly on the property of the tunneling interface. For an ideal contact between a metal and a superconductor, the interface can be approximately treated as an elastic scattering barrier. In this case, the tunneling effect can be understood by studying an elastic scattering problem of electrons on the interface. However, the scattering at the surface of a superconductor is more complicated in comparison with that at the surface of a normal metal. Besides the conventional reflection and transmission of normal electrons, there are also reflection of normal holes and transmission of hole quasiparticles generated by an off-diagonal scattering potential in the superconducting state. The reflection of holes is a peculiar feature of reflection on the surface of superconductors that was first pointed out by Andreev. This kind of reflection is called the Andreev reflection [74]. It is important when the surface tunneling barrier is low. When the tunneling barrier is high, the Andreev reflection is suppressed and the normal reflection of electrons becomes more important.

In the study of the Andreev reflection and transmission of holes at the surface of superconductor, the semi-classical WKB approximation is usually adopted [75, 76, 77]. This approximation simplifies the steps in solving the self-consistent gap equation around the surface. However, it is not necessary to take this approximation if the self-consistency of the gap function is not strictly required.

Let us investigate the electron scattering on the surface of superconductor using the Hamiltonian defined by Eq. (3.53). We assume that the interface is perpendicular to the x-direction, and the scattering potential $U(x)$ depends only on x and is finite just in the vicinity of the interface. In addition, we assume that the system is translation invariant in the direction parallel to the interface and the momentum parallel to the interface is conserved.

The Andreev reflection and transmission of holes on the surface of a superconductor results from the superconducting condensation, which breaks the number conservation of normal electrons in the superconducting state. In the BCS theory, up-spin electrons are hybridized with down-spin holes. Thus both electrons and holes with opposite spins can be scattered on the superconductor surface. Fig. 5.1 shows schematically the reflection and transmission wave vectors for both electrons and holes.

The Andreev reflection differs from the normal reflection of electrons, and

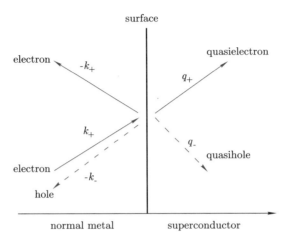

Figure 5.1   The electron and hole reflection and transmission at the superconductor surface. The incident electron is from the normal metal side, and the reflection and transmission waves can be either electron- or hole-like. The directions of the probability currents for the electron and hole are marked with arrows. The electric current directions are opposite to those of the probability currents of the hole-type reflection and transmission waves.

leads to different physical effects. If the incident electron has an energy higher than the superconducting gap, it transmits almost completely through the surface barrier and becomes a quasiparticle propagating in the superconductor. In this case, the Andreev reflection is very weak and affects weakly on the transmission current. In contrast, if the energy of the incident electron is smaller than the superconducting gap, the electron cannot transmit through the barrier and propagate as a quasiparticle. Nevertheless, if the incident electron captures another electron with opposite momentum to form a Cooper pair at the interface, they can cotunnel into the superconductor and become a pair of condensed electrons, doubling the transmission current. Thus the transmission wave contains a pair of electrons with zero total momentum. Considering the charge and momentum conservation, this implies that there must be a reflection wave of holes with a momentum nearly opposite to the incident wavevector on the

normal metal side. Hence the Andreev reflection arises from the superconducting pairing of the transmitting electron with another electron at the interface.
The ordinary electron reflection suppresses the transmission probability, while
the Andreev reflection enhances the transmission probability because it corresponds the transmission of a Cooper pair.

In the ordinary reflection, the momenta of incident and reflection electrons
are conserved along the direction parallel to the interface, and take opposite
values along the direction normal to the interface. In the Andreev reflection,
however, the momentum of the reflected hole parallel to the interface equals the
corresponding momentum of the incident electron but in the opposite direction.
The momenta of the incident electron and the reflected hole normal to the
interface have opposite sign, but their magnitudes are not equal to each other.
Thus in the Andreev reflection, the momentum of the refection hole is different
from the incident electron. But their energies with respect to the Fermi surface
are the same, which is a consequence of elastic scattering.

Let us denote the wavevectors of the incident electron and the Andreev
reflection hole along the x-axis by k_+ and k_-, respectively. If the incident
momentum is $(k_+, k_\parallel)$ with $k_\parallel$ the component parallel to the interface, then the
momentum of the Andreev reflection hole is $(-k_-, -k_\parallel)$. Given the values of the
energy $E > 0$ and the momentum parallel to the interface $k_\parallel$ for the incident
electron, the wavevectors of the incident electron and the Andreev reflection
hole along the x direction in the limit $x \to -\infty$, i.e. $k_\pm$, are determined by the
equation

$$E_e = E_h = E, \tag{5.1}$$

where

$$E_e = \frac{\hbar^2}{2m}\left(k_+^2 + k_\parallel^2\right) - \mu, \tag{5.2}$$

$$E_h = \mu - \frac{\hbar^2}{2m}\left(k_-^2 + k_\parallel^2\right), \tag{5.3}$$

are the excitation energies of electrons and holes in the normal metal. Substituting these expressions into Eq. (5.1), we find that

$$k_\pm = \sqrt{\frac{2m(\mu \pm E)}{\hbar^2} - k_\parallel^2}. \tag{5.4}$$

The wavefunction on the normal metal side is a superposition of the plane waves for the incident electron, the reflected electron, and the Andeev reflection hole. If an up-spin electron is emitted from $x = -\infty$ to the interface, the wavefunction on the normal metal side is then given by

$$\psi\left(r\right) = e^{ik_\parallel \cdot r_\parallel} \begin{pmatrix} e^{ik_+ x} + be^{-ik_+ x} \\ ae^{ik_- x} \end{pmatrix}. \tag{5.5}$$

The coefficient of the incident electron wavefunction is normalized to 1. a and b are the coefficients of the Andreev hole and normal electron reflections, respectively. This wavefunction can be also written in a second quantized form as

$$\left|\psi\right\rangle = \left(c^\dagger_{k_+,k_\parallel,\uparrow} + bc^\dagger_{-k_+,k_\parallel,\uparrow} + ac_{-k_-,-k_\parallel,\downarrow}\right)\left|0\right\rangle, \tag{5.6}$$

where $\left|0\right\rangle$ is the Fermi sea. In the coordinate representation, $\left|\psi\right\rangle$ becomes

$$\psi\left(r\right) = e^{ik_\parallel \cdot r_\parallel} \left[\left(e^{ik_+ x} + be^{-ik_+ x}\right) c^\dagger_{r\uparrow} + ae^{ik_- x}c_{r\downarrow}\right]\left|0\right\rangle. \tag{5.7}$$

Creating a spin-up electron with momentum k in the Fermi sea is equivalent to annihilating a spin-down electron with momentum $-k$.

On the superconducting side, the wavefunction of electrons is a plane wave:

$$\psi(r) = e^{iq \cdot r} \begin{pmatrix} u_q \\ v_q \end{pmatrix}, \tag{5.8}$$

where q is the wavevector of superconducting quasiparticles. Since the momentum parallel to the interface is conserved, we have $q_\parallel = k_\parallel$. The coefficients u_q and v_q are determined by the BCS mean-field equations

$$\begin{pmatrix} \xi_q & \Delta_q \\ \Delta_q & -\xi_q \end{pmatrix} \begin{pmatrix} u_q \\ v_q \end{pmatrix} = E \begin{pmatrix} u_q \\ v_q \end{pmatrix}, \tag{5.9}$$

where

$$\xi_q = \frac{\hbar^2}{2m}q^2 - \mu,$$

and Δ_q is the gap function.

Given E, q is determined by the eigen energy of Eq. (5.9)

$$\xi_q^2 + \Delta_q^2 = E^2, \tag{5.10}$$

and (u_q, v_q) is given by

$$\begin{cases} u_q = \sqrt{\dfrac{1}{2} + \dfrac{\xi_q}{2E}}, \\[3mm] v_q = -\mathrm{sgn}(\Delta_q)\sqrt{\dfrac{1}{2} - \dfrac{\xi_q}{2E}}. \end{cases} \tag{5.11}$$

The associated creation operator of Bogoliubov quasiparticle is defined by

$$\gamma_{1,q}^\dagger = u_q c_{q\uparrow}^\dagger + v_q c_{-q\downarrow}. \tag{5.12}$$

Applying $\gamma_{1,q}$ to the BCS ground state creates a quasiparticle excitation of energy E. The matrix on the left-hand side of Eq. (5.9) has another eigenvalue $-E$, and the corresponding annihilation operator of superconducting quasiparticles is defined by

$$\gamma_{2,q} = -v_q c_{q\uparrow}^\dagger + u_q c_{-q\downarrow}. \tag{5.13}$$

Applying $\gamma_{2,q}^\dagger$ to the BCS ground state also creates a quasiparticle excitation of energy E. But the spin of $\gamma_{2,q}^\dagger$ is orthogonal to the wave function of the incident electron, and does not need to be considered.

In the discussion below, we consider the system with $|\Delta(q_x, k_\parallel)| = |\Delta(-q_x, k_\parallel)|$. In this case, if $q_x > 0$ is a solution to the equation, so is $-q_x$. Similar as in the normal metal side, for each energy E, $|q_x|$ can take two different values, which are larger and smaller than the Fermi wavevector, respectively. These two values of $|q_x|$, if denoted by q_+ and q_-, satisfy the inequalities $q_+^2 > 2m\mu/\hbar^2 - k_\parallel^2$ and $q_-^2 < 2m\mu/\hbar^2 - k_\parallel^2$, respectively. Since the solutions corresponding to $\pm q_x$ are degenerate, there are four solutions for a given energy E. The solutions of $q_x = \pm q_+$ correspond to the electron-type excitations with the probability flux along $\pm x$ direction, and those of $q_x = \pm q_-$ correspond to the hole-type excitations with the probability flux along $\mp x$ direction. It should be emphasized that the direction of the probability flux of hole-type excitations is anti-parallel to its electric current. For the scattering problem discussed here, it is sufficient just to consider the solution with the probability flux along the $+x$ direction.

Thus inside the superconductor, the transmission wave is a superposition of two solutions of $q = (q_+, k_\parallel)$ and $q = (-q_-, k_\parallel)$ and can be cast into the form

$$\psi(r) = e^{ik_\parallel \cdot r_\parallel} \left[ce^{iq_+x} \begin{pmatrix} u_+ \\ v_+ \end{pmatrix} + de^{-iq_-x} \begin{pmatrix} u_- \\ v_- \end{pmatrix} \right], \tag{5.14}$$

where c and d are the coefficients of two transmission waves.

If Δ_q does not depend on the value of $q_\pm$, such as in an isotropic s-wave superconductor, we have

$$\begin{pmatrix} u_- \\ v_- \end{pmatrix} = \begin{pmatrix} -v_+ \\ -u_+ \end{pmatrix}.$$

On the other hand, if $\Delta(-q_-, k_\parallel) = -\Delta(q_+, k_\parallel)$, we have

$$\begin{pmatrix} u_- \\ v_- \end{pmatrix} = \begin{pmatrix} -v_+ \\ u_+ \end{pmatrix}.$$

In order to determine the reflection and transmission coefficients, we need to know the detailed shape of the scattering potential. For an arbitrary scattering potential, it is difficult to obtain an analytic solution. However, if the scattering potential $U(x)$ is simply a δ-function, we can then obtain a relatively simple analytic solution.

5.2 Tunneling Conductance

In order to calculate the tunneling current of electrons at the metal-superconductor interface, let us evaluate the current contributed by the incident and reflected electrons and the Andreev reflection holes on the normal metal side. The tunneling current of normal electrons is determined by the difference between the currents with and without applying an external bias voltage to the tunneling junction.

In the normal metal, the electric current operator is defined by

$$\hat{J}_x = -e \sum_{k\sigma} v_x(k) c_{k\sigma}^\dagger c_{k\sigma}, \tag{5.15}$$

where $v_x(k) = \hbar k_x/m$ is the velocity of electrons along the x-axis. The wavefunction of an electron which is incident from $x \rightarrow -\infty$ and reflected on the superconductor surface is described by Eq. (5.6). Its contribution to the electric current is given by

$$J_x(k_+, k_\parallel) = \langle\psi|\hat{J}_x|\psi\rangle = -\frac{e\hbar}{m}\left[k_+\left(1 - |b|^2\right) + k_-|a|^2\right]. \qquad (5.16)$$

Here the vacuum is the electron Fermi sea and $|\psi\rangle$ must be normalized according to the incident electron wavefunction according to the boundary condition of the tunneling problem.

In the absence of external voltage, an electron or hole can transmit from the normal metal side to the superconductor side, and vice versa. Thus the net current vanishes. However, after applying a voltage V, the motion of electrons in the two directions are no longer balanced, and the tunneling current become finite.

Upon applying an external voltage V, the chemical potential of the normal metal is increased by eV in comparison with that of the superconductor. If the quasiparticle eigenstates in the superconductor side is not affected by the applied voltage, the contribution from the electrons or holes transmitted from the superconducting state to the current in the normal metal side is not changed before and after applying the voltage. The net current is therefore determined by the difference between the electric current in the normal metal with and without the external voltage, described by the formula

$$I_x = \frac{2e\hbar}{m}\sum_k{}'\left[f(E - eV) - f(E)\right]\left[k_+\left(1 - |b|^2\right) + k_-|a|^2\right], \qquad (5.17)$$

where the factor 2 results from the spin degeneracy, and $\sum'$ represents the summation over permitted incident and reflection angles. The above analysis neglects the correction of the applied voltage to the microscopic electronic structure, which is a non-linear effect. Hence Eq. (5.17) is just a result of linear approximation.

Eq. (5.17) is consistent with the results obtained by Blonder-Tinkham-Klapwijk (BTK)[78]. But the derivation here is simpler. It allows us to see more clearly how good and reliable this formula can be used in the analysis of tunneling effect. In BTK's original derivation, the assumption of $k_+ = k_- = k_F$ is made. Eq. (5.17) holds more generally than the formula derived by BTK.

Eq. (5.17) is a general formula for the tunneling current. It is valid independent on the shape of the interface scattering potential, provided the scattering
is elastic which preserves the energy as well as the momenta parallel to the
interface. However, if the surface barrier is very high or the scattering potential is very strong, it may not always be convenient and necessary to solve the
Schrödinger equation to obtain the reflection coefficients a and b. In this case,
as will be discussed in 5.5, it is more convenient to treat approximately the
scattering potential by an energy-independent tunneling matrix.

The differential conductance of the tunneling junction is given by the derivative of the tunneling current with respect to the applied bias:

$$G(V) = \frac{\mathrm{d}I_x}{\mathrm{d}V} = -\frac{2e^2\hbar}{m} \sum_k{}' \frac{\mathrm{d}f(E - eV)}{\mathrm{d}E} \left[k_+ \left(1 - |b|^2\right) + k_-|a|^2 \right]. \qquad (5.18)$$

At zero temperature, it becomes

$$G(V) = \frac{2e^2}{(2\pi)^D \hbar} \int{}' \mathrm{d}k_\parallel \left[1 - |b(eV)|^2 + \frac{k_-(eV)}{k_+(eV)}|a(eV)|^2 \right], \qquad (5.19)$$

where the scattering coefficients and $k_\pm$ take values at $E = eV$. $\int'$ represents
integration only over permitted incoming and reflecting angles. D is the dimension of the system.

In order to compare with the tunneling conductance in a metal-metal tunneling junction, we define the normalized differential conductance

$$\sigma(V) = \frac{G(V)}{G(V \to \infty)}. \qquad (5.20)$$

In the high bias limit $V \to \infty$, the effect of superconducting pairing potential
can be neglected, and $G(\infty)$ contributes purely from normal electrons. The
deviation of $\sigma(V)$ from 1 results from the effect of superconductivity.

5.3 Scattering from the δ-Function Interface Potential

If the scattering potential $U(x)$ is a δ-function,

$$V(x) = U\delta(x),$$

the boundary condition is relatively easy to handle and an analytic solution for the scattering coefficients can be obtained. From the solution, a great deal of useful information about the tunneling effect can be extracted, allowing us to understand more transparently the scattering effect. The conclusion such obtained can be also applied qualitatively to other tunneling systems with more general scattering potentials.

Around the interface, the superconducting gap parameter gradually decays to zero from the superconductor to the normal metal side, within a characteristic length scale of the order of superconducting coherence length ξ. In order to rigorously solve the scattering problem of electrons at the superconducting surface, we need to solve the Bogoliubov-de Gennes equation self-consistently. This is difficult and generally can be done only numerically. Nevertheless, if the coherence length ξ is very short, a natural approximation is to take the limit $\xi \to 0$. In this limit, we can neglect the small variation of the superconducting gap function at the interface, and assume the gap function to be entirely zero in the normal metal side and finite and uniform in the superconductor side. This is a common approximation used in the study of tunneling effect of superconducting quasiparticles, which can greatly simplify calculations.

The electron wavefunctions in the normal metal and superconductor sides are given by Eq. (5.5) and (5.14), respectively. At the interface, the continuity conditions for the wavefunction and its derivative are determined by the equations

$$\psi\left(x = 0^{+}\right) - \psi\left(x = 0^{-}\right) = 0,$$

$$\frac{\hbar^2}{2m}\psi'_x(x = 0^-) - \frac{\hbar^2}{2m}\psi'_x(x = 0^+) + U\psi(x = 0) = 0.$$

Based on these equations, we can further obtain the equations that determine the scattering coefficients:

$$
\begin{pmatrix}
0 & -1 & u_+ & u_- \\
-1 & 0 & v_+ & v_- \\
0 & iw + k_+ & q_+ u_+ & -q_- u_- \\
iw - k_- & 0 & q_+ v_+ & -q_- v_-
\end{pmatrix}
\begin{pmatrix}
a \\
b \\
c \\
d
\end{pmatrix}
=
\begin{pmatrix}
1 \\
0 \\
k_+ - iw \\
0
\end{pmatrix}.
\tag{5.21}
$$

The solution for this set of linear equations is

$$
\begin{pmatrix} a \\ b \\ c \\ d \end{pmatrix} = \frac{1}{p} \begin{pmatrix} -2k_+ (q_+ + q_-)\, v_+ v_- \\ 2k_+ (w_- v_- u_+ - w_+ v_+ u_-) - p \\ 2k_+ w_- v_- \\ -2k_+ w_+ v_+ \end{pmatrix},
\tag{5.22}
$$

where

$$
w = \frac{2mU}{\hbar^2},
$$

$$
w_+ = iw - k_- + q_+,
$$

$$
w_- = iw - k_- - q_-,
$$

$$
p = w_-(iw + k_+ + q_+)u_+ v_- - w_+(iw + k_+ - q_-)v_+ u_-.
$$

From Eq. (5.17), it is clear that the tunneling current depends strongly on the incident direction. The contribution to the tunneling current from incident electrons with momenta almost parallel to the interface is very small, while that from incident electrons perpendicular to the interface is large. Thus for most problems related to the electron tunneling, we only need to consider the situation in which the incident electron kinetic energy $\hbar^2 k_+^2 / 2m \gg E$. In this case, the deviations of the x-components of the momenta of the reflection and transmission electrons from the Fermi momentum are small, and can be approximately expressed as

$$
k_\pm \approx k_{F_x}(1 \pm \delta_k),
\tag{5.23}
$$

$$
q_\pm \approx k_{F_x}(1 \pm \delta_q),
\tag{5.24}
$$

where

$$
\delta_k = \frac{mE}{\hbar^2 k_{F_x}^2},
$$

$$
\delta_q = \frac{m\sqrt{E^2 - \Delta_{k_F}^2}}{\hbar^2 k_{F_x}^2},
$$

and Δ_{k_F} is the gap value at $k = k_F = (k_{F_x}, k_{\parallel})$. Using these expressions, the scattering coefficients can be approximately written as

$$
\begin{pmatrix} a \\ b \\ c \\ d \end{pmatrix} \approx \frac{1}{\gamma} \begin{pmatrix} 4(1+\delta_k)v_+v_- \\ \delta_k u_+v_- - (2\delta_k - \eta + 2)\eta(u_+v_- - u_-v_+) \\ 2(2-\eta)(1+\delta_k)v_- \\ 2\eta(1+\delta_k)v_+ \end{pmatrix}, \tag{5.25}
$$

where

$$
Z = \frac{2mU}{\hbar^2 k_F},
$$
$$
\gamma = 4u_+v_- - \eta^2(u_+v_- - u_-v_+),
$$
$$
\eta = iZ + \delta_q + \delta_k.
$$

If the incident direction is nearly normal to the superconductor surface, k_{F_x} is almost equal to the Fermi wavevector k_F. In this case, δ_k and δ_q can be neglected, and the above result can be further simplified as

$$
\begin{pmatrix} a \\ b \\ c \\ d \end{pmatrix} \approx \frac{1}{\gamma} \begin{pmatrix} 4v_+v_- \\ -(Z^2 + 2iZ)(u_+v_- - u_-v_+) \\ 2(2-iZ)v_- \\ 2iZv_+ \end{pmatrix}, \tag{5.26}
$$

where $\gamma \approx 4u_+v_- + Z^2(u_+v_- - u_-v_+)$. This is just the result obtained by utilizing the semi-classical WKB approximation [76, 77] for the Andreev reflection. It is simply to verify that this solution satisfies the equation

$$
|b|^2 + |c|^2 - |a|^2 - |d|^2 = 1,
$$

which is just the current conservation equation along the x-direction. This is because the current is proportional to the product of the wavevector and the magnitude square of the scattering coefficients, and under the approximation $k_{Fx} \approx k_F$, the x-direction wavevectors for the reflection and transmission waves are all equal to k_{F_x}.

In the limit that the δ-function potential vanishes, i.e. $U = 0$, the scattering coefficients determined by Eq. (5.26) possess the following two features: (1) $b =$

$d = 0$, namely there are only Andreev hole reflection and electron transmission, but no electron reflection and hole transmission; (2) The Andreev hole reflection coefficient has a simple form, $a = v_+/u_+$. Furthermore, when $E < |\Delta_{q+}|$, $\xi_{q+} = \sqrt{E^2 - \Delta_{q+}^2}$ is purely imaginary. As u_+ equals the complex conjugate of v_+, we have $|a| = 1$. In this case, the Andreev reflection doubles the electric current. Of course this is only an ideal case. When $U \neq 0$, the enhancement of the electric current due to the Andreev reflection can be either smaller or larger than 1.

The differential conductance of the metal-superconductor tunneling junction can be evaluated using Eq. (5.18) and the scattering coefficients derived above. To simplify the calculation, we only consider the limit that the incident direction is normal to the superconductor surface and the Fermi wavevector $k_F \to \infty$. In this case, the scattering coefficients are given by Eq. (5.25) and the zero temperature differential conductance can be expressed as

$$G(V) = \frac{2e^2}{(2\pi)^D \hbar} \int' dk_\parallel g(eV), \tag{5.27}$$

where

$$g(E) = 1 - |b(E)|^2 + |a(E)|^2. \tag{5.28}$$

In the limit $E \to \infty$, the effect of superconducting pairing is very small and it can be shown that

$$g(\infty) = \frac{4}{4 + Z^2} \tag{5.29}$$

does not depend on the momentum and energy of scattering electrons, neither on the pairing symmetry.

For an isotropic s-wave superconductor, $\Delta_q = \Delta$ and $v_\pm = -u_\mp$. $u_\pm$ and $v_\pm$ depend only on E, but not on the component of the incident momentum parallel to the surface $k_\parallel$. In this case, it can be shown that $g(E)$ is also independent on momentum $k_\parallel$, and the tunneling conductance is proportional to $g(eV)$, i.e. $G(V) \propto g(eV)$.

The energy dependence of $g(E)$ is given by

$$g(E) = \begin{cases} \dfrac{4E}{2E + (2 + Z^2)\sqrt{E^2 - \Delta^2}}, & \text{if } E > \Delta, \\[4mm] \dfrac{8\Delta^2}{4E^2 + (2 + Z^2)^2(\Delta^2 - E^2)}, & \text{if } E < \Delta. \end{cases} \qquad (5.30)$$

From this expression, the bias dependence of the differential conductance can be obtained. Fig. 5.2 shows the normalized differential conductance as a function of applied bias voltage at several different scattering potentials for the isotropic s-wave superconductor. In the case $Z = 0$, the conductance is independent on the bias voltage at $eV < \Delta$ and the tunneling current is twice of that in a metal-metal junction. This enhancement of electric currents in the small voltage is a characteristic feature of the Andreev reflection, resulting from the superconducting pairing of transmission electrons. With the increase of scattering potential, the transmission probability decreases and the normalized zero-bias differential conductance is given by

$$\sigma(0) = \frac{8 + 2Z^2}{(2 + Z^2)^2}. \qquad (5.31)$$

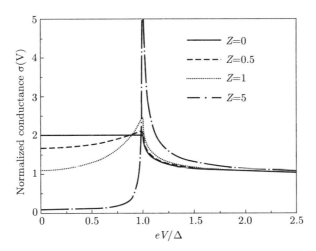

Figure 5.2   Normalized differential conductance $\sigma(V)$ for an isotropic s-wave superconductor tunneling junction.

$\sigma(0)$ vanishes as $2/Z^2$ in the limit of $Z \to \infty$. When the bias voltage equals the gap value, σ increases with increasing Z,

$$\sigma(V = \Delta/e) = \frac{4 + Z^2}{2}, \qquad (5.32)$$

which diverges in the limit $Z \to \infty$.

For a d-wave superconductor, the tunneling current depends on the orientation of the metal-superconductor interface. In the following, we only discuss the tunneling conductance when the x-axis is parallel to the nodal direction of the superconducting gap function. When the x-axis is parallel to the anti-nodal direction, i.e. the direction of the maximal gap, the differential conductance is qualitatively the same as for an isotropic s-wave superconductor.

When the x-axis is parallel to the nodal direction of the d-wave superconductor, the gap function is given by

$$\Delta_k = \Delta \tilde{k}_x \tilde{k}_y,$$

where $\tilde{k}_{x,y} = k_{x,y}/k_F$ is the normalized momentum. When the incident direction is nearly parallel to the x-axis, $|\tilde{k}_x| \approx 1$, the gap parameter is $\Delta_{q+} \approx \Delta \tilde{k}_y$ for the quasielectron along the transmission direction and $\Delta_{q-} \approx -\Delta \tilde{k}_y$ for the quasihole. Based on this observation, we have $v_\pm = \mp u_\pm$, and

$$g(E) = \begin{cases} \dfrac{8E^2 + 4E(2 + Z^2)\sqrt{E^2 - \Delta^2 \tilde{k}_y^2}}{\left[(2 + Z^2)E + 2\sqrt{E^2 - \Delta^2 \tilde{k}_y^2}\right]^2}, & \text{if } E > |\Delta \tilde{k}_y|, \\[4mm] \dfrac{8\Delta^2 \tilde{k}_y^2}{(2 + Z^2)^2 E^2 + 4(\Delta^2 \tilde{k}_y^2 - E^2)}, & \text{if } E < |\Delta \tilde{k}_y|. \end{cases} \qquad (5.33)$$

Fig. 5.3 shows the normalized differential conductance as a function of the bias voltage V for the d-wave superconductor at several different scattering potentials. Unlike in an s-wave superconductor, $\sigma(V)$ increases as Z increases in the low bias limit. Furthermore, at $E = 0$, $g(0) = 2$ independent of $\tilde{k}_y$, thus the differential conductance at zero bias is given by

$$\sigma(V = 0) = \frac{4 + Z^2}{2}, \qquad (5.34)$$

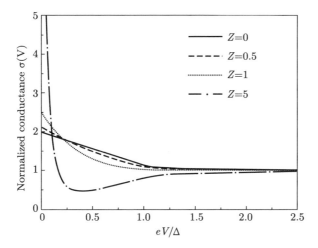

Figure 5.3   Normalized differential conductance $\sigma(V)$ for a d-wave superconductor tunneling junction with the nodal direction parallel to the x axis.

which diverges in the limit $Z \to \infty$. This divergence is a manifestation of the surface resonance states of the d-wave superconductor. It provides a criterion to judge by experimental measurements whether there are gap nodes along the direction normal to the superconductor surface. The zero-energy resonance state appears when the denominator of the scattering coefficient given in Eq. (5.25) becomes zero, namely

$$(2 + Z^2)E + 2\sqrt{E^2 - \Delta^2 \tilde{k}_y^2} = 0. \tag{5.35}$$

This equation does not have real solutions. But it has a complex solution, given by

$$E = -\frac{2i\Delta \tilde{k}_y}{\sqrt{Z^4 + 4Z^2}}. \tag{5.36}$$

The real part of this complex solution is zero, thus the resonance state appears exactly at zero energy. The inverse of the imaginary part is the lifetime of the surface resonance state. When $Z \to \infty$, the lifetime becomes infinite.

When $eV = \Delta$, σ does not diverge. Instead, it decreases with increasing Z and approaches $\pi/4$ as $Z \to \infty$, different from that in an s-wave superconductor.

The tunneling measurement results of high-T_c superconductors agree qualitatively with the theoretical predictions. Fig. 5.4 shows the differential conductance along the direction parallel to the CuO_2 plane for $Bi_2Sr_2CaCu_2O_8$ [79].
It verifies the existence of the zero energy resonance peak in high-T_c superconductors. Furthermore, it was also found that this zero energy resonance peak
exists only in the superconducting phase [80], and disappears in the normal
phase. The zero-bias conductance peaks in $YBa_2Cu_3O_{6+x}$ and other high-T_c
cuprates have also been found by experimental measurements [81, 82, 83, 84].
These tunneling measurements provide strong support to the *d*-wave pairing
symmetry of high-T_c superconductivity.

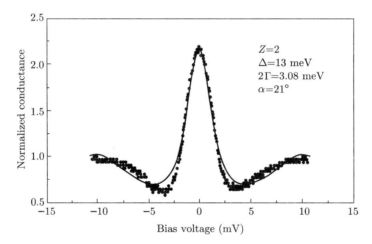

Figure 5.4   Tunneling differential conductance of $Bi_2Sr_2CaCu_2O_8$ along the direction of
CuO_2 planes. The solid line is a fitting curve obtained based on the extended BTK formula.
Γ is the inverse of the effective quasiparticle lifetime. α is the angle between the tunneling
current and the anti-nodal direction. Reprinted with permission from [79]. Copyright (1998)
by the American Physical Society.

5.4 The Surface Bound State

The above discussion shows that in a tunneling junction of *d*-wave superconductor, if the interface is perpendicular to the nodal direction and the tunneling

probability is very small $(Z \to \infty)$, there exists a zero-energy resonance at the interface. This resonance state leads to to a sharp zero-energy peak in the tunneling differential conductance, which can be measured by tunneling experiments. Naturally, one may ask how strongly the existence of such zero-energy resonance state depends on the detailed structure of the metal-superconductor interface, and whether they can be detected by other experimental techniques, such as photoelectron spectroscopy, Raman scattering, etc. This is of particular interesting from the experimental point of view. In order to address this question, we study a d-wave superconductor which contacts directly with the vacuum or with an insulator. We will show that the zero-energy resonance state exists in all these cases and is an intrinsic property of d-wave superconductors.

To explore this problem, let us first consider a system of superconductor covered by a metallic film with a thickness L. We then investigate how the zero-energy bound state emerges at the interface by setting L to zero. The reason for adopting this approach is that the boundary condition can be more easily handled in a finite L system [75].

Let us assume that the interface between the metallic film and the superconductor is located at $x = 0$, and the other surface of the film is at $x = -L$. What we are interested in is the bound state on the surface of superconductor. For this system, there is no change in the wavefunction of electrons on the superconductor side in comparison with the metal-superconductor tunneling junction previously studied: there is a transmission wave along the $+x$ direction but no incident wave, as given by Eq. (5.14). However, for the bound state problem, $q_\pm$ must be complex, and the wavefunction should decay along the $+x$ direction. If the states other than the bound states are considered, the contribution from the incident waves should also be considered in the superconductor. Apparently, the electron wavefunction in the metallic film is changed. It is now the superposition of the wavefunctions for all four possible momenta $k = (\pm k_\pm, k_\parallel)$. As the wavefunction vanishes at $x = -L$, it should take the form

$$\psi(r) = e^{ik_\parallel \cdot r_\parallel} \begin{pmatrix} \alpha \sin k_+(x+L) \\ \beta \sin k_-(x+L) \end{pmatrix} \qquad (-L < x < 0). \qquad (5.37)$$

From the continuity of the wavefunction and its first order derivative at the metal-superconductor interface, we find that the coefficients α, β, c, and d satisfy the equations

$$
\begin{cases}
\alpha \sin k_+ L = c u_+ + d u_-, \\
\beta \sin k_- L = c v_+ + d v_-, \\
\alpha k_+ \cos k_+ L = i q_+ u_+ c - i q_- u_- d, \\
\beta k_- \cos k_- L = i q_+ v_+ c - i q_- v_- d.
\end{cases}
\tag{5.38}
$$

These equations have solutions if and only if

$$
\frac{(i q_+ \tan k_+ L - k_+) u_+}{(i q_+ \tan k_+ L - k_-) v_+} = \frac{(i q_- \tan k_+ L + k_+) u_-}{(i q_- \tan k_+ L + k_-) v_-}.
\tag{5.39}
$$

In the limit $L \to 0$, it becomes

$$
\frac{u_+}{v_+} = \frac{u_-}{v_-}.
\tag{5.40}
$$

For the s-wave superconductor, Eq. (5.40) is valid when $E = \Delta$. Thus the s-wave superconductor may have an interface bound state, but not at zero energy.

For the d-wave superconductor, if the nodal direction is perpendicular to the superconductor surface, then $v_\pm = \mp u_\mp$. Eq. (5.40) holds in the limit $E \to 0$. In this case,

$$
u_\pm^2 \to \pm \frac{i \Delta_k}{E},
$$

$$
q_\pm = \sqrt{k_{F_x}^2 \pm \frac{i 2 m \Delta_k}{\hbar^2}} \approx k_F \pm \frac{i m \Delta_k}{\hbar^2 k_F}.
$$

This is a bound state solution. The imaginary part of $q_\pm$ represents the characteristic decay wavevector and the corresponding decay length is

$$
l \approx \frac{\hbar^2 k_F}{m \Delta_k} = \frac{2 \varepsilon_F}{\Delta_k} k_F^{-1}.
\tag{5.41}
$$

Usually $\varepsilon_F \gg \Delta_k$, and thus the size of the bound state is very large. l is of the order of the superconducting coherence length.

5.5 Tunneling Hamiltonian

When the metal-superconductor junction is separated by an insulating layer not ideally contacted, the scattering of electrons at the interface may not be elastic. In this case, the electron tunneling cannot be treated as an elastic process, and the theoretical study becomes rather complicated. A frequently used approach is to treat the tunneling matrix element as a phenomenological parameter under proper approximations. This phenomenological approach does not require the conservation of electron energy and momentum during the tunneling process. In comparison with the previous approach which assumes the tunneling process is elastic, the application range of this phenomenological approach is broader. This approach is often used to study the tunneling problem in combination with the Green's function method. However, it is difficult to calculate the tunneling matrix elements from microscopic models and to know how good the approximation is.

If the interface potential is not a δ-function, the scattering coefficients usually can only be obtained numerically. If the insulating layer is thick and the transmission coefficient is small, the tunneling junction can be approximately described by an effective tunneling Hamiltonian as

$$H = H_L + H_R + H_T. \tag{5.42}$$

H_L and H_R are the Hamiltonians for the left- and right-hand side of the junction, respectively. The tunneling Hamiltonian H_T is defined by

$$H_T = \sum_{m,n}(T_{m,n}c^\dagger_{R,n}c_{L,m} + h.c.). \tag{5.43}$$

It describes the tunneling process of an electron from state m on the left side of the junction to state n on the right side of the junction. $T_{m,n}$ is the corresponding tunneling matrix element. The quantum states on the two sides of the junction can be very different.

If both sides of the tunneling junction are uniform conductors, the wavevector k and spin σ are good quantum numbers. In this case, $m = (k, \sigma)$ and $T_{m,n} = T_{k\sigma,k'\sigma'}$ is the tunneling matrix element of electrons from the state

(k, σ) to (k', σ'). If the scattering potential is spin-independent and the charac-teristic energy scales, including temperature and frequency, are much smaller than the Fermi energies of the normal metal and the superconductor, we can approximate the tunneling matrix elements as a state-independent constant:

$$T_{m,n} \approx T_{k_F \sigma, k'_F \sigma'} = T.$$

This approximation is broadly and successfully used in the study of tunneling problems, especially in the analysis of experimental results. Of course, this approximation is valid only when the tunneling matrix element is not sensitive to the momentum of electron. Otherwise, it needs to be modified.

The tunneling Hamiltonian Eq. (5.42) treats the two sides of the junction as two relatively independent systems connected by the tunneling matrix, similar to the tight-binding approximation in the band theory. This model neglects the interplay between H_T and $H_{L,R}$, and is applicable only to the system with small tunneling probability.

The tunneling Hamiltonian was first established in the study of elastic scat-tering at interface. It was extended to inelastic scattering systems by simply phenomenological consideration. For an elastic scattering system, the tunnel-ing matrix elements $T_{m,n}$ can be represented in terms of the interface barrier potential. However, for an inelastic scattering system, the tunneling matrix elements $T_{m,n}$ have to be taken as phenomenological fitting parameters.

Bardeen carried out the first microscopic investigation on the tunneling matrix elements [85]. Later, Harrison derived an explicit expression of $T_{m,n}$ under the WKB approximation [86]. Their works established a microscopic framework for the tunneling Hamiltonian, and provided a theoretical picture to understand the physical meaning of the tunneling Hamiltonian (5.42) and its application.

Let us consider an elastic scattering system with the interface lying in $x_a < x < x_b$. We assume that the scattering potential $U(x)$ depends only on x and the momentum parallel to the interface is conserved. In this case, we only need to consider the motion of electrons along the x-direction. The Hamiltonian is defined as

$$H = \begin{cases} H_L(x), & x < x_a, \\ -\dfrac{\hbar^2}{2m}\partial_x^2 + U(x), & x_a < x < x_b, \\ H_R(x), & x > x_b. \end{cases} \qquad (5.44)$$

It is not necessary to know the detailed expressions of $H_{L,R}(x)$. But we assume that the interactions among electrons or superconducting quasiparticles are very weak and the tunneling can be treated as a single-particle problem.

To discuss the tunneling problem, we need first understand the physical meaning of the electron operators $c_{L,m}$ and $c_{R,m}$ in the tunneling Hamiltonian. For this purpose, we assume that $\psi_{L,m}(x)$ is the wavefunction created by $c^\dagger_{L,m}$, which is defined in the entire space. Within the metal as well as the insulating layer side, $\psi_{L,m}$ is the eigen wavefunction of the Hamiltonian:

$$H\psi_{L,m} = E_m^L \psi_{L,m}, \qquad x < x_b. \qquad (5.45)$$

Physically, we are interested in the scattering problem with energy $E_m^L \ll U(x)$. In this case, $\psi_{L,m}$ would decay exponentially in the insulating region $x_a < x < x_b$. When $x > x_b$, $\psi_{L,m}$ is not the eigen solution of H_R. We do not need to know the concrete form of $\psi_{L,m}$. Nevertheless, the wavefunction needs to be continuous at $x = x_b$. Furthermore, $\psi_{L,m}$ decays fast in the region $x > x_b$, and vanishes in the limit $x \to +\infty$.

The state created by $c^\dagger_{R,m}$ is similarly defined. When $x > x_a$, $\psi_{R,m}$ is the solution to the eigen equation

$$H\psi_{R,m} = E_m^R \psi_{R,m}. \qquad (5.46)$$

It decays exponentially with decreasing x in the range $x_a < x < x_b$. In the region $x < x_a$, $\psi_{R,m}$ is not the solution to H_L, but it satisfies the continuity condition at $x = x_a$. Again, $\psi_{R,m}$ decays fast in the region $x < x_a$, and becomes zero as $x \to -\infty$.

$\psi_{L,m}$ and $\psi_{R,m}$ together form a set of non-orthogonal but complete bases. Any eigenstate $\psi(x)$ of H can be represented as linear superpositions of $\psi_{L,m}$ and $\psi_{R,m}$ as

$$\psi = \sum_{\alpha,m} a_{\alpha,m}\psi_{\alpha,m}. \qquad (5.47)$$

From the eigen equation of the Hamiltonian,

$$H\psi = E\psi, \tag{5.48}$$

we find that the eigen energy E is determined by the equation

$$\det A_{\beta,n;\alpha,m} = 0, \tag{5.49}$$

where

$$A_{\beta,n;\alpha,m} = \langle \psi_{\beta,n} | H - E | \psi_{\alpha,m} \rangle. \tag{5.50}$$

In Eq. (5.50), the off-diagonal terms of $\langle \psi_{\beta,n} | H | \psi_{\alpha,m} \rangle$ and $\langle \psi_{\beta,n} | \psi_{\alpha,m} \rangle$ are small. The eigenvalues are predominately determined by the diagonal terms such that E is approximately equal to E_m^α. This implies that the tunneling takes place just between states with $E_m^L \approx E_n^R \approx E$. The corresponding tunneling matrix element is then given by

$$T_{m,n} = A_{L,m;R,n} \approx \langle \psi_{L,m} | H - E_n^R | \psi_{R,n} \rangle = \int_{-\infty}^{x_1} dx \psi_{L,m}^* (H - E_n^R) \psi_{R,n}, \tag{5.51}$$

where x_1 can take any value between x_a and x_b. Using the equation

$$\int_{-\infty}^{x_1} dx \psi_{R,n} (H - E_m^L) \psi_{L,m}^* = 0, \tag{5.52}$$

we find that the tunneling matrix element can be further expressed as

$$T_{m,n} \approx \int_{-\infty}^{x_1} dx \left(\psi_{L,m}^* H \psi_{R,n} - \psi_{R,n} H \psi_{L,m}^* \right), \tag{5.53}$$

where the approximation $E_m^L \approx E_n^R$ is used. If we assume

$$H_L(x) = -\frac{\hbar^2}{2m} \partial_x^2, \tag{5.54}$$

Eq. (5.53) can be further simplified as

$$\begin{aligned} T_{m,n} &\approx -\frac{\hbar^2}{2m} \int_{-\infty}^{x_1} dx \left(\psi_{L,m}^* \partial_x^2 \psi_{R,n} - \psi_{R,n} \partial_x^2 \psi_{L,m}^* \right) \\ &= -\frac{\hbar^2}{2m} \left[\psi_{L,m}^*(x_1) \partial_x \psi_{R,n}(x_1) - \psi_{R,n}(x_1) \partial_x \psi_{L,m}^*(x_1) \right]. \end{aligned} \tag{5.55}$$

This is the formula that was first obtained by Bardeen [85]. Eq. (5.55) depends on $\psi_{R,n}(x)$ and $\psi_{L,m}(x)$, but not explicitly on $H_R(x)$.

Now let us derive the tunneling matrix elements between two normal metals using Eq. (5.55). For simplicity, we assume H_R to have the form

$$H_R(x) = -\frac{\hbar^2}{2m}\partial_x^2 + V_R, \tag{5.56}$$

where V_R is a constant.

$H_{L,R}(x)$ describe free electrons. Their eigenstates can be represented by wavevectors. In the left conductor, since the tunneling probability is small, the reflection probability is close to 1. Therefore, the electron wavefunction is approximately a superposition of incident and reflection waves with an equal weight, which can be represented as

$$\psi_{L,k_x} = c_L \sin(k_x x + \gamma_L), \qquad x < x_a, \tag{5.57}$$

where c_L is the normalization constant independent on k_x, γ_L is the phase shift due to the scattering potential and

$$k_x = \sqrt{\frac{2mE_x}{\hbar^2}}. \tag{5.58}$$

In the scattering region, it is difficult to obtain a rigorous solution. However, if $U(x)$ is a slowly varying potential, the WKB approximation can be used, which gives

$$\psi_{L,k_x} = c_L \sqrt{\frac{k_x}{2p(x)}} \exp\left[-\int_{x_a}^{x} p(x)\mathrm{d}x\right], \qquad x_a < x < x_b, \tag{5.59}$$

where

$$p(x) = \sqrt{\frac{2m[V(x) - E_x]}{\hbar^2}}. \tag{5.60}$$

Similarly, the electron wavefunction on the right conductor can be obtained using the WKB approximation,

$$\psi_{R,q_x} = \begin{cases} c_R \sin(q_x x + \gamma_R), & x > x_b, \\ c_R \sqrt{\dfrac{q_x}{2p(x)}} \exp\left[-\int_{x}^{x_b} p(x)\mathrm{d}x\right], & x_a < x < x_b, \end{cases} \tag{5.61}$$

where c_R is the normalization constant, γ_R is the scattering phase shift and

$$q_x = \sqrt{\frac{2m(E_x - V_R)}{\hbar^2}}. \tag{5.62}$$

Substituting these results into Eq. (5.55), we find that the tunneling matrix elements can be expressed as [86]

$$
\begin{aligned}
T_{k_x,q_x} &= \frac{\hbar^2 c_L c_R \sqrt{k_x q_x}}{2m} \exp\left[-\int_{x_a}^{x_b} p(x)dx\right] \\
&= \frac{\hbar c_L c_R \sqrt{v_{L,x} v_{R,x}}}{2} \exp\left[-\int_{x_a}^{x_b} p(x)dx\right],
\end{aligned} \tag{5.63}
$$

where $v_{L,x} = \hbar k_x/m$ and $v_{R,x} = \hbar q_x/m$ are the velocities of electrons in the left and right conductors, respectively.

Eq. (5.63) shows that the tunneling probability is not only closely related to the scattering potential, but also depends on the velocities of the incident and transmission electrons. This result is obtained under the assumption that both sides of the tunneling junction are normal metals. In the case the tunneling probability is low, the hole reflection and transmission rates induced by the Cooper pairing are also small and negligible. This suggests that Eq. (5.63) can be also applied to a tunneling system with one or both sides of the junction being superconductors.

5.6 Tunneling Current

In an external bias, the tunneling circuit is actually a non-equilibrium system. To evaluate the tunneling current, one has to use the method of closed-time-path Green's functions which is generally difficult to implement. However, if the bias is small, the non-linear effect is negligible and the tunneling current can be evaluated using the conventional perturbation theory based on equilibrium states [9].

We assume that the eigenstates in both the metallic and superconducting sides of the junction can be labelled by the momenta of electrons and the tunneling Hamiltonian H_T, defined by Eq. (5.64), can be expressed as

$$H_T = \sum_{k,q} \left(T_{k,q} c_{Lk\sigma}^\dagger c_{Rq\sigma} + h.c.\right). \tag{5.64}$$

The corresponding tunneling current operator is defined by

$$\hat{I} = -\frac{2e}{\hbar}\text{Im}\sum_{k,q} T_{k,q} c_{Lk\sigma}^{\dagger} c_{Rq\sigma}. \tag{5.65}$$

In many textbooks, the tunneling current operator is defined through the time derivative of the particle numbers in the left or the right side of the junction [9]. These two kinds of definitions are equivalent if H_L commutes with the particle number in the left conductor,

$$N_L = \sum_{k,\sigma} c_{Lk\sigma}^{\dagger} c_{Lk,\sigma}. \tag{5.66}$$

However, if the left junction is also a superconductor, then under the BCS mean-field theory, H_L contains the BCS pairing interaction which does not preserve the particle number:

$$H_L = H_L^{(0)} + \sum_k \left(\Delta_{L,k} c_{Lk\uparrow}^{\dagger} c_{L-k\downarrow}^{\dagger} + \Delta_{L,k} c_{L-k\downarrow} c_{Lk\uparrow}\right),$$

$$0 = \left[H_L^{(0)}, N_L\right].$$

In this case, the electric current operator defined through the time-derivative of N_L is

$$-e\frac{dN_L}{dt} = \frac{-ie}{\hbar}[H, N_L]$$

$$= \frac{4e}{\hbar}\text{Im}\sum_k \Delta_{L,k} c_{L-k\downarrow} c_{Lk\uparrow} - \frac{2e}{\hbar}\text{Im}\sum_{k,q} T_{k,q} c_{Lk\sigma}^{\dagger} c_{Rq\sigma}. \tag{5.67}$$

Compared with Eq. (5.65), there is an extra term in the electric current operator contributed by paired electrons. In an equilibrium system, the average value of this term is zero. Thus the currents evaluated using Eq. (5.67) and Eq. (5.65) are equal to each other.

We will use H_T as a perturbation and treat the sum of H_L and H_R, $H_0 = H_L + H_R$, as the unperturbed Hamiltonian. In the interaction picture, the time-evolution of operators is defined by

$$B(t) = e^{iH_0 t} B e^{-iH_0 t}. \tag{5.68}$$

The left and right sides of the junction are in different equilibrium states. The thermodynamic average of this time-dependent operator is determined by the formula

$$\langle B \rangle_0 = \frac{\text{Tr} e^{-\beta K_0} B}{\text{Tr} e^{-\beta K_0}}, \tag{5.69}$$

where

$$K_0 = H_0 - \mu_L N_L - \mu_R N_R. \tag{5.70}$$

The chemical potential difference between the left and right conductors equals the external voltage:

$$eV = \mu_L - \mu_R.$$

Thus in order to study the time-evolution of operators in this non-equilibrium system using the approach based on the equilibrium states, one should replace H_0 by K_0. This is equivalent to replacing B by a new operator $\tilde{B}$, defined by

$$B(t) = e^{iK_0 t} \tilde{B} e^{-iK_0 t} = e^{iH_0 t} B e^{-iH_0 t},$$
$$\tilde{B} = e^{-iK_0 t} e^{iH_0 t} B e^{-iH_0 t} e^{iK_0 t}. \tag{5.71}$$

In the case H_L and H_R commute with their corresponding particle number operators, $\tilde{B}$ is simply given by

$$\tilde{B} = e^{i(\mu_L N_L + \mu_R N_R)} B e^{-i(\mu_L N_L + \mu_R N_R)}. \tag{5.72}$$

However, for the BCS mean-field Hamiltonian, H_L and H_R do not commute with the particle number operators. In this case the relationship between $\tilde{B}$ and B becomes much more complicated, and has not been explicitly studied in literature. Nevertheless, in the steady state, H_L and H_R commute with the particle number operators on thermal average, and Eq. (5.72) holds approximately. Thus for the BCS Hamiltonian, Eq. (5.72) is still approximately valid, although it is not a rigorous operator identity.

Using Eq. (5.72), the tunneling Hamiltonian in the interaction picture can be expressed as

$$H_T(t) = e^{iK_0 t} e^{i(\mu_L N_L - \mu_R N_R)t} H_T e^{-i(\mu_L N_L - \mu_R N_R)t} e^{-iK_0 t}$$

$$= e^{iK_0 t} \sum_{k,q,\sigma} \left(T_{k,q} e^{ieVt} c^\dagger_{Lk\sigma} c_{Rq\sigma} + T^*_{k,q} e^{-ieVt} c^\dagger_{Rq\sigma} c_{Lk\sigma} \right) e^{-iK_0 t}$$

$$= e^{ieVt} A(t) + e^{-ieVt} A^\dagger(t), \tag{5.73}$$

where

$$A(t) = \sum_{k,q,\sigma} T_{k,q} c^\dagger_{Lk\sigma}(t) c_{Rq\sigma}(t), \tag{5.74}$$

$$c_{\alpha k\sigma}(t) = e^{-iK_0 t} c_{\alpha k\sigma} e^{iK_0 t}. \tag{5.75}$$

Similarly, the tunneling current operator in the interaction picture is defined by

$$\hat{I}(t) = \frac{ie}{\hbar} \left[e^{ieVt} A(t) - e^{-ieVt} A^\dagger(t) \right]. \tag{5.76}$$

Up to the first order in H_T, the expectation value of the tunneling current can be expressed as

$$I(t) = \left\langle \hat{I}(t) \right\rangle_0 - \frac{i}{\hbar} \int_{-\infty}^{t} dt' \left\langle \left[\hat{I}(t), H_T(t') \right] \right\rangle_0. \tag{5.77}$$

There is no tunneling current in the absence of perturbation,

$$\left\langle \hat{I}(t) \right\rangle_0 = 0. \tag{5.78}$$

Thus $I(t)$ is completely determined by the second term in Eq. (5.77).

Substituting the expressions of $\hat{I}$ and $H_T(t)$ into Eq. (5.77), we find that $I(t)$ contains both the normal tunneling current due to quasiparticle tunneling and the Josephson current due to the tunneling of Cooper pairs. If we use $I_Q(t)$ and $I_J(t)$ to represent respectively these two kinds of currents, then

$$I(t) = I_Q(t) + I_J(t), \tag{5.79}$$

$$I_Q(t) = -\frac{2e}{\hbar^2} \mathrm{Im} \int_{-\infty}^{\infty} dt' e^{ieV(t-t')} X_{ret}(t - t'), \tag{5.80}$$

$$I_J(t) = \frac{2e}{\hbar^2} \mathrm{Im} \int_{-\infty}^{\infty} dt' e^{-ieV(t+t')} Y_{ret}(t - t'), \tag{5.81}$$

where

$$X_{ret}(t - t') = -i\theta(t - t')\left\langle\left[A(t), A^\dagger(t')\right]\right\rangle_0,$$ (5.82)

$$Y_{ret}(t - t') = -i\theta(t - t')\left\langle\left[A^\dagger(t), A^\dagger(t')\right]\right\rangle_0,$$ (5.83)

$$A(t) = \sum_{k,q,\sigma} T_{k,q} c_{Lk\sigma}^\dagger(t) c_{Rq\sigma}(t).$$ (5.84)

I_J exists only in the superconducting junction. There is no Josephson current if either side of the junction is a normal metal, i.e. $Y_{ret} = 0$.

Below we discuss the property of tunneling current of normal electrons, and leave the discussion on the Josephson tunneling current to Chapter 6.

5.7 Tunneling Current of Quasiparticles

The tunneling current of quasiparticles, Eq. (5.80), can be also expressed as

$$I_Q = -\frac{2e}{\hbar^2}\mathrm{Im}X_{ret}(eV),$$ (5.85)

where $X_{ret}(eV)$ is the Fourier transform of $X_{ret}(t)$ in the frequency space:

$$X_{ret}(eV) = \int_{-\infty}^{\infty} dt e^{ieVt} X_{ret}(t).$$ (5.86)

$X_{ret}(eV)$ can be calculated using the finite-temperature Green's function theory. The Matsubara Green's function corresponding to $X_{ret}(t)$ is defined by

$$X(\tau) = -\left\langle T_\tau A(\tau) A^\dagger(0)\right\rangle_0$$
$$= \sum_{k,q}|T_{k,q}|^2[G_{L,11}(k, -\tau)G_{R,11}(q, \tau) + G_{L,22}(k, \tau)G_{R,22}(\mathbf{q}, -\tau)].$$ (5.87)

In the imaginary frequency space, it becomes

$$X(i\omega_n) = \int_0^\beta d\tau e^{i\omega_n\tau} X(\tau) = X_1(i\omega_n) + X_2(-i\omega_n),$$ (5.88)

where

$$X_\alpha(i\omega_n) = \frac{1}{\beta}\sum_{k,q,p_n}|T_{k,q}|^2 G_{L,\alpha\alpha}(k, p_n)G_{R,\alpha\alpha}(q, p_n + \omega_n).$$ (5.89)

Using the spectral representation of the Green's functions, Eq. (5.89) can be represented using the retarded Green's function as

$$X_\alpha(i\omega_n) = \sum_{k,q} \frac{|T_{k,q}|^2}{\pi^2} \int d\omega_1 d\omega_2 \mathrm{Im}G^R_{L,\alpha\alpha}(k,\omega_1)$$

$$\mathrm{Im}G^R_{R,\alpha\alpha}(q,\omega_2)\frac{f(\omega_1)-f(\omega_2)}{i\omega_n+\omega_1-\omega_2}. \tag{5.90}$$

Substituting Eq. (5.90) into Eq. (5.88) and taking the analytical continuation, i.e. $i\omega_n \to \omega + i0^+$, we have

$$I_Q(eV) = \frac{2e}{\pi\hbar^2}\sum_{k,q}|T_{k,q}|^2\left[j_1(k,q,eV)-j_2(k,q,-eV)\right], \tag{5.91}$$

where

$$j_\alpha(k,q,\omega) = \int d\omega_1 \mathrm{Im}G^R_{L,\alpha\alpha}(k,\omega_1)\mathrm{Im}G^R_{R,\alpha\alpha}(q,\omega_1+\omega)\left[f(\omega_1)-f(\omega_1+\omega)\right]. \tag{5.92}$$

If the tunneling matrix element $T_{k,q}$ does not depend on k and q, i.e. $T_{k,q} = T_0$, then from the definition of quasiparticle density of states,

$$\rho(\omega) = -\frac{1}{\pi}\sum_k \mathrm{Im}G^R_{11}(k,\omega) = -\frac{1}{\pi}\sum_k \mathrm{Im}G^R_{22}(k,-\omega), \tag{5.93}$$

we can reexpress $I_Q(eV)$ as

$$I_Q(eV) = \frac{4\pi e|T_0|^2}{\hbar^2}\int d\omega \rho_L(\omega)\rho_R(\omega+eV)\left[f(\omega)-f(\omega+eV)\right]. \tag{5.94}$$

Hence the normal tunneling current is determined by the convolution of the density of states on the two sides of the junction.

Eq. (5.94) is a formula commonly used in the analysis of tunneling measurement results. At zero temperature, it reduces to

$$I_Q(eV) = \frac{4\pi e|T_0|^2}{\hbar^2}\int_0^{eV} d\omega \rho_L(\omega+eV)\rho_R(\omega). \tag{5.95}$$

If both sides of the junction are superconductors, the above integral with respect to ω is typically an elliptic integral and there is no analytic solution. Nevertheless, it can be shown that the derivative of I_Q with respect to V, i.e.

the differential conductance, reaches the maximum at $eV = \Delta_0^L + \Delta_0^R$, where Δ_0^L and Δ_0^R are the maximal gap values. If both sides are d-wave superconductors, the low energy density of states is linear, then at low voltage,

$$
\begin{aligned}
I_Q(eV) &\approx \frac{4\pi e|T_0|^2 N_{L,F} N_{R,F}}{\hbar^2 \Delta_0^L \Delta_0^R} \int_0^{eV} d\omega(\omega + eV)\omega \\
&= \frac{10\pi e^4|T_0|^2 N_{L,F} N_{R,F}}{3\hbar^2 \Delta_0^L \Delta_0^R} V^3,
\end{aligned}
\tag{5.96}
$$

which varies cubically with V. Thus the low voltage differential conductance is proportional to V^2. The quadratic power here is a consequence of the convolution of two linear density of states. If one side of the junction is an s-wave superconductor, and the other side is a d-wave superconductor, then the tunneling current is zero if eV is smaller than the s-wave superconducting gap Δ_s.

If one end of the tunneling junction is a superconductor and the other, say, the left end, is a normal metal, then the low energy electron density of states in the left end can be approximated by the value at the Fermi energy $N_{L,F}$. In this case, the tunneling current at low bias becomes

$$
I_Q(eV) = \frac{4\pi e|T_0|^2 N_{L,F}}{\hbar^2} \int_0^{eV} d\omega \rho_R(\omega),
\tag{5.97}
$$

and the corresponding differential conductance is given by

$$
\frac{dI_Q}{dV} = \frac{4\pi e^2|T_0|^2 N_{L,F}}{\hbar^2} \rho_R(eV),
\tag{5.98}
$$

which is proportional to the density of states of quasiparticles in the superconductor. Hence the density of states of quasiparticles in superconductors can be probed through the measurement of the tunneling current in a superconductor-insulator-metal junction.

The above analysis is valid when the tunneling matrix elements are momentum independent. However, in high-T_c cuprates, T_{kq} cannot be treated as a constant since the electron velocity along the c-axis depends strongly on the in-plane momentum of electrons in the CuO_2 plane if the tunneling current is along the c-axis. If the left-hand side of the junction is a high-T_c superconductor, and the right-hand side is a normal metal conductor, then

$$v_c(k) \propto \cos^2(2\varphi_L), \tag{5.99}$$

$$|T_{k,q}|^2 = |T_0|^2 \cos^2(2\varphi_L), \tag{5.100}$$

where φ_L is the azimuthal angle of the in-plane component of the momentum k, and T_0 is approximately a constant. In this case, the zero temperature tunneling current is determined by the formula

$$I_Q = -\frac{2e\pi N_F^R|T_0|^2}{\hbar^2} \sum_k \cos^2 2\varphi_L \int_0^{eV} d\omega \mathrm{Im}\left[G_{L,11}(k,\omega-eV) + G_{L,22}(k,\omega)\right]. \tag{5.101}$$

For an ideal d-wave superconductor, Eq. (5.101) can be simplified as

$$I_Q = \frac{2e\pi^2 N_F^R|T_0|^2}{\hbar^2} \sum_k \cos^2 2\varphi_L \int_0^{eV} d\omega \delta\left(\omega - E_k^L\right), \tag{5.102}$$

where $E_k^L = \sqrt{\varepsilon_{L,k}^2 + \Delta_{L,k}^2}$ is the quasiparticle spectrum. By changing the momentum summation into an integral over ε_k and setting $x = \cos(2\varphi_L)$, the above expression can be simplified as

$$I_Q = \frac{4e\pi N_F^R N_F^L|T_0|^2}{\hbar^2} \int_0^1 \frac{dx x^2}{\sqrt{1-x^2}} \sqrt{(eV)^2 - (\Delta_{L,0}x)^2}. \tag{5.103}$$

The right-hand side is an elliptic integral. When $eV \ll \Delta_{L,0}$, Eq. (5.103) can be approximately integrated out, which gives

$$I_Q \approx \frac{\pi^2 N_F^R N_F^L|T_0|^2 e^5 V^4}{4\hbar^2 \Delta_{L,0}^3}. \tag{5.104}$$

The corresponding differential conductance [87] is given by

$$\frac{dI_Q}{dV} \approx \frac{\pi^2 N_F^R N_F^L|T_0|^2 e^5 V^3}{\hbar^2 \Delta_{L,0}^3}. \tag{5.105}$$

This V^3-dependence of the differential conductance is clearly different from the result obtained in the system where the tunneling matrix elements $T_{k,q}$ are momentum independent. In the latter case, the differential conductance is linearly proportional to the density of states and varies linearly with V. The difference shows that the tunneling matrix elements have strong effects on the tunneling current and should be seriously considered in the analysis of tunneling experimental results in high-T_c superconductors. In fact, the cubic voltage dependence of the differential conductance can be obtained simply through dimensional analysis. The low energy density of states in the d-wave superconductors is linear, hence proportional to the energy. The function $\cos(2\varphi_L)$ in the tunneling matrix element has the same momentum dependence as the gap function, and thus has the same dimension as the energy. It implies that the effective dimension of $\cos^2(2\varphi_L)$ is 2. This, in combination with the linear density of states, leads to the cubic power of V in the differential conductance.

On the other hand, if both sides of the tunneling junction are high-T_c superconductors, then the tunneling matrix elements along the c-axis can be expressed as

$$|T_{k,q}|^2 = |T_0|^2 \cos^2(2\varphi_L)\cos^2(2\varphi_R). \tag{5.106}$$

Since the contribution to the quasiparticle tunneling current from the two sides of the junction is independent, it is simple to show that the differential conductance at low voltage is proportional to V^6 based on the dimensional analysis. This is confirmed by more sophisticated calculations. The zero temperature tunneling current is determined by the formula

$$I_Q = \frac{\pi e|T_0|^2}{\hbar^2} \sum_{k,q} \cos^2(2\varphi_L)\cos^2(2\varphi_R) \int_{-eV}^{0} \delta(\omega + E_k^L)\delta(\omega + eV - E_q^R). \tag{5.107}$$

After integrating over energy, it becomes

$$I_Q = \frac{e|T_0|^2 N_F^L N_F^R}{4\pi\hbar^2} \int_0^{eV} d\omega\,\omega(eV - \omega) \int_0^{2\pi} d\varphi_L d\varphi_R \mathrm{Re}$$

$$\frac{\cos^2(2\varphi_L)}{\sqrt{(eV-\omega)^2 - \Delta_{L,0}^2 \cos^2(2\varphi_L)}} \frac{\cos^2(2\varphi_R)}{\sqrt{\omega^2 - \Delta_{R,0}^2 \cos^2(2\varphi_R)}}. \tag{5.108}$$

At low bias, only low energy quasiparticles contribute to the tunneling current. The integrals over φ_L and φ_R can be calculated approximately, which yields

$$I_Q \approx \frac{\pi|T_0|^2 N_F^L N_F^R e^8 V^7}{560\hbar^2 \Delta_{L,0}^3 \Delta_{R,0}^3}. \tag{5.109}$$

The corresponding differential conductance is

$$\frac{\mathrm{d}I_Q}{\mathrm{d}V} = \frac{\pi|T_0|^2 N_F^L N_F^R e^8 V^6}{80\hbar^2 \Delta_{L,0}^3 \Delta_{R,0}^3}. \qquad (5.110)$$

Hence, as predicted, $\mathrm{d}I_Q/\mathrm{d}V$ is proportional to V^6 at small bias, different from the result $\mathrm{d}I_Q/\mathrm{d}V \propto V^2$ in the system where $T_{k,q}$ is a constant.

It should be emphasized that Eqs. (5.105) and (5.110) are valid only for ideal superconductors with infinite quasiparticle lifetime. If the superconductor is affected by disorder potentials or other effects, the quasiparticle lifetime becomes finite. If the energy scale of the inverse lifetime is larger than or equal to eV, the effect of the angular factor $\cos^2(2\varphi)$ will be smeared out. The tunneling current will become the same as that obtained based on the assumption that $T_{k,q}$ is a constant. Hence when applying these formulae to analyze the tunneling experiments of high-T_c superconductors, we should take full account of the scattering effect induced by impurities or other elementary excitations.

Chapter 6
Josephson Effect

6.1 Josephson Tunneling Current

Between two superconductors, there is not only the normal quasi-particle tunneling current, but also the so-called Josephson current contributed by the tunneling of Cooper pairs. This is also called the Josephson effect.

The Josephson current, defined by Eq. (5.81), can be also written as

$$I_J(t) = \frac{2e}{\hbar^2} \mathrm{Im} \left[e^{-2ieVt} Y_{ret}(eV) \right], \tag{6.1}$$

where

$$Y_{ret}(\omega) = \int_{-\infty}^{\infty} dt' e^{i\omega t} Y_{ret}(t) \tag{6.2}$$

is the Fourier transform of $Y_{ret}(t)$. Unlike I_Q, I_J is time-dependent. It indicates that a constant voltage can generate a time-dependent tunneling current. This peculiar property of the Josephson effect is a typical manifestation of quantum phase coherence.

In the imaginary time representation, the Matsubara function corresponding to $Y_{ret}(t)$ is defined by

$$
\begin{aligned}
Y(\tau) &= -\left\langle T_\tau A^\dagger(\tau) A^\dagger(0) \right\rangle_0 \\
&= -\sum_{k,q} T^*_{k,q} T^*_{-k,-q} \left[G_{L,12}(k,\tau) G_{R,21}(q,-\tau) + G_{L,12}(k,-\tau) G_{R,21}(q,\tau) \right].
\end{aligned}
\tag{6.3}
$$

The corresponding Fourier transform is

$$Y(i\omega) = \int_0^\beta d\tau e^{i\omega\tau} Y(\tau) = \tilde{Y}(i\omega) + \tilde{Y}(-i\omega), \tag{6.4}$$

where

$$\tilde{Y}(i\omega) = -\frac{1}{\beta} \sum_{k,q,p_n} T^*_{k,q} T^*_{-k,-q} G^L_{12}(k, p_n + \omega) G^R_{21}(q, p_n). \tag{6.5}$$

If one or both ends of the junction are $d_{x^2-y^2}$-superconductors and the tunneling matrix element $T^*_{k,q} T^*_{-k,-q}$ is invariant under $\pi/2$ rotation in momentum space, then

$$\sum_k T^*_{k,q} T^*_{-k,-q} G_{L,12}(k, p_n) = 0. \tag{6.6}$$

Consequently, the Josephson current vanishes:

$$I_J(t) = 0. \tag{6.7}$$

This is a consequence of d-wave pairing gap whose average over the Fermi surface is zero. It is also a common feature for all non-s-wave superconductors.

The absence of the Josephson tunneling current in a junction with one of the ends being a d-wave superconductors is a consequence of the first order perturbation. The contribution from higher order perturbations to the tunneling current is generally finite and should be evaluated in order to find the Josephson current of d-wave superconductors. Needless to say, the Josephson tunneling current between a s- and a d-wave superconductor is much smaller than that between two s-wave superconductors with the same tunneling matrix elements. This is a difference between these two kind of junctions.

However, if a Josephson junction is formed by coupling two d-wave superconductors through a weak link along a direction parallel to the ab-plane, $T^*_{k,q} T^*_{-k,-q}$ is not invariant under $\pi/2$ rotation in space. In this case, the Josephson current is finite even in the first order perturbation. When both sides of the junctions are d-wave superconductors, a physically interesting case is shown in Fig. 6.1, where the crystalline axes on one side are different from those in the other side. For this kind of Josephson junction, a simple calculation based on Eq. (5.63) and (6.5) yields

$$\tilde{Y}(i\omega) \propto -\frac{1}{\beta} \sum_{p_n} \chi(\theta_L, p_n + \omega) \chi(\theta_R, p_n) \Delta_{L,0} \Delta^*_{R,0}, \tag{6.8}$$

where

$$\chi(\theta_i, p_n) = \int_{\theta_i - \pi/2}^{\theta_i + \pi/2} d\phi \frac{\cos(2\phi)\cos(\phi - \theta_i)}{(ip_n)^2 - \varepsilon_k^2 - |\Delta_{i,0}|^2 \cos^2(2\phi)} \qquad (i = L, R).$$

It can be shown that under the $\pi/2$ rotation of θ_i, $\chi(\theta_i, p_n)$ changes sign:

$$\chi(\theta_i + \frac{\pi}{2}, p_n) = -\chi(\theta_i, p_n),$$

as a consequence of d-wave pairing. If $\theta_R = \theta_L + \pi/2$, then $\chi(\theta_L, p_n)\chi(\theta_R, p_n) < 0$ at $i\omega = 0$. This generates a π-phase shift in the Josephson current.

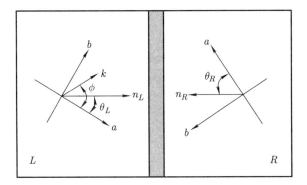

Figure 6.1   A Josephson junction of two superconductors connected along the ab-plane but with different crystalline orientations. θ_L and θ_R are the azimuthal angles between the a-axes and the normal direction on the left- and right-hand sides of the junction, respectively.

In general, if φ_L and φ_R are the superconducting phases on the left- and right-hand sides of the junction, $\Delta_{L,0} = |\Delta_{L,0}| \exp(i\varphi_L)$ and $\Delta_{R,0} = |\Delta_{R,0}| \exp(i\varphi_R)$, $\tilde{Y}(i\omega)$ can then be expressed as

$$\tilde{Y}(i\omega) \propto C(\theta_L, \theta_R, i\omega)e^{-i(\varphi_L - \varphi_R)}. \qquad (6.9)$$

In the absence of an applied bias or when the bias voltage is very low, if we neglect the frequency dependence in $C(\theta_L, \theta_R, i\omega)$, the above expression then becomes

$$\tilde{Y}(i\omega) \propto C(\theta_L, \theta_R)e^{i(\varphi_L - \varphi_R)}. \qquad (6.10)$$

Substituting it into Eq. (6.4) and performing an analytic continuation, we obtain immediately the formula of Josephson current between two d-wave superconductors:

$$I_J = I_0 C(\theta_L, \theta_R) \sin(2eVt + \varphi_L - \varphi_R). \tag{6.11}$$

This formula is a natural manifestation of the gauge invariance. By applying an external magnetic field, instead of an electric field, to the junction, the above formula becomes

$$I_J = I_0 C(\theta_L, \theta_R) \sin(\varphi_L - \varphi_R - \int_L^R \frac{2e}{\hbar} A \cdot \mathrm{d}l), \tag{6.12}$$

where A is the vector potential.

Compared with the corresponding formula for an s-wave superconductor, an extra geometrical factor $C(\theta_L, \theta_R)$, which is related to the orientation of the crystalline axes of the d-wave superconductor, emerges in Eq. (6.12). When the sign of $C(\theta_L, \theta_R)$ is changed, the phase of the Josephson current changes π. This effect is absent in the Josephson junction of s-wave superconductors. In a circuit containing one or more Josephson junctions of d-wave superconductors, it is possible to generate spontaneously a half-quantum magnetic flux. Detection of this spontaneously generated half-quantum flux allows us to determine whether the superconductor in the circuit has the d-wave pairing symmetry. This will be discussed in the next section.

Similar to $\chi(\theta_i, p_n)$, $C(\theta_L, \theta_R)$ changes sign when θ_L or θ_R is rotated by $\pi/2$:

$$C(\theta_L + \frac{\pi}{2}, \theta_R) = -C(\theta_L, \theta_R),$$

$$C(\theta_L, \theta_R + \frac{\pi}{2}) = -C(\theta_L, \theta_R).$$

Using this symmetry property, $C(\theta_L, \theta_R)$ can be expanded as [88]

$$C(\theta_L, \theta_R) = \sum_{nn'} \{C_{n,n'} \cos[(4n+2)\theta_L] \cos[(4n'+2)\theta_R]$$

$$+ D_{n,n'} \sin[(4n+2)\theta_L] \sin[(4n'+2)\theta_R]\}.$$

Both $C_{n,n'}$ and $D_{n,n'}$ depend on the band structures. However, in most cases, only the zeroth order terms ($n = n' = 0$) are important. Thus we have

$$C(\theta_L, \theta_R) \approx C_{0,0} \cos(2\theta_L) \cos(2\theta_R) + D_{0,0} \sin(2\theta_L) \sin(2\theta_R),$$

where the first term is that originally obtained by Sigrist and Rice [89]. If we further assume $D_{0,0} = -C_{0,0}$, the above expression then becomes the formula first obtained by Tsuei [90] for an extremely disordered Josephson junction:

$$C(\theta_L, \theta_R) \approx C_{0,0} \cos[2(\theta_L + \theta_R)].$$

6.2 Spontaneous Magnetic Flux Quantization

In a superconducting ring composed of a single or multiple Josephson junctions, for example, a superconducting quantum interference device (SQUID), if all the ends of the junctions are s-wave superconductors, then the flux trapped inside the ring is quantized and the minimal flux quantum is given by

$$\Phi_0 = \frac{h}{2e} = 2 \times 10^{-15} Wb = 2.07 \times 10^{-7} \text{Gs} \cdot \text{cm}^2. \tag{6.13}$$

This flux quantization is a consequence of superconducting phase coherence, which is a macroscopic quantum phenomenon that can be generated by applying an external magnetic field or a supercurrent. However, if one or more of the superconductors in the ring are d-wave superconductors, then the condition for the flux quantization inside the ring is modified and a half-quantum flux may emerge spontaneously. This can be understood from the Ginzburg-Landau supercurrent formula (1.57)

$$J_s = \frac{ie\hbar}{2m} \left(\psi^* \nabla \psi - \psi \nabla \psi^* \right) + \frac{2e^2}{m} A\psi^* \psi.$$

In the superconducting state, the fluctuation of the pairing amplitude is weak, and only the spatial variation of the superconducting phase ϕ needs to be considered. In this case, the supercurrent formula is simplified to

$$J_s = \frac{e\hbar}{m} |\psi|^2 \nabla \varphi - \frac{2e^2 |\psi|^2}{m} A. \tag{6.14}$$

It can also be expressed as

$$A + \frac{m}{2e^2 |\psi|^2} J_s = \frac{\Phi_0}{2\pi} \nabla \varphi. \tag{6.15}$$

Taking a loop integration for the above equation around the ring, the left-hand side just equals the sum of the magnetic flux generated by the external magnetic field inside the ring, $\Phi_a = HS$ (S is the area enclosed by the loop), and the flux generated by the supercurrent around the ring, $\Phi_s = LI_s$ (L is inductance of the ring):

$$\oint_C dl \cdot \left(A + \frac{m}{2e^2|\psi|^2} J \right) = \Phi_a + \Phi_s. \tag{6.16}$$

On the right-hand side, the loop integral of $\nabla\varphi$, up to a multiple of 2π, is equal to the sum of the phase difference across each junction, hence

$$\Phi_a + \Phi_s = \frac{\Phi_0}{2\pi} \sum_{\langle ij \rangle} \phi_{ij} + n\Phi_0. \tag{6.17}$$

$\langle ij \rangle$ represents the Josephson junction formed by two neighboring superconductors i and j. ϕ_{ij} is the phase difference across the junction, and n is an integer.

In the absence of the external voltage, the supercurrent inside the ring can be expressed in terms of the junction parameters according to Eq. (6.11) as

$$I_s = |I_{ij}^0| \sin(\phi_{ij} + \delta_{ij}), \tag{6.18}$$

where δ_{ij} equals either 0 or π, depending on the sign of the coefficient $C(\theta_L, \theta_R)$ in Eq. (6.11). If $\delta_{ij} = \pi$, this junction is called a π-junction.

As far as the spontaneous flux quantization is concerned, the supercurrent I_s around the ring is generally very small and the magnetic flux generated by I_s is just of the order of Φ_0. However, the critical current I_{ij}^0 at each junction is generally much larger than I_s, i.e. $|I_{ij}^0| \gg I_s$. This implies that $L|I_{ij}^0| \gg \Phi_0$ and $|\sin(\phi_{ij} + \delta_{ij})| \ll 1$. Thus the sine function in Eq. (6.18) can be expanded up to the leading order as

$$\frac{I_s}{|I_{ij}^0|} \approx \phi_{ij} + \delta_{ij} + 2\pi m_{ij}, \tag{6.19}$$

where m_{ij} is an integer satisfying the inequality

$$|\phi_{ij} + \delta_{ij} + 2\pi m_{ij}| \ll \pi.$$

Substituting Eq.(6.19) into Eq. (6.17), we have

$$2\pi \frac{\Phi_a + \Phi_s}{\Phi_0} + \sum_{\langle ij \rangle} \frac{I_s}{|I_{ij}^0|} \approx \sum_{\langle ij \rangle} \delta_{ij} + 2\pi m, \tag{6.20}$$

where m is an integer. After neglecting the second term on the left-hand side, this equation becomes

$$\Phi_a + \Phi_s \approx \left(\sum_{\langle ij \rangle} \delta_{ij} + 2\pi m \right) \frac{\Phi_0}{2\pi}. \tag{6.21}$$

If the number of π-junctions inside the ring is odd, the sum over δ_{ij} is equal to an odd number times of π. The flux quantization condition then becomes

$$\Phi_a + \Phi_s = \left(m + \frac{1}{2} \right) \Phi_0, \tag{6.22}$$

which is fundamentally different from the system without π-junctions. It should be emphasized that the half-quantum flux relies on the phase change of the pairing order parameter. Only when the pairing phase is momentum dependent, for example, in a d-wave superconductor, the half-quantum flux can emerge. It does not appear in a ring with only isotropic s-wave superconductors. Therefore, the detection of the half-quantum flux can be used to judge *decisively* whether the superconducting pairing is of d-wave symmetry. Furthermore, in the absence of an external field, $\Phi_a = 0$, Eq. (6.22) shows that the ring encloses a finite flux whose minimal value equals half of the flux quantum Φ_0. This is a spontaneous generation of the half-quantum flux in a ring consisting of odd number of π-junctions. It reveals a fundamental difference between s and d-wave superconductors.

On the other hand, if the number of π-junctions is even, the sum over the phase difference δ_{ij} equals an integer multiple of 2π. In this case, the magnetic flux quantization condition is the same as for a ring formed purely by s-wave superconductors

$$\Phi_a + \Phi_s = m\Phi_0. \tag{6.23}$$

In the absence of an external magnetic field, the state with $m = 0$ generally has a minimal energy, and the system has no spontaneously generated magnetic flux.

6.3 The Phase-Sensitive Experiments

Pairing symmetry of superconducting electrons can be detected by the measurement of quantum interference effects of Josephson junctions. Unlike thermodynamic or electromagnetic transport measurements, this class of experiments depend on the phases of superconducting gap functions, but not on their magnitudes. In other words, they are phase-sensitive, and The measurement results depend entirely on the macroscopic interference between Josephson junctions, regardless of the microscopic detail of specific materials.

The phase-sensitive experiments of high-T_c superconductors focus mainly on the measurement of quantum interference effects of single or double Josephson junctions and the detection of spontaneous magnetic fluxes in a circuit of Josephson junctions. A detailed discussion on the experimental results can be found from Refs. [91, 92].

6.3.1   Quantum Interference Effect of Josephson Junctions

One of the most important applications of the Josephson effect is to fabricate superconducting quantum interference devices (SQUID). The simplest SQUID is a loop circuit composed of two parallel connected Josephson junctions. The supercurrent in the circuit is very sensitive to the magnetic flux enclosed by the loop, providing an ideal tool to probe weak magnetic fields. Furthermore, the quantum interference effect with two Josephson junctions can be used to detect pairing symmetry. It plays an important role in the study of high-T_c superconductivity.

A typical SQUID used in high-T_c quantum interference experiments, as depicted in Fig. 6.2(a), is constructed by connecting a high-T_c superconductor, say YBCO, with an s-wave superconductor through a weak link. It is called a corner SQUID, since the two junctions are located on the two edges of a corner, touching the ac and bc surfaces of a YBCO crystal, respectively. We denote these two junctions as a and b, and the corresponding phase differences across the junctions as ϕ_a and ϕ_b, respectively. The supercurrent in the system is the sum of the Josephson tunneling currents through these two junctions:

$$I_s = I_{s,a} + I_{s,b},$$
$$I_{s,a} = |I_a| \sin(\phi_a + \delta_a), \qquad (6.24)$$
$$I_{s,b} = |I_b| \sin(\phi_b + \delta_b),$$

where $\delta_{a,b}$ equals zero for a zero junction, or π for a π-junction.

Taking the integral over Eq. (6.15) along the loop of this SQUID, it can be shown that the phase difference between ϕ_a and ϕ_b satisfies the equation

$$\phi_a - \phi_b = 2\pi \frac{\Phi_a}{\Phi_0} + 2\pi \frac{I_{s,a}L_a - I_{s,b}L_b}{\Phi_0} + 2\pi n, \qquad (6.25)$$

where L_a and L_b are the self-inductances of the a and b junctions, respectively. For a symmetric SQUID, $I_{s,a}L_a = I_{s,b}L_b$, the second term on the right-hand side of Eq. (6.25) vanishes and the above equation becomes

$$\phi_a - \phi_b = 2\pi \frac{\Phi_a}{\Phi_0} + 2\pi n. \qquad (6.26)$$

If the maximal tunneling currents of these two junctions are equal to each other, i.e. $|I_a| = |I_b|$, then the total supercurrent is

$$I_s = 2|I_a| \sin \gamma_0 \cos \frac{1}{2}(\phi_a + \delta_a - \phi_b - \delta_b), \qquad (6.27)$$

where γ_0 is an adjustable phase factor,

$$\gamma_0 = \frac{1}{2}(\phi_a + \delta_a + \phi_b + \delta_b).$$

The maximal value of I_s takes place at $\gamma_0 = \pi/2$ and is given by

$$I_{max} = 2|I_a| \cos \left(\frac{\pi \Phi_a}{\Phi_0} + \frac{\delta_a - \delta_b}{2} \right). \qquad (6.28)$$

For both a- and b-junctions, if one end of the junction is an s-wave superconductor and the other is a $d_{x^2-y^2}$-wave superconductor, and the tunneling directions are along the two principle axes of the $d_{x^2-y^2}$-wave superconductor, then $|\delta_a - \delta_b| = \pi$ as shown in Fig. 6.2(a). In this case, $|I_{max}|$ reaches the maxima and minima when Φ_a equals the half-integer and integer flux quanta, respectively. On the other hand, if the two junctions are formed by connecting the s-wave and the $d_{x^2-y^2}$ superconductors on the same surface through weak links, as shown in Fig. 6.2(b), then $|\delta_a - \delta_b| = 0$. Now $|I_{max}|$ reaches maxima when Φ_a equals zero or an integer flux quantum, and reaches minima when Φ_a equals an half-integer flux quantum.

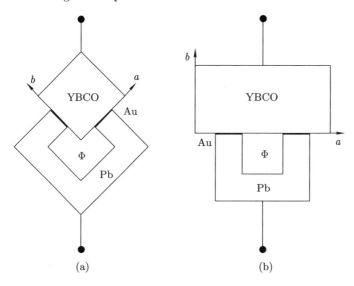

(a)   (b)

Figure 6.2   A SQUID consisting of an s-wave superconductor, Pb, and a high T_c superconductor, YBCO. The two superconductors are linked by the gold film. The two tunneling junctions are located respectively on the ac and bc-surfaces of YBCO for the corner SQUID (a), and on the same ac-surface of YBCO for the edge SQUID (b).

The above discussion indicates that, in a weak external magnetic field, the quantum interference current in the SQUID composed of a s- and a d-wave superconductor depends on the geometric configuration of the junctions. When the two junctions lie on the same edge of the d-wave superconductor, the maximal supercurrent appears when the external magnetic flux is zero. In contrast, when the two junctions lie on the two adjacent edges, the maximal supercurrent appears when the external flux equals $\pm\Phi_0/2$. Clearly, this property can be used to detect the pairing symmetry.

Based on the above discussions, Wollman $et.$ $al.$ carried out the first phase-sensitive measurement for high-T_c superconductors [93]. They measured the maximal bias current as a function of the external flux in two different SQUID shown in Fig. 6.2. The results are shown in Fig. 6.3. In their experiments, what they measured is the periodic modulation of the resistance with an applied magnetic field in the fluctuation regime of the critical current. The variation of the minimal resistance with the external flux exhibits a phase shift at each given bias current. For the two SQUID shown in Fig. 6.2 (a) and (b), it can be shown that the phase shifts equal $\Phi_0/2$ and 0 in the limit of zero bias current, respectively. In order to determine this phase shift, they extrapolated the data to the limit of zero bias current and found that the resulting shift is around $0.3 \sim 0.6$ Φ_0 for the corner-SQUID shown in Fig. 6.2(a). On the other hand, for the edge-SQUID shown in Fig. 6.2 (b), they found that the phase shift in the zero bias current limit is around zero, which is qualitatively different from the previous case. Their results do not ensure that these two phase shifts are precisely located at 0 and $\Phi_0/2$, but the trend it revealed provides strong evidence in support of the *d*-wave pairing symmetry of high-T_c superconductivity.

In Fig. 6.3, the extrapolated results exhibit a considerable variation for different samples. The variance may arise from two effects. First, the SQUID may not be as symmetric as expected due to the twin crystal structure or the orthogonal distortion of the YBCO crystals, which leads to the difference between the tunneling matrix elements and structures of the two junctions. Second, there may exist residual fluxes inside the loop. Both of them can cause the phase shifts to deviate from their ideal values.

In order to eliminate the uncertainty resulting from the asymmetry as well as the residual flux in the SQUIDs, Wollman *et. al.* proposed to use a single Josephson junction to detect the paring symmetry [93, 94]. Similar to the SQUID shown in Fig. 6.2, they probe the pairing symmetry by measuring the interference effect caused by the phase difference between the tunneling currents from the two adjacent edges. The structure of this Josephson tunneling junction is illustrated in Fig. 6.4. It is a single Josephson tunneling junction because the same junction touches both *ac* and *bc* surfaces of YBCO, unlike

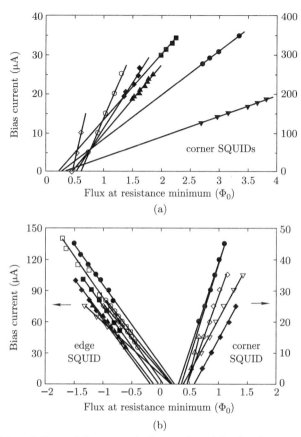

Figure 6.3   Extrapolations of the magnetic flux at the minimal resistance as a function of the bias current for the two SQUID shown in Fig. 6.2. Different symbols stand for different samples. For the corner SQUID shown in figure 6.2(a), the flux in the zero current limit should take the half-quantum value $\Phi_0/2$ for the d-wave superconductor. In contrast, for the edge SQUID shown in Fig. 6.2(b), the flux in the zero current limit should occur at zero flux regardless of the pairing symmetry. Reprinted with permission from [93]. Copyright (1993) by the American Physical Society.

the SQUID shown in Fig. 6.2(a).

The Josephson junction shown in Fig. 6.4(b) is not much different from the usual one. For this kind of junctions, the phase difference between the two superconductors can be adjusted by an external magnetic field. The tunneling

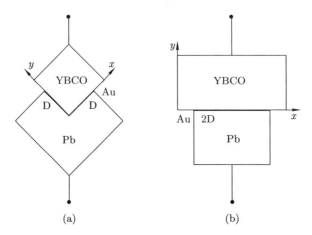

Figure 6.4   Josephson junction consisting of an *s*-wave superconductor and a *d*-wave super-conductor. (a) a corner sharing junction; (b) an edge-sharing junction.

currents vary in a pattern very similar to the single slit Fraunhofer diffraction in optics. For a perpendicular magnetic field, if we assume the length of the junction 2D is smaller than the London penetration depth, then within this length scale of the junction, the magnetic field distribution is nearly uniform. In the London gauge, the vector potential reads

$$A_x = 0, \qquad A_y = B_0 x.$$

Under this gauge, the gauge invariant phase difference varies with x and can be expressed as

$$\phi_a - \phi_b - \frac{2e}{\hbar} \int_{-l_a - D/2}^{l_b + D/2} B_0 x = \phi_a - \phi_b - \frac{2e L B_0 x}{\hbar},$$

where $L = l_a + l_b + D$ equals the summation of the penetration lengths λ_a and λ_b of superconductors on the two adjacent sides, and the width of the junction D. According to Eq. (6.12), the total Josephson tunneling current is

$$I_s = j_0 \int_{-D}^{D} dx \sin\left(\phi_a - \phi_b - \frac{2e}{\hbar} B_0 L x\right)$$

$$= 2D j_0 \sin(\phi_a - \phi_b) \frac{\sin(\pi\Phi/\Phi_0)}{\pi\Phi/\Phi_0}, \qquad (6.29)$$

where j_0 is the Josephson current density per unit length. The dependence of the critical tunneling current on the magnetic flux is

$$I_{max} = I_0 \left| \frac{\sin(\pi\Phi/\Phi_0)}{\pi\Phi/\Phi_0} \right|, \tag{6.30}$$

where $\Phi = 2B_0LD$. I_{max} reaches the maximum when $\Phi_a = 0$.

For the Josephson junction shown in Fig. 6.4(a), an s-wave superconductor is connected to the two adjacent surfaces, i.e. the ac and bc surfaces, of a high-T_c superconductor. For the convenience in the calculation, we take different gauges for the vector potentials on the two sides of the junction, and calculate the tunneling currents on the ac- and bc-surfaces separately. The total current is simply given by the sum of these two tunneling currents. It should be noted that the directions of currents from the two surfaces are orthogonal to each other, and only the components along the vertical direction contribute to the total current. If the superconducting electrons are s-wave paired, then the total tunneling current is given by the formula

$$\begin{aligned} I_s &= \frac{1}{\sqrt{2}} j_0 \int_0^D dx \sin\left(\phi_a - \phi_b - \frac{2e}{\hbar} B_0 Lx \right) \\ &+ \frac{1}{\sqrt{2}} j_0 \int_0^D dy \sin\left(\phi_a - \phi_b + \frac{2e}{\hbar} B_0 Ly \right) \\ &= \sqrt{2} j_0 D \sin(\phi_a - \phi_b) \frac{\sin(\pi\Phi/\Phi_0)}{\pi\Phi/\Phi_0}. \end{aligned} \tag{6.31}$$

Except an extra factor $\sqrt{2}$, this result is the same as for the Josephson junction shown in Fig 6.4(b).

On the other hand, if the superconductor possesses the $d_{x^2-y^2}$-wave pairing symmetry, the tunneling currents from the ac and bc surfaces exhibit a π-phase difference and the total tunneling current becomes

$$\begin{aligned} I_s &= \frac{1}{\sqrt{2}} j_0 \int_0^D dx \sin\left(\phi_a - \phi_b - \frac{2e}{\hbar} B_0 Lx \right) \\ &- \frac{1}{\sqrt{2}} j_0 \int_0^D dy \sin\left(\phi_a - \phi_b + \frac{2e}{\hbar} B_0 Ly \right) \\ &= -\sqrt{2} j_0 D \cos(\phi_a - \phi_b) \frac{\sin^2(\pi\Phi/2\Phi_0)}{\pi\Phi/\Phi_0}. \end{aligned} \tag{6.32}$$

It reaches the maximum when $\cos(\phi_a - \phi_b) = 1$,

$$I_{max} = I_0 \frac{\sin^2(\pi\Phi/2\Phi_0)}{\pi\Phi/\Phi_0}. \tag{6.33}$$

Different from the Josephson junction formed by two *s*-wave superconductors, $I_{max} = 0$, rather than taking the maximal value, at $\Phi = 0$. From the derivative, it can be shown that the maximum of I_{max} is determined by the equation

$$\pi\Phi = \tan(\pi\Phi/2\Phi_0).$$

Solving this equation numerically, we find that the maximum of I_{max} occurs at $\Phi \approx 0.742\Phi_0$.

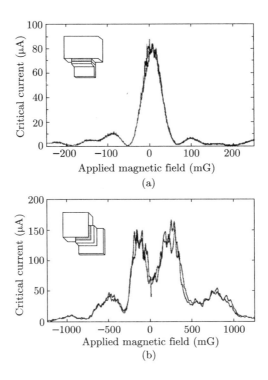

Figure 6.5   Magnetic field dependence of the tunneling current for the two Josephson junctions shown in Fig. 6.4. Reprinted with permission from [94]. Copyright (1995) by the American Physical Society.

Wollman *et. al.* measured the field dependence of the critical tunneling currents for these two Josephson junctions. Fig. 6.5 shows their results [94]. For the edge Josephson junction illustrated in Fig. 6.4(b), the maximal critical current occurs at zero field. In contrast, for the corner Josephson junction in Fig. 6.4(a), the critical current at zero field is a local minimum. It does not reach zero, probably because the junction is not completely symmetric: YBCO does not possess the tetragonal symmetry due to the existence of the CuO chains. This suggests that the pairing symmetry in YBCO cannot be purely d-wave, it should also contain a small s-wave component, which contributes a finite critical current at zero field. This set of experimental results agree with the theoretical prediction for d-wave superconductors. It lends strong support to the theory that the high-T_c electrons are d-wave paired.

6.3.2   Spontaneous Quantized Flux

In a superconducting quantum interference ring consisting of an odd number of π-junctions, there exists a spontaneously generated half-quantum flux, which can be used to judge whether high-T_c electrons have the d-wave pairing symmetry or not. The appearance of a spontaneous quantized flux relies only on the sign change of the superconducting gap function along different directions, rather than on the gap amplitude. This is again a phase-sensitive probe for the pairing gap, similar to the measurement of the interference effect in a Josephson junction or SQUID. It provides a powerful tool to determine qualitatively the pairing symmetry. Furthermore, the presence of the spontaneous half-quantum flux is not affected by the symmetry in the tunneling parameters of Josephson junctions or SQUID. If the orientation of the crystalline axes can be delicately designed to generate a spontaneous half-quantum magnetic flux, then not only the sign change of the gap function under the spatial rotation, but also the direction of the nodal lines can be accurately determined.

Tsuei and Kirtley from the IBM carried out the first experimental measurement for the spontaneous half-quantum flux. They grew epitaxially a high quality high-T_c film of YBCO on a tricrystal substrate of SiTrO$_3$ with three delicately designed orientations. Four Josephson rings were etched on the film as shown in Fig. 6.6. Because of the difference in the orientations of the crystalline axes, the phase differences at different crystal interfaces are different. The central ring crosses three interfaces, whose phase changes π for the *d*-wave superconductor. Hence it is a π-junction. For the other three rings, two of them cross the same interface twice, and the third one does not cross any interface. The phase changes around these rings are zero, and thus there are no spontaneously generated magnetic fluxes.

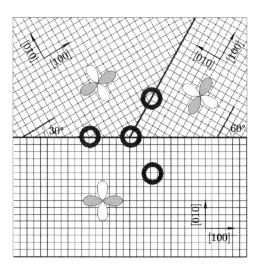

Figure 6.6   YBCO film epitaxially grown on the (100) surface of SiTiO$_3$: orientations of the crystalline axes and the phase patterns of the $d_{x^2-y^2}$ order parameters. The tricrystal substrate SiTiO$_3$ is composed of three domains with different crystalline orientations. The crystalline axes of the YBCO film are aligned by the crystalline axes of SiTiO$_3$. Tunneling junctions naturally formed at the interfaces exhibit different phase differences. The thickness of the film is 1200 Å. The four etched rings each has an inner radius 48 μm and a width 10 μm. If YBCO is a *d*-wave superconductor, the central ring should be a π-ring and contains a spontaneously generated half-quantum flux. For the other three rings, the accumulated phase differences are zero and there are no spontaneously generated magnetic fluxes. Reprinted from [95], with permission from Elsevier.

Tsuei and Kirtly measured the fluxes in the four rings using the scanning

SQUID microscope. Their results are shown in Fig. 6.7 [90, 92]. In the absence
of external magnetic field, they found that the flux enclosed by the central
tri-junction ring equals $\Phi_0/2$ within the experimental error, while the fluxes
through the other three junctions are zero. Their results are fully consistent
with the theoretical prediction of d-wave superconductors. They measured systematically how the magnetic fluxes change using different tri-crystal films by
varying the crystalline orientations of the substrate. All the results they obtained support YBCO to have the d-wave pairing symmetry.

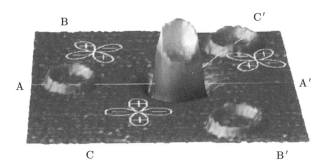

Figure 6.7   The scanning SQUID image for the tricrystal Josephson rings shown in Fig.
6.6. Within the experimental error, the flux through the central three-junction ring equals
$\Phi_0/2$, while for other three rings the fluxes are zero. Reprinted with permission from [92].
Copyright (2000) by the American Physical Society.

They have also investigated systematically properties of spontaneously generated half-quantum fluxes in other high-T_c superconductors, including both
hole- and electron-doped ones[92, 95, 96, 97, 98, 99, 100]. Their results are all
consistent with the d-wave pairing theory, and suggest that the d-wave pairing
is a universal feature of high-T_c superconductivity.

6.4 Paramagnetic Meissner Effect

The preceding discussion shows that, there are spontaneously generated orbital currents or half-quantum fluxes in a Josephson junction ring composed
of d-wave superconductors. This spontaneously generated flux can also exist in

granular superconductors. It may induce the so-called paramagnetic Meissner effect, or the anti-Meissner effect, leading to a positive magnetic susceptibility in a weak magnetic field [89, 101].

The paramagnetic Meissner effect arises from the spontaneous fluxes generated in the granular superconductor. It depends on the granular structure. Two adjacent grains can be regarded as a weakly coupled Josephson junctions. A large number of weakly coupled grains form a Josephson junction network which contains numerous loops. In a *d*-wave superconductor, some of the loops contain zero or even number of π-junction and there are no spontaneously generated magnetic fluxes. But the other loops contain odd number of π-junctions and have finite magnetic fluxes. In the absence of an applied magnetic field, these spontaneous magnetic fluxes (orbital moments) are randomly oriented and the net flux is zero. However, an applied magnetic field can polarize these fluxes. If the polarized magnetic moments from the π-loops surpass the contribution from the superconducting diamagnetic current, the granular superconductor becomes paramagnetic.

However, the magnetic moments of these spontaneously generated fluxes are generally very small. A weak magnetic field can completely polarize them. Hence the paramagnetic susceptibility is very weak. It becomes visible only in the vicinity of the superconducting transition temperature and in the weak field limit. Moreover, the paramagnetic susceptibility decreases with increasing field. It drops to zero when the paramagnetic moments contributed from the π-loops become fully polarized. On the other hand, the superconducting diamagnetic current increases with the applied magnetic field. When the magnetic field is above a threshold, the diamagnetic moments dominate and the granular superconductor becomes diamagnetic. This transition field from the paramagnetic to diamagnetic phases is very small, typically less than 1 Gauss. Thus the magnetic field used to measure the paramagnetic Meissner effect should be as weak as possible, provided that the background contribution to the fluctuating magnetic moment can be well screened.

To measure the paramagnetic Meissner effect, one needs to cool down the granular system from the normal phase to the superconducting phase. There are two situations that should be distinguished. One is the so-called field cooling. It is to apply a magnetic field before cooling down. The other is the zero-field cooling. It is to apply a magnetic field after cooling down to the superconducting phase.

In the case of field cooling, the spontaneous fluxes generated by the Josephson loops just below the superconducting transition temperature are aligned with the applied field. In a weak applied field, the superconducting diamagnetic current is small, and the paramagnetic response is stronger than the diamagnetic one. The system is paramagnetic. With the increase of the field, the diamagnetic effect becomes stronger and stronger, and the system eventually becomes diamagnetic. Fig. 6.8 shows the susceptibility of the granular $Bi_2Sr_2CaCu_2O_8$ superconductor under the field cooling as a function of temperature at several different external magnetic fields. As expected, the magnetic susceptibility is positive, or paramagnetic, in the weak fields. It decreases with increasing field and becomes diamagnetic after the magnetic field exceeds a threshold.

In the case of zero field cooling, the magnetic field is applied after the temperature falls far below the superconducting transition temperature and the susceptibility is measured with increasing temperature. As the measurement time is usually very short, there is no enough time for the orbital moments spontaneously generated in the π-loops to relax towards the direction of the applied magnetic field. The orbital current in each π-loop is randomly oriented as in the zero field and the orbital moments cancel each other, yielding an extremely small net paramagnetic moment. As a result, the diamagnetic effect dominates and the susceptibility is always negative. Nevertheless, the difference between the susceptibility under high and low magnetic fields,

$$\delta\chi(T, H < H^*) = \frac{M(T, H)}{H} - \frac{M(T, H^*)}{H^*}, \tag{6.34}$$

still contains information about the paramagnetism in the π-Josephson loops. $M(T, H)$ is the total magnetic moment of the system. In the experiment, H^* is typically around a few Gauss, much higher than the field to fully polarize the paramagnetic orbital moments so that the nonlinear effect of paramagnetic

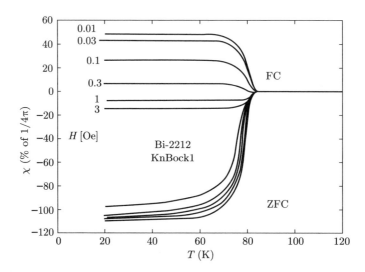

Figure 6.8   Temperature dependence of the magnetic susceptibility at various magnetic fields under the field cooling (FC) and the zero field cooling (ZFC) for $Bi_2Sr_2CaCu_2O_8$ granular superconductors. Reprinted with permission from [102]. Copyright (1992) by the American Physical Society.

moments can be ignored. Thus $M(T, H^*)/H^*$ contributes almost entirely by the diamagnetic current. The relative susceptibility $\delta\chi(T, H < H^*)$ defined above subtracts the contribution from the nearly field-independent diamagnetic susceptibility. It represents the paramagnetic response from the π-rings in the weak external magnetic field.

Fig. 6.9 shows how the relative susceptibility $\delta\chi(T, H < H^*)$ of Bi_2Sr_2 $CaCu_2O_8$ granular superconductors varies with the temperature and the applied magnetic field. A peak in $\delta\chi(T, H < H^*)$ appears below T_c and its height grows with decreasing H. The presence of this peak is not difficult to understand physically. At low temperatures, the spontaneously generated Josephson currents are frozen and difficult to be flipped by a weak magnetic field. Thus the paramagnetic response is vanishingly small and $\delta\chi$ should decrease with decreasing temperature. Close to T_c, the spontaneously generated magnetic moments decrease and become zero exactly at T_c. Thus $\delta\chi$ should decrease with increasing temperature just below T_c. This indicates that there must be a peak in $\delta\chi(T, H < H^*)$ between low temperature and T_c. The decrease of the peak height is a natural consequence of the decrease in the spontaneous magnetization of Josephson orbital moments in high fields.

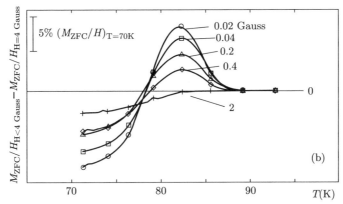

Figure 6.9   Temperature dependence of the relative susceptibility $\delta\chi(T, H < H^*)$ for $Bi_2Sr_2CaCu_2O_8$ granular superconductors under the zero field cooling. H^*=4 Gauss. Reprinted from [103], with permission from Elsevier.

It should be pointed out that the paramagnetic Meissner effect occurs not just in granular superconductors with d- or other non-s-wave pairing symmetry. It was also observed in some bulk s-wave superconductors. For example, weak paramagnetic Meissner effects were observed in Nb [104] and Al [105] superconductors. In these materials, the paramagnetic Meissner effect is likely to result from the sample inhomogeneity, especially on their surfaces. Compared with the paramagnetic Meissner effect in the d-wave granular superconductor [106], the s-wave counterpart is much weaker, and the observed paramagnetic susceptibility is also much smaller than that of the BSCCO granular superconductor. In addition, the paramagnetic Meissner effect in the d-wave granular superconductor is stable. It is an equilibrium property. However, in s-wave superconductors, the paramagnetic Meissner effect is a property of metastable state. It decays with time. These two major differences can be used to determine whether the paramagnetic Meissner effect is due to the spontaneous magnetization of Josephson π-loops.

Chapter 7

Single Impurity Scattering

7.1 Non-Magnetic Impurity Scattering

Impurity scattering is an ubiquitous and important physical effect in real materials. In most circumstances, for example, in the preparation of high quality samples, impurity is a factor that we do not want but cannot avoid. Vacancies, defects, and dislocations in materials are typical sources of impurity scattering. These impurities with strong disorder characteristics are not produced on purpose. In fact, they emerge randomly and it is difficult to precisely control them. On the other hand, it is also a common practice to intentionally change the material property by systematic doping impurities. The intrinsic physical properties can be understood by measuring and analyzing the effects induced by impurities. Historically, the impurity scattering has played a crucial role in the study of semiconductors. It has also been an important and active subject of high-T_c study.

In the study of high-T_c superconductivity, it is often to use zinc, nickel, or other dopants to partially replace some of the atoms in high-T_c materials. For different chemical and crystal structures and different dopants, the substituted atoms and their locations are different. Their effects on superconductivity are also different. Apparently, the effect is stronger if it is the copper atoms in the CuO_2 planes that are substituted. Otherwise, the effect is weaker. Which atom is substituted by a dopant depends on how close the ionic radii and chemical valences of two kinds of atoms are. Doping of zinc or nickel atoms is mainly to replace copper atoms in high-T_c cuprates since they are all divalent cations and have similar ionic radii. Their effects on the superconducting properties are significant and in a certain sense universal. They can be detected by experimental measurements. Besides there are also many investigations on lithium, magnesium, and aluminum doping. These atoms substitute the copper atoms in the CuO_2 planes as well.

The Zn^{2+} cation exhibits a fully occupied valence electron configuration $3d^{10}$, and is a non-magnetic ion. Because of the strong antiferromagnetic fluctuations in high-T_c superconductors, a non-magnetic impurity is often viewed as a magnetic one. However, this viewpoint lacks of convincing theoretical justification. It is based on an implicit assumption that the zinc atom only replaces a spin but has no effect on surrounding antiferromagnetic correlations. It might be correct in the low hole doping limit, but fails in general. Moreover, it cannot explain why zinc has qualitatively the same effect on physical properties in both underdoped and overdoped high-T_c superconductors, while antiferromagnetic fluctuations in the overdoped regime are much weaker than those in the underdoped regime. The valence electron configuration of Ni^{2+} is $3d^8$, which has 8 electrons in the 3d orbitals. Due to the Hund's rule coupling, these 3d electrons are bounded into a spin 1 cation. As a result, Ni^{2+} is magnetic. In order to study the effect of the zinc and nickel impurities, one should start with a low energy effective Hamiltonian derived in Sect. 2.8, and consider antiferromagnetic fluctuations and the impurity scattering effect in a unified way.

In an s-wave superconductor, the non-magnetic impurity scattering may eliminate the gap anisotropy and slightly reduce the superconducting transition temperature T_c. But non-magnetic impurity scattering itself does not change T_c significantly, neither other physical properties. This is just the statement of the Anderson theorem [107] on the non-magnetic impurity scattering in an s-wave superconductor. A proof of this theorem with the condition for its validity is given in Appendix D. In contrast, a magnetic impurity serves as a pair-breaker in the superconducting state. It flips spins and breaks Cooper pairs, affecting strongly on the superconducting energy gap.

For a d-wave superconductor, the Anderson theorem is no longer valid. Similar to a magnetic impurity in an s-wave superconductor, a non-magnetic impurity is also a pair-breaker for d-wave paired electrons, and a thorough investigation on its effects can assist us to understand physical properties of *d*-wave superconductors. In fact, both magnetic and non-magnetic impurity scattering effects have played important roles in the study of high-T_c supercon- ductors, especially in the determination of pairing symmetry. In some cases, the single-impurity problem can be exactly solved. In this chapter, the prob- lem of single-impurity scattering in *d*-wave superconductors will be discussed. This discussion is also useful to the understanding of many-impurity scattering problems to be addressed in the next chapter.

In a superconductor, the electronic structure of an impurity is different from the surrounding atoms. The interacting potential between the impurity and the surrounding conduction electrons takes the Coulomb form:

$$H_{\text{imp}} = \sum_\sigma \int dr c_\sigma^\dagger(r) U(r) c_\sigma(r).$$

Due to the screening of conduction electrons, the range of the impurity poten- tial is usually confined within one lattice constant. As a result, the impurity potential can be assumed completely local, i.e. $U(r) = V_0 \delta(r - r_0)$, where r_0 is the impurity coordinate. In the Nambu spinor representation, the scattering potential is expressed as

$$U(r) = V_0 \sigma_3 \delta(r). \tag{7.1}$$

For such a δ-potential, there is only the *s*-wave scattering, and all other par- tial waves vanish. In this case, the non-magnetic single-impurity problem in a superconductor can be solved exactly. On the other hand, if the range of the impurity scattering potential is finite [108], the scattering channels with nonzero angular momenta contribute as well. This may cause difficulty in theo- retical calculations. In the discussion below, we will only consider the impurity scattering problem of δ-function potential.

Let us use the Green's function theory to analyze the single-impurity scat- tering problem. Fig 7.1 shows the Feynman diagrams associated with a single- impurity scattering, which are free of the crossing diagrams and simple to com- pute. The single-particle Green's function satisfies the self-consistent Dyson equation, given by

$$G(r, r', \omega) = G^{(0)}(r - r', \omega) + G^{(0)}(r, \omega) V_0 \sigma_3 G(0, r', \omega). \tag{7.2}$$

Set $r = 0$, the above equation becomes

$$G(0, r', \omega) = G^{(0)}(-r', \omega) + G^{(0)}(0, \omega)V_0\sigma_3 G(0, r', \omega). \qquad (7.3)$$

From this equation, we find that

$$G(0, r', \omega) = \frac{c}{c - G_0(\omega)\sigma_3} G^{(0)}(-r', \omega), \qquad (7.4)$$

where

$$G_0(\omega) = \frac{1}{\pi N_F} G^{(0)}(r = 0, \omega) = \frac{1}{\pi N_F V} \sum_k G^{(0)}(k, \omega). \qquad (7.5)$$

V is the volume of the system, and N_F is the density of states in the normal state on the Fermi surface. c is a parameter characterizing the scattering potential:

$$c = \frac{1}{\pi N_F V_0}. \qquad (7.6)$$

It is related to the s-wave scattering phase shift δ_0 through the equation

$$c = \cot \delta_0. \qquad (7.7)$$

Substituting Eq. (7.4) into Eq. (7.2), $G(r, r', \omega)$ is obtained as

$$G(r, r', \omega) = G^{(0)}(r - r', \omega) + G^{(0)}(r, \omega)T(\omega)G^{(0)}(-r', \omega), \qquad (7.8)$$

where

$$T(\omega) = V_0\sigma_3 \frac{c}{c - G_0(\omega)\sigma_3} \qquad (7.9)$$

is the T-matrix describing the impurity scattering. T is a function of ω, independent of the wavevector. This is a consequence of the δ-function scattering potential.

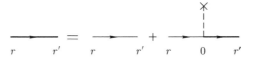

Figure 7.1   Feynman diagrams for the single-impurity scattering problem.

For the d-wave superconductor, we assume that $\Delta_k = \Delta_\varphi = \Delta_0 \cos 2\varphi$ depends only on the azimuthal angle φ of the wavevector, and ξ_k depends only on the amplitude of k but not on its angle φ. Now the radial integral of $G_0(\omega)$ over k defined in Eq. (7.5) can be evaluated, and the result is

$$
\begin{aligned}
G_0(\omega) &= \frac{1}{2\pi^2} \int_0^{2\pi} d\varphi \int_{-\infty}^{\infty} d\varepsilon \frac{\omega + \varepsilon\sigma_3 + \Delta_\varphi \sigma_1}{\omega^2 - \varepsilon^2 - \Delta_\varphi^2} \\
&= -\frac{1}{2\pi} \int_0^{2\pi} d\varphi \frac{\omega\theta(\mathrm{Re}\sqrt{\Delta_\varphi^2 - \omega^2})}{\sqrt{\Delta_\varphi^2 - \omega^2}},
\end{aligned}
\tag{7.10}
$$

where $\theta(x)$ is the standard Heaviside step function. It should be emphasized that in Eq. (7.10), the real part of $\sqrt{\Delta_\varphi^2 - \omega^2}$ must be non-negative in order to avoid the ambiguity in the evaluation of the square root of this complex number. In deriving Eq. (7.10), we have used the property that the average of Δ_φ on the Fermi surface is zero. The particle-hole symmetry and the wide band-width assumption are also used in the integration of ε. By adding an infinitesimal imaginary part to ω, Eq. (7.10) then becomes

$$
G_0(\omega + i0^+) = -\frac{1}{2\pi} \int_0^{2\pi} d\varphi \left[\mathrm{Re}\frac{\omega}{\sqrt{\Delta_\varphi^2 - \omega^2}} + i\mathrm{Re}\frac{|\omega|}{\sqrt{\omega^2 - \Delta_\varphi^2}} \right].
\tag{7.11}
$$

The right-hand side of Eq. (7.10) is an elliptic integral. When $0 < |\omega| \ll \Delta_0$,

$$
G_0(\omega) \approx -\frac{2\omega}{\pi\Delta_0} \ln\frac{2\Delta_0}{-i\omega}.
\tag{7.12}
$$

In the limit the imaginary part of ω approaching zero, the corresponding retarded Green's function

$$
G_0(\omega + i0^+) \approx -\frac{2\omega}{\pi\Delta_0} \ln\frac{2\Delta_0}{|\omega|} - i\frac{|\omega|}{\Delta_0}.
\tag{7.13}
$$

In a d-wave superconductor, $G_0(\omega)$ depends only on ω because Δ_k averages to zero over the Fermi surface. In this case, T is a diagonal matrix, given by

$$
T(\omega) = T_0 + T_3\sigma_3,
\tag{7.14}
$$

where

$$
T_0 = \frac{1}{\pi N_F} \frac{G_0(\omega)}{c^2 - G_0^2(\omega)},
\tag{7.15}
$$

$$
T_3 = \frac{1}{\pi N_F} \frac{c}{c^2 - G_0^2(\omega)}.
\tag{7.16}
$$

In the discussion of non-magnetic impurity scattering, two different limits should be considered. One is the strong scattering limit, which is also called the resonance scattering or the unitary scattering limit. The other is the weak scattering limit, which is also called the Born scattering limit. The d-wave superconductors under these two limits exhibit different behaviors, both qualitatively and quantitatively. The strong scattering limit corresponds to the limit $c \to 0$ with the phase shift $\delta_0 \to \pi/2$. The weak scattering Born limit corresponds to $c \to \infty$ or $\delta_0 \to 0$.

7.2   The Resonance State

The pole of the T-matrix corresponds to the frequency of a resonance state induced by the impurity scattering. It is determined by

$$G_0(\Omega) = \pm c. \tag{7.17}$$

The pole Ω has no real solution. It has only complex solutions when $c \ll 1$. Eq. (7.17) is difficult to solve exactly in general. But it can be solved approximately when $\mathrm{Re}\,\Omega \ll \Delta_0$. Substituting the expression of $G_0(\omega)$ given by Eq. (7.12) into Eq. (7.17), we have

$$\bar{\Omega}\left(\ln\frac{4}{\pi c} \pm i\frac{\pi}{2} - \ln\bar{\Omega}\right) = 1, \tag{7.18}$$

where $\bar{\Omega} = \pm 2\Omega/\pi\Delta_0 c$. After taking the logarithm on both sides, it becomes

$$\ln\bar{\Omega} + \ln\left(\ln\frac{4}{\pi c} \pm i\frac{\pi}{2} - \ln\bar{\Omega}\right) = 0. \tag{7.19}$$

Suppose $\ln\bar{\Omega}$ is very small, Ω can be solved by expanding the left-hand side with respect to $\ln\bar{\Omega}$. Up to the first order terms in $\ln\bar{\Omega}$, the solution is given by

$$\Omega \approx \pm\frac{\pi c\Delta_0}{2}\exp\left(-\frac{a}{a-1}\ln a\right), \tag{7.20}$$

$$a = \ln\frac{4}{\pi c} \pm i\frac{\pi}{2}.$$

In the limit $|c| \to 0$, $|a| \gg 1$, Eq. (7.20) is approximately given by

$$\Omega \approx \pm\Omega_0 - i\Gamma_0, \tag{7.21}$$

where

$$\Omega_0 = \frac{2\pi|c|\Delta_0 \ln(4/\pi|c|)}{4\ln^2(4/\pi|c|) + \pi^2}, \tag{7.22}$$

$$\Gamma_0 = \frac{\pi^2|c|\Delta_0}{4\ln^2(4/\pi|c|) + \pi^2}. \tag{7.23}$$

The retarded Green's function $G(r, r', \omega)$ is analytic in the upper half complex plane of ω. The pole of the T-matrix only exists in the lower half-plane.

The complex solution of Ω indicates that the pole of the T-matrix corresponds to a resonance state, not a bound state. The resonance frequency decreases with increasing c. In the unitary scattering limit $c \to 0$, both the resonance frequency and Γ_0 become 0, and the resonance state becomes a bound state. It should be noted that the resonance state only exists in the strong scattering regime. In the limit of weak scattering ($c \to \infty$), there is no resonance state. In fact, when $c \sim 1$, $\Omega_0 \sim \Delta_0$ and $\Gamma_0 \sim \Omega_0$, it is no longer meaningful to interpret the pole of the T-matrix as a resonance state.

In a superconductor, there are two characteristic length scales. One is the Fermi wavelength l_F related to the Fermi wave vector $l_F = 2\pi/k_F$. The other is the Cooper pair coherence length determined by the Fermi velocity v_F and the energy gap Δ_0 through $\xi_0 = \hbar v_F/\Delta_0$. In the presence of a resonance state, another characteristic length appears related to the impurity resonance state $l_0 = \hbar v_F/\Omega_0$. When the condition $l_F \ll \xi_0 \ll l_0$ is satisfied, the influence from the resonance state becomes important.

The appearance of the resonance state divides the energy space into three characteristic regimes. The first is the weak scattering regime, where $|G_0(\omega)| \ll |c|$, $|\omega| \ll \Omega_0$, and $T_{11} \approx -T_{22}$. The second is the high frequency regime, where $|\omega| \gg \Omega_0$, $G_0(\omega)| \gg |c|$, and $T_{11} \approx T_{22}$. The third is the resonance regime $|\omega| \sim \Omega_0$, where $T_{\pm}$ changes rapidly with ω and the resonance state appears at $|\omega| = \Omega_0$. The impurity scattering effect is relatively weak in the Born limit. But in the resonance regime, the impurity potential scattering plays an important role. Thus in the discussion of a specific problem, in order to capture the key points with a simple and correct method, it is crucial to know in which regime the system is.

In a high-T_c superconductor, the zinc impurity is a strong scattering center, as already explained in Sect. 2.8. The zinc impurity is expected to generate a sharp peak near the Fermi surface. This resonance peak was observed experimentally [109]. Fig. 7.2 shows the scanning tunneling microscope spectroscopy above the zinc impurity in $Bi_2Si_2CaCu_2O_8$. The resonance state energy generated by the zinc impurity is 1.5meV, much smaller than the superconducting gap $\Delta_0 \sim 48$meV. According to Eq. (7.22), the scattering parameter is found to be $c = \cot \delta_0 = 0.07$. It corresponds to a scattering phase shift $\delta_0 = 0.48\pi$. This phase shift is very close to $\pi/2$, indicating that the zinc impurity scatter-

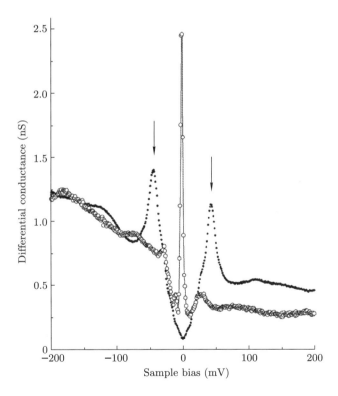

Figure 7.2   Scanning tunneling differential conductance measured at (open circles) and far away (solid dots) from the Zn impurity site in the $Bi_2Si_2CaCu_2O_{8+\delta}$ high T_c superconductor. Reprinted by permission from [109].

ing potential is indeed in the unitary limit. According to Eq. (7.6), the zinc scattering potential can be expressed using c as

$$V_0 = \frac{1}{\pi c N_F} \sim \frac{5}{N_F},$$

where the density of states N_F is inversely proportional to the band width. For the $Bi_2Si_2CaCu_2O_8$ superconductor, N_F is estimated to be $1.5(eV)^{-1}$ from the specific heat measurement [110]. Correspondingly, the zinc scattering potential strength is estimated as $V_0 \sim 3.3eV$. This value, as discussed in Sect. 2.8, is just the energy of the Zhang-Rice singlet, much larger than the band width.

7.3 Correction to the Quasiparticle Density of States

The density of states of superconducting electrons is determined by the imaginary part of the retarded Green's function

$$\rho(r,\omega) = -\frac{1}{\pi} \mathrm{Im} G_{11}(r, r, \omega + i0^+) = \rho_0(\omega) + \delta\rho(r, \omega), \qquad (7.24)$$

where

$$\rho_0(\omega) = -\frac{1}{\pi} \mathrm{Im} G_{11}^{(0)}(r = 0, \omega + i0^+), \qquad (7.25)$$

$$\delta\rho(r,\omega) = -\frac{1}{\pi} \mathrm{Im} G_{1\alpha}^{(0)}(r, \omega + i0^+) T_{\alpha\alpha}(\omega + i0^+) G_{\alpha 1}^{(0)}(-r, \omega + i0^+). \qquad (7.26)$$

ρ_0 is the electron density of states in the absence of impurity, independent of the spatial coordinate. $\delta\rho(r, \omega)$ is the correction to the density of states induced by the impurity. The potential scattering does not change the total number of states at each site, the summation of $\delta\rho(r, \omega)$ over frequency is 0, satisfying the sum rule ([111])

$$\int d\omega \delta\rho(r, \omega) = 0. \qquad (7.27)$$

If the impurity scattering is in the Born limit, $c \to \infty$, the correction to the density of states is inversely proportional to c and can be neglected. In the presence of resonance states, the correction to the density of states remains small in the low and high energy regimes, but it becomes significant in the resonance energy regime.

In the following, we concentrate to discuss the properties of the impurity density of states in the limit of $c \to 0$, where the analytic behavior of the density of states can be relatively simple to derive. The results obtained in this limit, nevertheless, hold qualitatively even when c is not so small. It should be emphasized that in the analysis of the density of states in the zero energy limit, we should first take the limit $c \to 0$ and then set $\omega \to 0$. Otherwise, $\mathrm{Im}T(\omega \to 0) = 0$ at any finite $c \neq 0$, and the contribution of the resonance state is entirely ignored.

At the impurity site, the correction to the density of states is given by

$$\delta\rho(0,\omega) = -N_F \mathrm{Im} \frac{G_0^3(\omega)}{c^2 - G_0^2(\omega)}. \tag{7.28}$$

From the unperturbed Green's function at the resonance frequency Ω_0,

$$G_0(\Omega_0) \approx -|c| - \frac{i\pi|c|}{2\ln(4/\pi|c|)},$$

and the corresponding $\delta\rho$ is found to be

$$\delta\rho(0,\Omega_0) = \frac{N_F|c|\ln(4/\pi|c|)}{\pi}. \tag{7.29}$$

It shows that at $\omega = \pm\Omega_0$, $\delta\rho \sim |c|\ln^{-1}|c|$ has a weak peak at the resonance frequency. The weight of this peak (namely the area enclosed by the peak) is approximately given by

$$w(r=0) \sim \delta\rho(0,\Omega_0)\Gamma_0 \approx \frac{\pi N_F \Delta_0 c^2}{4} \ln^{-3}\frac{4}{\pi|c|}.$$

Fig. 7.3 shows the correction to the density of states at the impurity site. Two resonance peaks induced by the impurity emerge in the density of states. As shown by the figure, the smaller is c, the closer is the resonance energy to zero. A larger c gives rise to a higher resonance peak in the density of states. Note that $\delta\rho(0,\omega)/(cN_F)$ is normalized by c in this figure. Without this normalization, the resonance peak of $c = 0.01$ is almost 10 times higher than that of $c = 0.001$.

In the limit $c \to 0$ and $\omega \to 0$, the T-matrix exhibits a singularity. However, $G^{(0)}(r, \omega \to 0) \neq 0$ at $r \neq 0$, and $G^{(0)}(r, \omega)$ is regular as a function of ω. It

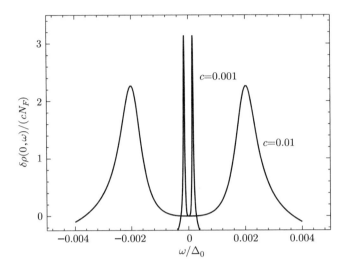

Figure 7.3   Impurity correction to the quasiparticle density of states $\delta\rho(0,\omega)/(cN_F)$ around the resonance energy in the unitary limit.

becomes real in the limit $\omega \to 0$,

$$\lim_{\omega \to 0} G^{(0)}(\pm r, \omega + i0^+) = -\frac{1}{V} \sum_k \frac{\varepsilon_k \sigma_3 + \Delta_k \sigma_1}{\varepsilon_k^2 + \Delta_k^2} \cos(k \cdot r). \qquad (7.30)$$

In this case,

$$\delta\rho(r,\omega) \approx -\frac{1}{\pi} G_{1\alpha}^{(0)}(r,0) \mathrm{Im} T_{\alpha\alpha}(\omega) G_{\alpha 1}^{(0)}(-r,0). \qquad (7.31)$$

At the resonance frequency Ω_0,

$$\mathrm{Im} T(\Omega_0) \approx -\frac{1}{\pi N_F c} \begin{pmatrix} \dfrac{\pi}{8 \ln(4/\pi c)} & 0 \\ 0 & \dfrac{2 \ln(4/\pi c)}{\pi} \end{pmatrix}.$$

Thus $\delta\rho(r, \Omega_0)$ diverges as $|c|^{-1}$, and the spectral weight of the resonance peak on the neighboring sites is c^{-2} times stronger than that on the impurity site. This means that the resonance effect is significantly stronger on the four neighboring sites of the impurity. To detect the resonance state experimentally, one should therefore concentrate to measure the density of states not just on the impurity site, but more on its neighboring sites.

In the limit $c \to 0$, $T_0 \approx 1/\pi N_F G_0(\omega)$, and T_3 is small compared to T_0. Therefore, when $\Delta_0 \gg \omega \gg \Omega_0$, $\delta\rho(r, \omega)$ is approximately given by

$$\delta\rho(r, \omega) = \frac{1}{\pi^2 N_F} \left[G_{11}^{(0)}(r, 0) \right]^2 \mathrm{Im} G_0^{-1}(\omega). \tag{7.32}$$

Hence the ω dependence of $\delta\rho(r, \omega)$ is determined by the imaginary part of $G_0^{-1}(\omega)$, and its r dependence is determined by the square of $G_{11}^{(0)}(r, 0)$. The imaginary part of G_0^{-1}, according to Eq. (7.13), is

$$\mathrm{Im} G_0^{-1}(\omega + i0^+) \approx \frac{\pi^2 \Delta_0}{|\omega| \left(4 \ln^2 \frac{2\Delta_0}{|\omega|} + \pi^2 \right)}. \tag{7.33}$$

The dependence of $G_{11}^{(0)}(r, 0)$ on r is more complicated. But when $\varepsilon \gg \Delta_0$ and $r > \xi_0$, $G_0(r, 0)$ in Eq.(7.30) contributes mainly from the momentum summation in the region of $k \perp r$, due to the rapid oscillation of $\cos(k \cdot r)$.

If r is along the anti-nodal direction,

$$G_{11}^{(0)}(r, 0) = 2N_F \int_0^1 dx \frac{1}{\sqrt{1 - x^2}} \exp\left[-\frac{\Delta_0 \bar{r}}{\varepsilon_F} |2x^2 - 1|x \right] \sin(\bar{r}x), \tag{7.34}$$

where $\bar{r} = k_F r$. On the other hand, if r is along the nodal direction,

$$G_{11}^{(0)}(r, 0) = 2N_F \int_0^1 dx \frac{1}{\sqrt{1 - x^2}} \exp\left(-\frac{2\Delta_0 \bar{r}}{\varepsilon_F} \sqrt{1 - x^2} x^2 \right) \sin(\bar{r}x). \tag{7.35}$$

Fig. 7.4 shows the behavior of $G_{11}^{(0)}(r, 0)$ as a function of r along these two directions under the condition $\varepsilon_F \gg \Delta_0$. $G_{11}^{(0)}(r, 0)$ takes a maximal value at $\bar{r} \approx 2$, and then decreases with increasing r. The decay along the anti-nodal direction is apparently faster than that along the nodal direction.

7.4 Tunneling Spectrum of the Zn-Impurity Resonance State

The impurity effect on the superconducting electron density of states can be measured by the scanning tunneling microscope (STM). From the result shown in Sect. 7.3, we know that the impurity correction to the density of states at

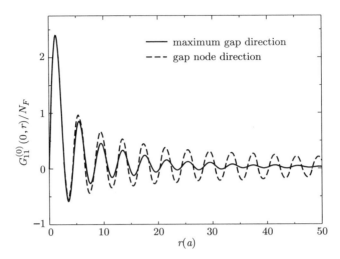

Figure 7.4 $G_{11}^{(0)}(r,0)/N_F$ as a function of r along the gap nodal (dashed line) and anti-nodal (solid line) directions, respectively. $k_F = \pi/2a$, $\xi_0 = 5\pi a$, and a is the lattice constant.

the resonance energy is much weaker on the impurity site than on its four nearest neighbors. This result is not consistent with the experimental observations around the zinc impurity in high-T_c superconductors. What the experiment found [109], as shown in Fig. 7.5, is that the impurity induced STM spectral weight is the strongest at the zinc impurity site and very weak on the four nearest neighboring sites. This dislocation of the spectral weight seems to contradict the theory which takes Zn as a strong non-magnetic impurity. However, if the effect of the anisotropy in the c-axis tunneling matrix element is taken into account, it can be shown that the spatial dislocation of the STM spectroscopy is in fact not a negative, but an affirmative evidence for the non-magnetic resonance impurity theory of Zn.

In general, the differential conductance measured by STM is proportional to the electron density of states. This is a basic assumption commonly used in the analysis of STM experimental data. But this assumption is not valid in all cases. In fact, the tunneling conductance depends not just on the electron density of states, but also on the tunneling matrix elements. If the tunneling matrix elements do not exhibit an apparent anisotropy (or momentum dependence) so

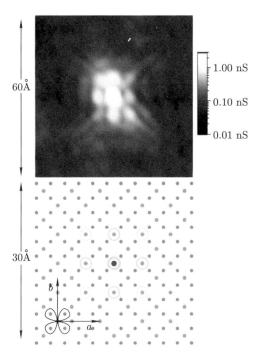

Figure 7.5   Spatial distribution of the differential tunneling conductance of the Zn resonance

state in $Bi_2Si_2CaCu_2O_{8+\delta}$. The differential conductance is largest at the zinc site (i.e. the

brightest spot), and then exhibits alternating dark and bright spots on the sites away from

the impurity. The tunneling differential conductance shows local minima at four nearest-

neighboring sites (dark spots), and local maxima at four next-nearest neighboring Cu sites

(bright spots). The a- and b-axes are rotated from the crystal axes by $\pi/4$. Reprinted by

permission from [109].

that it can be approximately treated as a constant, the differential tunneling
conductance is indeed determined purely by the density of states. On the contrary, if the tunneling matrix elements are strongly momentum dependent, the
tunneling conductance is strongly modified and no longer proportional to the
electron density of states. This is just the situation encountered in the STM
measurement of the zinc impurity in high-T_c superconductors.

High-T_c cuprates are quasi-two-dimensional materials with weak interlayer couplings. It is easier to cleave a high-T_c sample along a surface parallel to the CuO_2 planes. Most of the STM experiments place the probe tips perpendicular to the CuO_2 planes, and measure the tunneling currents along the c-axis direction. However, the c-axis hopping of electrons is highly anisotropic because the hopping of the oxygen holes in the $2p$-orbitals along the c-axis needs the assistance of the copper $4s$-orbital in cuprates. As the overlap between the oxygen anti-bonding p-orbitals and the copper $4s$-orbital possesses the $d_{x^2-y^2}$ symmetry, the electron hopping depends strongly on the in-plane momentum of electrons on the CuO_2 planes.

Following the discussion given in Sect. 2.7, one can show by the symmetry argument that the tunneling Hamiltonian along c-axis is given by [112, 53, 113],

$$H_t = \sum_i t_\perp c_i^\dagger D_i + h.c.,$$

where $t_\perp$ is the tunneling constant, and c_i is the electron operator in the metallic probe tip. D_i is an effective electron operator defined in the superconductor

$$D_i = \sum_j F_{i,j} \left(\delta_{j,i_0} + \frac{1}{\sqrt{2}} \delta_{j \neq i_0} \right) d_j,$$

where d_j is the annihilation operator of electron at site j. i_0 is the coordinate of the impurity site, and

$$F_{i,j} = \frac{1}{N} \sum_k \frac{\cos k_x - \cos k_y}{\sqrt{\cos^2(k_x/2) + \cos^2(k_y/2)}} e^{ik \cdot (i-j)}.$$

From the second order perturbation of H_t, it can be shown that the tunneling conductance g_i at site i is proportional to the density of states of the effective electron operator D_i,

$$g_i(V) \propto \left\langle D_i \delta(eV - H_{imp}) D_i^\dagger \right\rangle, \tag{7.36}$$

where V is the bias voltage and H_{imp} is the system Hamiltonian defined by Eq. (2.40) for a zinc impurity.

In Eq. (7.36), if $D_i = d_i$, $g_i(V)$ is proportional to the local density of states of electrons $\left\langle d_i \delta(eV - H_{imp}) d_i^\dagger \right\rangle$. However, in high-$T_c$ superconductors, $D_i \neq d_i$.

From the expression of $F_{i,j}$, it is simple to show that: (1) $F_{i,j} = 0$ if $i = j$; (2) $|F_{i,j}|$ reaches the maximum if i and j are nearest neighbors. These properties of $F_{i,j}$ lead to the following results: (1) The quasiparticle excitation at site i has no contribution to the tunneling conductance at that site; (2) The tunneling conductance at site i contributes mainly by the quasiparticle excitations on its four nearest neighboring sites. Based on these properties, it is not difficult to understand why the spatial dislocation occurs in the STM spectroscopy of the impurity resonance state. The tunneling current on the impurity site comes mainly from the density of states on its four nearest neighbor sites, giving rise to the strongest STM spectra. In contrast, on the four nearest neighboring sites of the zinc impurity, the STM spectroscopy comes mainly from their nearest neighbors on which the weights of the resonance states are very small. Thus the spatial dislocation of the zinc impurity resonance state is caused by the anisotropic tunneling matrix elements along the c-axis ([113]). It is actually consistent with the theory that Zn is a strong non-magnetic impurity.

7.5   Comparison with the Anisotropic s-Wave Superconductors

The previous discussion indicates that the unitary non-magnetic impurity scattering has a strong impact on the d-wave pairing states. However, it is not clear whether this effect is due to the gap symmetry or simply due to the gap anisotropy. Can a strongly anisotropic s-wave superconductor exhibit the same effect? This is an important question that needs to be addressed in order to understand the effect of impurity scattering in high-T_c superconductors. To answer this question, let us compare the impurity scattering in a d-wave superconductor with that in a strongly anisotropic s-wave superconductor whose gap function is defined by

$$\Delta(k) = \Delta_0 |\cos(2\phi)|.$$

It has the same amplitude as the d-wave superconducting gap, but it does not change sign on the Fermi surface.

The quasiparticle Green's function and the Dyson equation for the anisotropic s-wave superconductor can be similarly derived as for the d-wave superconductor. The difference is that the summation of $G^{(0)}$ over momentum k on the right-hand side of Eq. (7.5) now has an extra off-diagonal term in additional to the $G_0(\omega)$ term previously obtained for the d-wave superconductor:

$$G_0^s(\omega) = \frac{1}{\pi N_F V} \sum_k G^{(0)}(k,\omega) = G_0(\omega) + G_1(\omega)\sigma_1, \qquad (7.37)$$

where

$$G_1(\omega) = \frac{1}{\pi N_F V} \sum_k \frac{\Delta_0 |\cos(2\varphi)|}{\omega^2 - \varepsilon_k^2 - \Delta_k^2}. \qquad (7.38)$$

Now the T-matrix becomes

$$T(\omega) = V_0 \sigma_3 \frac{c}{c - G_0^s(\omega)\sigma_3} = \frac{1}{\pi N_F} \frac{G_0(\omega) - c\sigma_3 - G_1(\omega)\sigma_1}{c^2 + G_1^2(\omega) - G_0^2(\omega)}, \qquad (7.39)$$

and the poles of the T-matrix are determined by the equation

$$G_0^2(\omega) = c^2 + G_1^2(\omega). \qquad (7.40)$$

In comparison with Eq. (7.17), the resonance states previously discussed for the d-wave superconductor can be straightforwardly generalized to the s-wave case. We only need to replace c^2 in the d-wave case by $c^2 + G_1^2(\omega)$. Such a replacement indicates that the impurity scattering effect is weakened by the G_1 term in the anisotropic s-wave superconductor. In particular, the smaller is c, the stronger is the correction from the G_1 term. This correction becomes dominant in the unitary scattering limit.

In the limit $\omega \ll \Delta_0$,

$$G_1(\omega) \approx -\frac{2}{\pi} \left(\sqrt{1 - \frac{\omega^2}{\Delta_0^2}} + i\frac{|\omega|}{\Delta_0} \right). \qquad (7.41)$$

Since $G_1(\omega \to 0) \to -2/\pi$ is finite, it suggests that in the low energy limit, the effective scattering parameter c in the anisotropic s-wave superconductor is equivalent to $\sqrt{c^2 + 4/\pi^2}$ in the d-wave superconductor. Thus no matter how strong the scattering potential is, the impurity scattering in the s-wave superconductor is not in the resonance scattering limit and there is not any low energy resonance state. This is just a consequence of the Anderson theorem ([107]).

The above analysis indicates that the low energy resonance state is caused by the gap symmetry, rather than the gap anisotropy. In other words, if the non-magnetic impurity induced resonance state is observed in a superconductor, the superconducting electrons are not s-wave paired no matter whether the gap function is isotropic or anisotropic. Therefore, the experimental observation of the zinc impurity induced resonance state is a direct evidence in support of non-s-wave pairing in high-T_c superconductors.

7.6 The Classical Spin Scattering

In a conventional metal, a classical spin acts like a potential scatterer. However, in a singlet pairing superconductor, the scattering effect of a classical spin is different. This problem was first studied by Lu Yu. He investigated the effect of a paramagnetic spin in an s-wave superconductor, and predicted that the magnetic impurity induces a bound state inside the superconducting gap [114]. Later on this problem was further explored by Shiba [115], who provided a unified description for both the weak and strong scattering limits.

The interaction between a classical spin and an electron is defined by the Hamiltonian

$$H_{int} = J \int dr \delta(r) c_r^+ \frac{\sigma}{2} c_r \cdot S. \tag{7.42}$$

It includes a spin-flip term. In the treatment of this spin-flip term, it is more convenient to double the dimension of the Nambu spinor from 2 to 4 by defining the following spinor operators:

$$c_k = \begin{pmatrix} c_{k\uparrow} \\ c_{k\downarrow} \end{pmatrix}, \qquad d_k = \begin{pmatrix} c_k \\ c_{-k}^\dagger \end{pmatrix}. \tag{7.43}$$

In this new representation, the BCS Hamiltonian (1.11) and the free Green's function (4.3) become

$$H_{BCS} = \frac{1}{2} \sum_k d_k^\dagger \begin{pmatrix} \xi_k & i\Delta_k\sigma_2 \\ -i\Delta_k\sigma_2 & -\xi_k \end{pmatrix} d_k, \tag{7.44}$$

$$G^{(0)}(k,\omega) = \left[\omega - \begin{pmatrix} \xi_k & i\Delta_k\sigma_2 \\ -i\Delta_k\sigma_2 & -\xi_k \end{pmatrix} \right]^{-1}, \tag{7.45}$$

where

$$\sigma_2 = \begin{pmatrix} 0 & -i \\ i & 0 \end{pmatrix},$$

and the Pauli matrices σ_α act in spin space. Using the identity

$$c_r^\dagger \sigma_\alpha c_r = c_r \sigma_2 \sigma_\alpha \sigma_2 c_r^\dagger,$$

the interaction Hamiltonian Eq. (7.42) can be expressed as

$$H_{int} = \frac{1}{4} J \int dr \delta(r) d_r^\dagger \mathcal{A} d_r, \tag{7.46}$$

where

$$\mathcal{A} = \begin{pmatrix} \sigma \cdot S & 0 \\ 0 & \sigma_2 \sigma \cdot S \sigma_2 \end{pmatrix}. \tag{7.47}$$

This classical spin scattering problem can be similarly solved as for a non-magnetic impurity. The Dyson equation of the single-particle Green's function is now given by

$$G(r, r', \omega) = G^{(0)}(r - r', \omega) + G^{(0)}(r, \omega) \frac{1}{4} J \mathcal{A} G(0, r', \omega). \tag{7.48}$$

From this equation, we find that

$$G(0, r', \omega) = \left[1 - \frac{1}{4} J F(\omega) \mathcal{A} \right]^{-1} G^{(0)}(-r', \omega), \tag{7.49}$$

$$F(\omega) = G^{(0)}(r = 0, \omega). \tag{7.50}$$

Substituting Eq. (7.49) into Eq. (7.48), we obtain

$$G(r, r', \omega) = G^{(0)}(r - r', \omega) + G^{(0)}(r, \omega) T(\omega) G^{(0)}(-r', \omega). \tag{7.51}$$

It has the same form as Eq. (7.8), but the T-matrix now becomes

$$T(\omega) = \frac{1}{4}J\mathcal{A}\left[1 - \frac{1}{4}JF(\omega)\mathcal{A}\right]^{-1}. \tag{7.52}$$

F and $\mathcal{A}$ satisfy the commutation relation

$$\mathcal{A}F(\omega) = F(\omega)\mathcal{A}.$$

Using this result and the identity $\mathcal{A}^2 = S^2 I$, the T-matrix can be simplified as

$$T(\omega) = \left[\frac{1}{4}J\mathcal{A} + \left(\frac{1}{4}JS\right)^2 F(\omega)\right]\frac{1}{1 - \left(\frac{1}{4}JS\right)^2 F^2(\omega)}. \tag{7.53}$$

As the average of $\mathcal{A}$ over the direction of the impurity spin is zero, i.e. $\langle\mathcal{A}\rangle = 0$, the spin averaged T-matrix is found to be

$$\langle T(\omega)\rangle = \left(\frac{1}{4}JS\right)^2 F(\omega)\frac{1}{1 - \left(\frac{1}{4}JS\right)^2 F^2(\omega)}, \tag{7.54}$$

independent on the pairing symmetry.

We now apply the above result to the magnetic impurity states and make comparison between the isotropic s-wave and the d-wave superconductors. In an isotropic s-wave superconductor, $\Delta_k = \Delta$,

$$F(\omega) = -\frac{\pi N_F}{\sqrt{\Delta^2 - \omega^2}}\left[\omega + i\Delta\sigma_2\begin{pmatrix} 0 & 1 \\ -1 & 0 \end{pmatrix}\right]. \tag{7.55}$$

From the eigenvalues of $F(\omega)$, which are given by $-\pi N_F(\omega \pm \Delta)/\sqrt{\Delta^2 - \omega^2}$, we find the poles of the T-matrix

$$\omega = \pm\frac{1 - \alpha}{1 + \alpha}\Delta < \Delta, \tag{7.56}$$

$$\alpha = \left(\frac{\pi JSN_F}{4}\right)^2. \tag{7.57}$$

This pair of poles fall on the real axis, and their absolute values are smaller than the pairing gap. Therefore, there is a pair of bound states generated by a classical paramagnetic impurity inside the s-wave superconducting gap. The binding energy depends on the density of states in the normal state and the coupling constant J. When J is very small, the bound state approaches the gap edge. With the increase of J, the bound state moves close to the Fermi level.

However, in a d-wave superconductor,

$$F(\omega) = \pi N_F G_0(\omega) \tag{7.58}$$

is a constant matrix, and the poles of the T-matrix are determined by the equation

$$\left(\frac{\pi J S N_F}{4}\right)^2 G_0^2(\omega) = 1. \tag{7.59}$$

If we define

$$c^{-1} = \frac{\pi J S N_F}{4}, \tag{7.60}$$

Eq. (7.59) then simply reduces to the equation for determining the poles in a non-magnetic impurity system, i.e. Eq. (7.17). The results previously obtained on the resonance state induced by a non-magnetic impurity can be directly used here. But the correction to the density of states by a magnetic impurity is not the same as that for a non-magnetic one. The T-matrix after average over the impurity spin is simple. It contains only the T_0 term, not the T_3 term.

7.7 The Kondo Effect

The interaction of a quantum magnetic impurity with electrons in a metal can induce the Fermi surface instability and change qualitatively the behaviors of resistivity, magnetic susceptibility, and other physical quantities. This is just the so-called Kondo effect [116]. The Kondo interaction is described by the Hamiltonian, defined by Eq. (7.42), but S is now a quantum spin operator rather than a classical one. The Kondo effect exists in a d-wave superconductor. It can also affect strongly physical properties of superconducting quasiparticles.

In a normal metal, the Kondo effect arises from the screening of the magnetic impurity by the conduction electrons. It generates a Kondo resonance state at the Fermi level. This effect was first discovered by Kondo through a third order perturbation calculation. It is actually a non-perturbative effect. A small Kondo coupling can lead to a qualitative change in the ground state. This problem was well studied in 1960s and 1970s through the poor-man scaling [117], numerical renormalization group [118] and other methods. In the normal metal, the Kondo system has two fixed points. One is the unstable fixed point at $J = 0$, and the other is the stable fixed point at $J \to \infty$. They correspond to the weak and strong coupling limits, or equivalently high or low temperature limits, respectively. In the weak coupling limit, the impurity spin is decoupled from the conduction electrons and behaves like a free magnetic moment without screening. It gives rise to a Curie-Weiss like impurity magnetic susceptibility. In the strong coupling limit, on the other hand, the coupling between the impurity and conduction electrons is strong. The impurity spin is completely screened and the system behaves like a normal Fermi liquid. The magnetic susceptibility is Pauli paramagnetic, just like in a normal metal. Moreover, the impurity scattering induces a resonance state on the Fermi surface which has a large impact on transport properties.

In a superconductor, the Kondo effect is greatly weakened by the superconducting energy gap. Nevertheless, the Kondo effect exists when the exchange coupling between the impurity spin and superconducting electrons is sufficiently strong. It can also screen the impurity spin and change the interaction between superconducting Cooper pairs. In particular, the Kondo effect can suppress the superconducting gap near the impurity. However, a self-consistent calculation for this screening effect is rather difficult. In most of calculations, the correction of the Kondo effect to the superconducting gap function is either ignored or just considered for the average gap in the whole space. These analyses are not based on the self-consistent solution. It is not clear to what extent they can be applied to real materials.

In a conventional isotropic s-wave superconductor, since the quasiparticle excitations on the Fermi surface are fully gapped, the Kondo effect is completely suppressed and the impurity magnetic moment is not screened, provided that the Kondo coupling is not significantly larger than the energy gap [119].

In a d-wave superconductor, the zero energy quasiparticle density of states is zero and the impurity moment is also unscreened in the weak coupling regime. However, as the quasiparticle density of states varies linearly with energy, the screening exists if the Kondo coupling is strong enough. When the characteristic Kondo temperature T_K is much smaller than the superconducting transition temperature T_c, the superconducting correlation is stronger than the Kondo screening effect and the system is in the weak coupling limit. On the other hand, if $T_K \gg T_c$, the Kondo screening takes place before the superconducting transition, and the screening survives even in the superconducting phase. This implies that there is a critical Kondo coupling J_c in a d-wave superconductor: the impurity moment is unscreened when J is below J_c and screened with a Kondo resonance state on the Fermi level when J is above J_c.

The existence of the Kondo effect above a critical coupling in a d-wave superconductor was confirmed by the mean-field calculations based on the large-N expansion [120, 121, 122]. Similar to the Kondo problem in a metal, the impurity magnetic moment behaves differently in the unscreened and screened phases. The transition from the unscreened phase to the screened one is a quantum phase transition. The critical coupling constant J_c depends on the superconducting gap Δ_0. It increases with increasing Δ_0. When the impurity concentration of the system becomes finite, the impurity induced quasiparticle density of states on the Fermi surface also becomes finite. This enhances the Kondo effect and drives J_c to zero [120].

In high-T_c superconductors, the magnetic impurity may exist either in or out of the CuO_2 planes. The Kondo coupling directly in the CuO_2 plane is generally stronger. It has higher possibility to induce the Kondo screening. The interaction between a magnetic impurity out of the CuO_2 planes and conduction electrons is relatively weak, and the impurity moment has less chance to be screened. When the Kondo screening happens, a Kondo resonance state emerges at the Fermi energy, which may account for the observed resonance state in the nickel or other magnetic impurity doped high-T_c superconductors. One can refer to Refs. [122, 123] for more detailed discussions on this topic.

Chapter 8

Many-Impurity Scattering

8.1 Scattering Potential and Disorder Average

Compared to the single-impurity scattering, the many-impurity one is considerably more difficult to handle. It is impossible to exactly solve the scattering problem for a many-impurity system with random impurity configurations. Usually two kinds of approximations are adopted in the analytic calculations. First, the impurity concentration is very low so that the interaction among impurities is negligible. Second, the screening from electrons to the impurity potential is strong, and the impurity potential is short-ranged and isotropic. Furthermore, it is assumed that the s-wave scattering plays the leading role, and the contribution from higher angular momentum channels can be neglected. The self-consistent T-matrix theory dealing with the impurity scattering in a d-wave superconductor is just established based on these approximations.

In the discussion of the many-impurity scattering problem, there are two important physical parameters: one is the scattering phase shift, and the other is the impurity concentration. Both are closely related to the two approximations mentioned above [108]. The s-wave scattering approximation neglects the anisotropy of the impurity state in a d-wave superconductor, especially the anisotropic decay of the impurity resonance state along the gap nodal and anti-nodal directions. This anisotropy can affect the interaction between impurities. A self-consistent study for this problem is rather difficult, and the current study on the many-impurity scattering is incomplete or even incorrect in some cases.

In this chapter, we assume the impurity scattering is described by a δ-

function potential:

$$U(r) = V_0 \delta(r). \tag{8.1}$$

This is equivalent to taking the *s*-wave scattering approximation for an arbitrary impurity scattering potential, namely if we start from an arbitrary scattering potential but take the *s*-wave approximation, we should obtain the same result with a properly adjusted V_0. But the derivation from the δ-function potential is more transparent. It avoids tedious and sometimes obscure discussions presented in literatures, and allows us to grasp more clearly the physics governing the many-impurity scattering problem. In fact, the *s*-wave scattering is an approximation that we have to take in order to treat analytically the impurity scattering problem. To go beyond this approximation, the vertex or other corrections to the Green's function need to be considered. These corrections are difficult to calculate. It is also difficult to gain intuitive physical insights from this tedious calculation.

Assuming the impurities are located at $\{R_l\} = \{R_1, ..., R_l, ...\}$, the interaction between impurities and electrons can be expressed as

$$H_i = \sum_{il} U(r_i - R_l) c_i^\dagger c_i$$

$$= \sum_q V_0 \rho_i(q) \left(c_{k+q,\uparrow}^\dagger \ c_{-k-q,\downarrow} \right) \tau_3 \begin{pmatrix} c_{k,\uparrow} \\ c_{-k,\downarrow}^\dagger \end{pmatrix}, \tag{8.2}$$

where

$$\rho_i(q) = \frac{1}{V} \sum_l e^{iq \cdot R_l} \tag{8.3}$$

is a function of the impurity configuration.

The effect of impurity scattering on a *d*-wave superconductor can be solved through the perturbation expansion. A key step is to take the random average for the impurity scattering potentials. At the *n*-th order of perturbation, the random average of the scattering potential is given by

$$f_n(q_1, q_2, ..., q_n) = \langle \rho_i(q_1) \rho_i(q_2)...\rho_i(q_n) \rangle_{imp}. \tag{8.4}$$

It is difficult to evaluate this average rigorously. However, in the limit the impurity concentration is very low and the total impurity number N_i is very large, the interaction among impurities is very small and negligible. A common approximation is to ignore the interference effect between different impurities and take

$$f_n(q_1, q_2, ..., q_n) \approx n_i \delta(\Sigma'q) + n_i^2 \delta(\Sigma'q)\delta(\Sigma'q) + n_i^3 \delta(\Sigma'q)\delta(\Sigma'q)\delta(\Sigma'q) + ..., \quad (8.5)$$

where $n_i = N_i/V$ is the impurity concentration. $\Sigma'q$ is to sum over all or part of q_i. More precisely, in the first term, $\Sigma'q$ is to sum over all q_i. In the second term, $\Sigma'q$ appears twice. In this case, all q_i are divided into two groups. The first $\Sigma'q$ is to sum over all the q_i in the first group, and the second $\Sigma'q$ is to sum over all the q_i in the second group. $(\Sigma'q)$'s in higher order terms should be similarly understood. This approximation neglects the interference effect between different impurities. It assumes that the scattering to electrons by different impurities is independent such that the momentum of an incident electron, after being scattered once or multiple times by one impurity, is not changed. This is equivalent to just keeping the reducible Feynman diagrams of impurity scattering, as shown in Fig. 8.1(a), and neglect all the irreducible diagrams, as shown in Fig. 8.1(b).

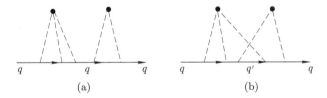

$$q \qquad q \qquad q \qquad\qquad q \qquad q' \qquad q$$
$$\text{(a)} \qquad\qquad\qquad \text{(b)}$$

Figure 8.1   (a) A reducible impurity scattering Feynman diagram in which the scattering of electrons by different impurities is independent. (b) An irreducible impurity scattering Feynman diagram, which contains the interaction between impurities. In the self-consistent T-matrix, or, the single-impurity approximation, this kind of diagrams are neglected.

Under this approximation, the many-impurity scattering problem is reduced effectively to a single-impurity one. The multi-impurity effect is reflected in the factor n_i associated with each single-impurity scattering term. But unlike the single-impurity problem discussed in the preceding chapter, the system becomes translation invariant after taking the disorder average. In addition, after the disorder average, the density of states divergence induced by the single-impurity resonance state in the unitary limit is smeared out. The divergence is replaced by a broadened and lowered peak. Although this result is obtained based on the T-matrix approximation, it indicates that it might be quite difficult to detect an impurity resonance state by measuring a macroscopic averaged quantity.

8.2 The Self-Energy Function

A major effect of disorder on superconductivity is to change the self-energy of quasiparticles. For a d-wave superconductor, the self-energy needs to be determined self-consistently from the Green's function of electrons. Under the approximations previously introduced, the Dyson equation of the single-particle Green's function, as illustrated in Fig. 8.2, is given by

$$G\left(q,\omega\right) = G^{(0)}\left(q,\omega\right) + G^{(0)}\left(q,\omega\right)\Sigma\left(\omega\right)G\left(q,\omega\right), \qquad (8.6)$$

where $\Sigma(\omega)$ is the self-energy function.   The self-energy is momentum-independent. This is due to the use of the δ-function potential. The solution to this equation is

$$G(q,\omega) = \frac{1}{\omega - \varepsilon_q\sigma_3 - \Delta_q\sigma_1 - \Sigma(\omega)}. \qquad (8.7)$$

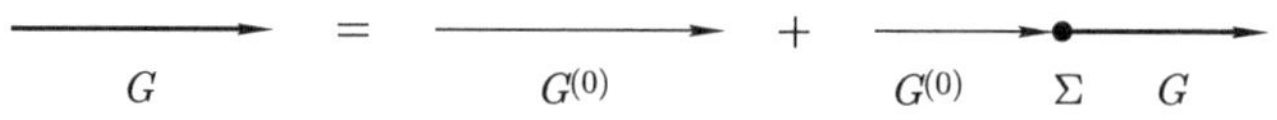

Figure 8.2   Diagrammatic representation of the Dyson equation.

In the single-impurity approximation, $\Sigma(\omega)$ is determined by the Feynman diagrams shown in Fig. 8.3. The first order correction to the self-energy is given by

$$\Sigma^{(1)}\left(\omega\right) = n_i V_0\sigma_3.$$

The second order correction is

$$\Sigma^{(2)}\left(\omega\right) = n_i V_0\sigma_3 c^{-1}\underline{G}\left(\omega\right)\sigma_3,$$

where $c = 1/\pi N_F V_0$ and

$$G(\omega) = \frac{1}{\pi N_F V} \sum_q G(q, \omega) \tag{8.8}$$

is the momentum averaged Green's function. G is similar to G_0 defined in Chapter 7, but G is determined by the full Green's function, not just $G^{(0)}$. c is related to the scattering phase shift by the equation

$$c = \cot \delta_0, \tag{8.9}$$

same as in the single-impurity system.

From the Feynman diagrams shown in Fig. 8.3, it is not difficult to show that the n-th order correction to the self-energy generally has the form

$$\Sigma^n(\omega) = n_i V_0 \sigma_3 \left(c^{-1} G(\omega) \sigma_3 \right)^{n-1}.$$

Therefore, the total self-energy is

$$\Sigma(\omega) = \Sigma^{(1)}(\omega) + \Sigma^{(2)}(\omega) + \Sigma^{(3)}(\omega) + \cdots = \frac{n_i V_0 c \sigma_3}{c - G(\omega) \sigma_3}. \tag{8.10}$$

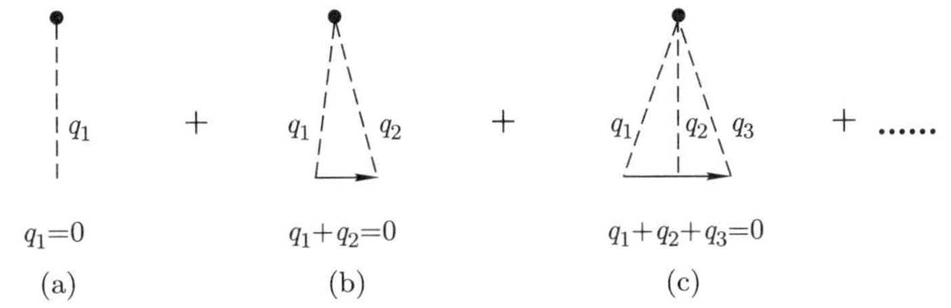

$q_1 = 0$ $q_1 + q_2 = 0$ $q_1 + q_2 + q_3 = 0$

(a)   (b)   (c)

Figure 8.3   The self-energy correction by the impurity scattering.

Eqs. (8.7), (8.8) and (8.10) form a set of self-consistent equations for determining the single-particle Green's function. The electron self-energy and G can generally be expressed as

$$G(\omega) = G_0(\omega) + G_1(\omega) \sigma_1 + G_2(\omega) \sigma_2 + G_3(\omega) \sigma_3, \tag{8.11}$$

$$\Sigma(\omega) = \Sigma_0(\omega) + \Sigma_1(\omega) \sigma_1 + \Sigma_2(\omega) \sigma_2 + \Sigma_3(\omega) \sigma_3. \tag{8.12}$$

From Eq. (8.10), we find that Σ's are determined by the equations

$$\Sigma_0(\omega) = -n_i c V_0 \underline{G}_0(\omega) A(\omega),$$

$$\Sigma_1(\omega) = -n_i c V_0 \underline{G}_1(\omega) A(\omega),$$

$$\Sigma_2(\omega) = -n_i c V_0 \underline{G}_2(\omega) A(\omega),$$

$$\Sigma_3(\omega) = n_i c V_0 [\underline{G}_3(\omega) - c] A(\omega),$$

where

$$A(\omega) = \frac{1}{\underline{G}_0^2(\omega) - \underline{G}_1^2(\omega) - \underline{G}_2^2(\omega) - [\underline{G}_3(\omega) - c]^2}.$$

Similarly, from Eq. (8.8), $\underline{G}$ can be solved as

$$\underline{G}_0(\omega) = [\omega - \Sigma_0(\omega)] \sum_q B(q,\omega),$$

$$\underline{G}_1(\omega) = \sum_q B(q,\omega) [\Delta_q + \Sigma_1(\omega)],$$

$$\underline{G}_2(\omega) = \Sigma_2(\omega) \sum_q B(q,\omega),$$

$$\underline{G}_3(\omega) = \sum_q [\varepsilon_q + \Sigma_3(\omega)] B(q,\omega),$$

where

$$B(q,\omega) = \frac{1}{\pi N_F V} \frac{1}{[\omega - \Sigma_0(\omega)]^2 - [\varepsilon_q + \Sigma_3(\omega)]^2 - [\Delta_q + \Sigma_1(\omega)]^2 - \Sigma_2^2(\omega)}.$$

For a d-wave superconductor, the momentum summation of the energy gap vanishes, i.e. $\sum_k \Delta_k = 0$. It is simple to show that $\Sigma_1 = \underline{G}_1 = \Sigma_2 = \underline{G}_2 = 0$ is a self-consistent solution to these equations [124]. In this case, $\underline{G}_3$ can be expressed as

$$\underline{G}_3(\omega) = \frac{1}{\pi N_F} \int \frac{d\varphi}{2\pi} \int_{-\infty}^{\infty} d\varepsilon \rho_N(\varepsilon) \frac{\varepsilon + \Sigma_3(\omega)}{[\omega - \Sigma_0(\omega)]^2 - [\varepsilon + \Sigma_3(\omega)]^2 - \Delta_\varphi^2}.$$
$$(8.13)$$

In obtaining this expression, we assumed that $\varepsilon(k)$ depends only on $|k|$ and $\Delta_k = \Delta_\varphi$ depends only on φ. If the system is particle-hole symmetric and $\rho_N(\varepsilon)$ does not strongly depend on ε, the above integral can be simplified as

$$\underline{G}_3(\omega) = \frac{1}{\pi} \int \frac{d\varphi}{2\pi} \int_{-\infty-\Sigma_3}^{\infty-\Sigma_3} d\varepsilon \frac{\varepsilon}{[\omega - \Sigma_0(\omega)]^2 - \varepsilon^2 - \Delta_\varphi^2}. \qquad (8.14)$$

Σ_3 is a complex number, the integral on the right-hand side is not necessary to be zero. Its value depends on the poles of the integrand

$$\varepsilon_\pm = \pm\sqrt{[\omega - \Sigma_0(\omega)]^2 - \Delta_\varphi^2}.$$

From the self-consistent equations, it can be shown that the absolute value of the imaginary part of $\varepsilon_\pm$ is larger than the absolute value of the imaginary part of Σ_3. Therefore, there are no poles in the complex ε-plane enclosed by the real axis and the line of $\mathrm{Im}\,\varepsilon = \mathrm{Im}\Sigma_3$. Since the integrand is an odd function of ε along the real axis, the integral along the real axis is zero. Consequently, $\underline{G}_3$ should also be zero [124].

The above result shows that $\underline{G} = \underline{G}_0$ is a constant matrix. Substituting this result into the self-consistent equations, we find that

$$\Sigma_0(\omega) = \frac{n_i V_0 c \underline{G}_0(\omega)}{c^2 - \underline{G}_0^2(\omega)}, \tag{8.15}$$

$$\Sigma_3(\omega) = \frac{n_i V_0 c^2}{c^2 - \underline{G}_0^2(\omega)}, \tag{8.16}$$

$$\underline{G}_0(\omega) = \frac{\omega - \Sigma_0(\omega)}{\pi N_F V} \sum_q \frac{1}{[\omega - \Sigma_0(\omega)]^2 - [\varepsilon_q + \Sigma_3(\omega)]^2 - \Delta_q^2}. \tag{8.17}$$

At $\omega = 0$, $\Sigma_0(0) \equiv -i\Gamma_0$ is purely an imaginary number, and $\Sigma_3(0)$ is real. Now we have

$$G_0(0) = -i\Gamma_0 \int \frac{d\varphi}{2\pi} \frac{1}{\sqrt{\Gamma_0^2 + \Delta_\varphi^2}}, \tag{8.18}$$

where Γ_0 is proportional to the quasiparticle scattering rate. If τ is the quasiparticle lifetime at the zero frequency, then

$$\tau^{-1} = 2\Gamma_0.$$

In the normal state, $\Delta_k = 0$, the integral in Eq. (8.17) can be integrated out rigorously. It can be shown that $\underline{G}_0(\omega)$ is frequency-independent:

$$\underline{G}_0(\omega) = \underline{G}(\omega) = -i. \tag{8.19}$$

Therefore the self-energy in the normal state is

$$\Sigma_N(\omega) = \frac{n_i V_0 c}{1 + c^2}(-i + c\sigma_3). \tag{8.20}$$

The electron scattering rate in the normal state is twice of the imaginary part of the self-energy:

$$\Gamma_N = -2\mathrm{Im}\Sigma_N(\omega) = \frac{2n_iV_0c}{1+c^2}. \tag{8.21}$$

In the superconducting state, Eqs. (8.15)~(8.17) can be solved numerically. In certain limit, this set of equations can be solved analytically but with approximations, which may help us to grasp the key physics governing the impurity scattering. This will be discussed in details in the following sections.

8.3 The Born Scattering Limit

In the discussion of physical properties of disordered *d*-wave superconductors, two scattering limits need more attention. One is the unitary limit, which is also called the resonance limit. The other is the Born scattering limit. They correspond to the strong and weak scattering limits, respectively. In the unitary limit, $c \to 0$, and the corresponding scattering phase shift $\delta_0 = \pi/2$. This is the largest phase shift that can be taken in the *s*-wave scattering channel. In the Born scattering limit, $c \to \infty$, the corresponding phase shift δ_0 is very small, close to zero. Many physical properties of *d*-wave superconductors are different, not just quantitatively, but also qualitatively, in these two limits.

A real system usually lies between the Born and unitary scattering limits. The effect of non-magnetic impurity scattering can be estimated, in principle, by interpolating between these two limits. Therefore, it is important to have a thorough understanding of these two scattering limits in order to understand more comprehensively the impurity effect in *d*-wave superconductors.

The problem of impurity scattering in a *d*-wave superconductor is difficult to solve analytically. A general solution to the Green's function in the whole frequency or temperature range can be obtained only through numerical calculations. However, in the unitary or Born scattering limit, the analysis is greatly simplified. The low energy or low temperature results can be obtained perturbatively through expansions with respect to frequency or temperature.

The zinc impurity has a large effect on high-T_c superconducting properties. The scattering induced by the zinc impurity is generally believed to be in the resonance scattering limit. Disorder effect induced by the structure inhomo-geneity or defects on the interface or CuO_2 planes is more complicated. It may lie in the resonance scattering limit, or the Born scattering limit, or somewhere between the two limits, depending on the interaction between impurities and superconducting quasiparticles.

In the Born scattering limit, $c \to \infty$, the electron self-energy is given by

$$\Sigma_0(\omega) = \frac{1}{2}\Gamma_N \underline{G}_0(\omega), \tag{8.22}$$

$$\Sigma_3(\omega) = n_i V_0. \tag{8.23}$$

Σ_3 is a constant independent on ω. It can be absorbed into the chemical po-tential by redefining ε_k. Thus only Σ_0 needs to be determined self-consistently. In this case,

$$\underline{G}_0(\omega) = \frac{\omega - \Sigma_0(\omega)}{\pi}\int\frac{d\varphi}{2\pi}\int_{-\infty}^{\infty}d\varepsilon\frac{1}{[\omega - \Sigma_0(\omega)]^2 - \varepsilon^2 - \Delta_\varphi^2}. \tag{8.24}$$

After integrating over ε, we obtain

$$\underline{G}_0(\omega) = -i\left[\omega - \Sigma_0(\omega)\right]\int\frac{d\varphi}{2\pi}\frac{\theta(\mathrm{Im}\sqrt{[\omega - \Sigma_0(\omega)]^2 - \Delta_\varphi^2})}{\sqrt{[\omega - \Sigma_0(\omega)]^2 - \Delta_\varphi^2}}. \tag{8.25}$$

Here the θ-function is introduced to avoid the uncertainty in taking the square root of a complex number. It means that the numerator of the integrand must have a positive imaginary part.

For the $d_{x^2-y^2}$ superconductor, $\Delta_\varphi = \Delta_0 \cos 2\varphi$, a useful variable substitu-tion formula for carrying out the integration over φ in Eq. (8.25) is

$$\int_0^{2\pi}\frac{d\varphi}{2\pi}f(\cos 2\varphi) = \frac{2}{\pi}\int_0^1\frac{dx}{\sqrt{1-x^2}}f(x), \tag{8.26}$$

where f is an arbitrary function. This formula is used to substitute all the integrals over φ below. After this substitution, Eq. (8.25) becomes

$$\underline{G}_0(\omega) = -\frac{i2\bar{\omega}}{\pi}\int_0^1\frac{dx}{\sqrt{1-x^2}}\frac{\theta\left(\mathrm{Im}\sqrt{\bar{\omega}^2 - x^2}\right)}{\sqrt{\bar{\omega}^2 - x^2}}, \tag{8.27}$$

where $\overline{\omega} = [\omega - \Sigma_0(\omega)]/\Delta_0$. This is an elliptic integral which cannot be represented by a simple elementary function.

The above self-consistent equations can be solved numerically. Fig. 8.4 and Fig. 8.5 show the real and imaginary parts of the self-energy as a function of frequency for three different values of Γ_N, respectively.

When $|\overline{\omega}| \ll 1$, the integral contributes mainly from the domain of $x \ll 1$ in which $1/\sqrt{1-x^2}$ can be approximately replaced by 1, we then have

$$\underline{G}_0(\omega) \approx -\frac{2\overline{\omega}}{\pi} \ln \frac{1 + \sqrt{1 - \overline{\omega}^2}}{\sqrt{-\overline{\omega}^2}} \approx -\frac{2\overline{\omega}}{\pi} \ln \frac{2}{\sqrt{-\overline{\omega}^2}} \qquad (|\overline{\omega}| \ll 1). \qquad (8.28)$$

The self-consistent equation of Σ_0 now becomes

$$\sqrt{-\overline{\omega}^2} = 2 \exp\left(\frac{\pi\Sigma_0}{\Gamma_N\overline{\omega}}\right). \qquad (8.29)$$

At the zero frequency, $\omega = 0$, the solution is by

$$\Sigma_0(0) = -i2\Delta_0 \exp\left(-\frac{\pi\Delta_0}{\Gamma_N}\right). \qquad (8.30)$$

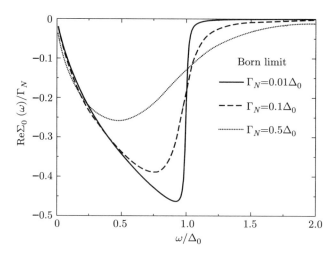

Figure 8.4   Frequency dependence of the real part of the self-energy $\Sigma_0(\omega)$ in the Born scattering limit $(c \to \infty)$.

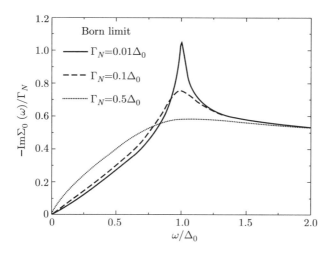

Figure 8.5   Frequency dependence of the imaginary part of the self-energy $\Sigma_0(\omega)$ in the Born scattering limit $(c \to \infty)$.

In the discussions below, we will encounter many integrals similar to Eq. (8.27). When the temperature or the energy scale is much smaller than the energy gap Δ_0, the contribution to this kind of integrals comes mainly from the domain of $x \ll 1$. Again, we can adopt the approximation $\sqrt{1 - x^2} \approx 1$. This is equivalent to taking the linear approximation for the quasiparticle energy spectrum or dispersion.

The low frequency behavior of $\Sigma_0(\omega)$ can be solved by the series expansion. Up to the 2nd order terms of ω,

$$\Sigma_0(\omega) \approx \Sigma_0(0) + \left(1 - \frac{\pi\Delta_0}{\Gamma_N}\right)\omega - \frac{\pi^2\Delta_0^2}{2\Gamma_N^2\Sigma_0(0)}\omega^2 + o(\omega^2). \tag{8.31}$$

In high-T_c superconductors, Γ_N is generally much smaller than Δ_0, and $\Sigma_0(0)$ is also very small. Therefore the disorder scattering only slightly changes the superconducting properties in this limit. Nevertheless, Σ_0 is finite. It can affect the conducting behavior of d-wave superconductors in low temperatures.

In the discussion for the disorder effect in d-wave superconductors, if the self-consistent condition is not enforced in the determination of the Green's function, then $\bar{\omega} = \omega/\Delta_0$ and in the low frequency limit $|\omega| \ll \Delta_0$,

$$\Sigma_0\left(\omega\right) \approx -\frac{\Gamma_N \omega}{\pi \Delta_0}\left(\ln\frac{2\Delta_0}{|\omega|} + i\frac{\pi}{2}\right). \tag{8.32}$$

Now $\Sigma_0(0) = 0$, which is different from the self-consistent solution. This differ-ence shows the importance of self-consistent treatment to the Green's function. However, in the frequency regime $\Gamma_N \ll |\omega| \ll \Delta_0$, the higher order corrections from the impurity scattering become less important. In this case, there is no much difference between the self-consistent and not-self-consistent results, and Eq. (8.32) is an approximate solution of $\Sigma_0(\omega)$.

8.4 The Resonant Scattering Limit

In the resonant scattering limit, $c \to 0$. The self-consistent equation of the self-energy becomes

$$\Sigma_0(\omega) = -\frac{n_i V_0 c}{\underline{G}_0(\omega)} = -\frac{\Gamma_N}{2\underline{G}_0(\omega)}, \tag{8.33}$$

$$\Sigma_3(\omega) = -\frac{n_i V_0 c^2}{\underline{G}_0^2(\omega)}. \tag{8.34}$$

Unlike in the Born scattering limit, $\underline{G}_0(\omega)$ now appears in the denominator. Compared to Σ_0, Σ_3 is a higher order small quantity and can be neglected. $\underline{G}_0(\omega)$ is still determined by Eq. (8.27).

Figs. 8.6 and 8.7 show the numeric solutions of the self-consistent equations. For small Γ_N, the real and imaginary parts of Σ_0 are very small, close to zero, at $\omega = \Delta_0$, different from the results obtained in the Born scattering limit.

In the limit $|\bar{\omega}| \ll 1$, the self-consistent equation is approximately given by

$$4\bar{\omega}\Sigma_0(\omega)\left(\ln 2 - \ln\sqrt{-\bar{\omega}^2}\right) \approx \pi \Gamma_N. \tag{8.35}$$

It is still difficult to solve this equation analytically. If we neglect the slowly varying logarithmic term, this equation is simplified as

$$[\omega - \Sigma_0(\omega)]\Sigma_0(\omega) \approx \frac{\pi \Gamma_N \Delta_0}{4\ln 2}. \tag{8.36}$$

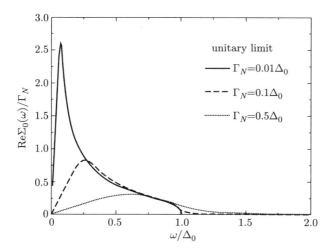

Figure 8.6   Frequency dependence of the real part of the self-energy $\Sigma_0(\omega)$ in the unitary scattering limit $(c \to 0)$.

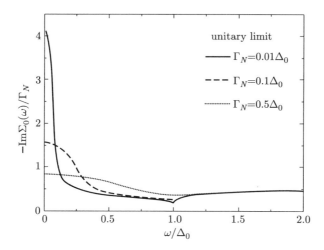

Figure 8.7   Frequency dependence of the imaginary part of the self-energy $\Sigma_0(\omega)$ in the unitary scattering limit $(c \to 0)$.

The solutions are

$$\Sigma_0(\omega) \approx \frac{\omega \pm \sqrt{\omega^2 - \pi \Gamma_N \Delta_0 / \ln 2}}{2}. \tag{8.37}$$

For these two solutions, the one with plus sign has a positive imaginary part corresponding to the advanced Green's function and does not need to be considered.

At low frequency, $\Sigma_0(\omega)$ can be expanded in terms of ω. The first three terms are

$$\Sigma_0(\omega) = \Sigma_0(0) + \frac{\omega}{2} + \frac{\omega^2}{8\Sigma_0(0)} + o(\omega^2), \tag{8.38}$$

where $\Sigma_0(0)$ is the self-energy at $\omega = 0$,

$$\Sigma_0(0) \approx -\frac{i\sqrt{\pi\Gamma_N\Delta_0}}{2\sqrt{\ln 2}}. \tag{8.39}$$

By comparison, we find that $|\Sigma_0(0)|$ drops with increasing Γ_N in the Born scattering limit, but increases with Γ_N in the resonance scattering limit. This indicates that the scattering effect is stronger in the resonant scattering limit than in the Born scattering limit in d-wave superconductors.

In the intermediate frequency regime, $\Delta_0 \gg \omega \gg |\Sigma_0(0)|$, the self-consistency in the self-energy becomes less important, similarly as in the Born scattering limit. In this case, $\bar{\omega} = \omega/\Delta_0$ and

$$\Sigma_0(\omega) \approx \frac{\pi\Delta_0\Gamma_N}{4\omega\left(\ln\dfrac{2\Delta_0}{|\omega|} + i\dfrac{\pi}{2}\right)}. \tag{8.40}$$

$\Sigma_0(\omega)$ exhibits different asymptotic behaviors at $\omega \ll |\Sigma_0(0)|$ and $\Delta_0 \gg \omega \gg |\Sigma_0(0)|$. This difference implies that the disorder scattering behaves quite differently for the energy smaller or larger than $|\Sigma_0(0)|$. If the energy is lower than $|\Sigma_0(0)|$, the impurity scattering affects strongly physical properties of d-wave superconductors, similar as in a gapless s-wave superconductor induced by magnetic impurities. On the other hand, if the energy is higher than $|\Sigma_0(0)|$, the impurity scattering is small, and the impurity correction to the superconducting properties is negligible.

In literature, an energy interval is regarded in a gapless region if $\omega \ll |\Sigma_0(0)|$, or in an intrinsic region if $\omega \gg |\Sigma_0(0)|$. The effect of impurity scattering is strong in the gapless region, but weak in the intrinsic region.

The single-particle Green's function can be represented using an unified formula in both the Born and unitary scattering limits. In the gapless region, $\omega \ll \Gamma_0$, if the self energy is expanded up to the first order in ω, the retarded Green's function in these two limits is found to be

$$G^R(k, \omega) = \frac{1}{a\omega + i\Gamma_0 - \varepsilon_k\sigma_3 - \Delta_k\sigma_1},\qquad(8.41)$$

where $a = 1 - \Sigma_0'(0)$. In the Born scattering limit,

$$a = \frac{\pi\Delta_0}{\Gamma_N},\qquad(8.42)$$

while in the unitary limit,

$$a = \frac{1}{2}.\qquad(8.43)$$

In the intrinsic region, $\Gamma_0 \ll \omega \ll \Delta_0$, after neglecting the impurity correction to ω, the Green's function is approximately given by

$$G^R(k, \omega) = \frac{1}{\omega + i\Gamma(\omega) - \varepsilon_k\sigma_3 - \Delta_k\sigma_1},\qquad(8.44)$$

where $\Gamma(\omega)$ is a frequency-dependent quasiparticle scattering rate. In the Born scattering limit

$$\Gamma(\omega) \approx \frac{\Gamma_N\omega}{2\Delta_0},\qquad(8.45)$$

while in the unitary limit,

$$\Gamma(\omega) \approx \frac{\pi^2\Delta_0\Gamma_N}{8\omega \ln^2(2\Delta_0/|\omega|)}.\qquad(8.46)$$

8.5   Correction to the Superconducting Critical Temperature

The impurity scattering does not affect the superconducting transition temperature T_c for an s-wave superconductor. This is a consequence of the Anderson theorem [107] (see Appendix D), which holds for all conventional metal-based superconductors. However, for d-wave superconductors, the Anderson theorem does not hold any more, and there is a finite correction to the critical transition temperature from the impurity scattering.

In order to study the impurity correction to the transition temperature T_c, it is more convenient to use the Matsubara Green's functions. Under the T-matrix approximation, the Green's function of electrons is given by

$$G(k, i\omega_n) = \frac{1}{i\tilde{\omega}_n - \tilde{\varepsilon}_k \sigma_3 - \Delta_k \sigma_1},$$ (8.47)

where

$$i\tilde{\omega}_n = i\omega_n - \Sigma_0(i\omega_n),$$

$$\tilde{\varepsilon}_k = \varepsilon_k + \Sigma_3(i\omega_n).$$

Substituting Eq. (8.47) into the gap equation,

$$\Delta_0 = -\frac{gk_BT}{2} \sum_{k,\omega_n} \phi_k \mathrm{Tr}\sigma_1 G(k, i\omega_n),$$ (8.48)

we obtain

$$1 = -gk_BT \sum_{k,\omega_n} \frac{\phi_k^2}{(i\tilde{\omega}_n)^2 - \Delta_k^2 - \tilde{\varepsilon}_k^2}.$$ (8.49)

At the critical point, $\Delta_k = 0$, from the expressions of $\Sigma_0(i\omega_N)$ and $\underline{G}_0(i\omega_n)$ given before, it can be shown that in the normal state,

$$\Sigma_{0,N}(i\omega_n) = -i\Gamma_N \mathrm{sgn}(\omega_n).$$

Note that $\Sigma_{0,N}(i\omega_n)$ differs from the corresponding expression in the real frequency, $\Sigma_{0,N}(\omega) = -i\Gamma_N$, which is ω independent. Substituting it into Eq. (8.49) and taking the average of ϕ_k^2 over the Fermi surface, we obtain

$$1 = \frac{gk_BT_cN_F}{2} \sum_{\omega_n} \int d\varepsilon \frac{1}{\left(|\omega_n| + \Gamma_N/2\right)^2 + \varepsilon^2} \equiv K(\Gamma_N).$$ (8.50)

The summation over ω_n or the integration over ε must have finite lower and upper bounds. Otherwise, the right hand side of the above equation diverges. This requirement is physically correct because electrons can form superconducting pairs only within a finite energy interval. The divergence is of the first order. It can be removed by imposing finite upper and lower bounds either to the summation over ω_n or to the integral over ε.

In the limit $\Gamma_N = 0$, it is more convenient to impose the restriction on the lower and upper bounds for the integral over ε, but not on the Matsubara frequency summation. In this case, the summation over ω_n can be solved analytically, which gives

$$K(0) = \frac{gN_F}{2} \int_0^{\omega_0} d\varepsilon \frac{1}{\varepsilon} \tanh \frac{\varepsilon}{k_B T_c}, \qquad (8.51)$$

where ω_0 is the characteristic frequency of electron pairing. The integral over ε can be completed using the method introduced in Sect. 3.1, which yields

$$\begin{aligned} K(0) &= \frac{gN_F}{2} \int_0^{\omega_0} d\varepsilon \frac{1}{\varepsilon} \tanh \frac{\varepsilon}{2k_B T_c} \\ &\approx \frac{gN_F}{2} \ln \frac{1.134\omega_0}{k_B T_c}. \end{aligned} \qquad (8.52)$$

In a pure system without impurities, the superconducting T_{c0} is determined by the equation

$$1 = \frac{gN_F}{2} \ln \frac{1.134\omega_0}{k_B T_{c0}}. \qquad (8.53)$$

Using this formula, Eq. (8.50) can be rewritten as

$$\ln \frac{T_{c0}}{T_c} + \frac{2}{gN_F} [K(\Gamma_N) - K(0)] = 0. \qquad (8.54)$$

The difference between $K(\Gamma_N)$ and $K(0)$, i.e. $K(\Gamma_N) - K(0)$, is a regular function of both ω and ω_n, because the divergent terms in both $K(0)$ and $K(\Gamma_N)$ are subtracted. Thus if we directly calculate this difference, both the upper and lower bounds in the integral over ε as well as in the summation over ω_n can be taken to infinity. After taking the integral over ε, the above equation becomes

$$\ln \frac{T_{c0}}{T_c} + \sum_{n \geqslant 0} \left(\frac{1}{n + \dfrac{1}{2} + \dfrac{\Gamma_N}{4\pi k_B T_c}} - \frac{1}{n + \dfrac{1}{2}} \right) = 0. \qquad (8.55)$$

The summation over n can be expressed using the digamma function $\psi(x)$ as

$$\sum_{n \geqslant 0} \left(\frac{1}{n + \dfrac{1}{2} + \dfrac{\Gamma_N}{4\pi k_B T_c}} - \frac{1}{n + \dfrac{1}{2}} \right) = \psi \left(\frac{1}{2} \right) - \psi \left(\frac{1}{2} + \frac{\Gamma_N}{4\pi k_B T_c} \right), \qquad (8.56)$$

where $\psi(x)$ is defined as

$$\psi(x) = -0.577 - \frac{1}{x} + \sum_{n=1}^{\infty} \left(\frac{1}{n} - \frac{1}{n+x} \right).$$

Thus the equation for determining the transition temperature T_c is given by

$$\ln \frac{T_{c0}}{T_c} = \psi \left(\frac{1}{2} + \frac{\Gamma_N}{4\pi k_B T_c} \right) - \psi \left(\frac{1}{2} \right). \tag{8.57}$$

Both the coupling constant g and the normal state density of states N_F do not appear explicitly in this equation. T_c is determined purely by T_{c0} and Γ_N. It implies that the impurity correction to the transition temperature is universal, determined just by Γ_N. Fig. 8.8 shows the numerical solution for Eq. (8.57). At $\Gamma_N \approx 1.764 T_{c0}$, the impurity scattering suppresses T_c completely.

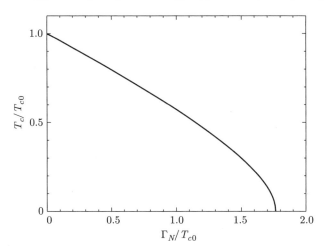

Figure 8.8   Correction to T_c by the impurity scattering: A universal plot of T_c/T_{c0} as a function of Γ_N/T_{c0}.

Fig. 8.9 shows the experimental results of the transition temperature T_c for $YBa_2(Cu_{1-z}M_z)_3O_7$ and other high-T_c superconductors as a function of the nickel or zinc impurity concentration z. For all the cases shown in the figure, T_c decreases linearly with z within the measurement errors. But the slopes are different for different superconductors. The experimental results agree with the

theoretical prediction for the impurity correction to T_c for d-wave superconductors.

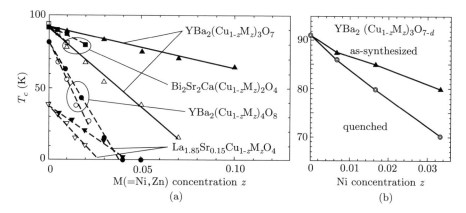

Figure 8.9 T_c versus the impurity concentration z for $YBa_2(Cu_{1-z}M_z)_3O_7$ and other high-
T_c superconductors, $M = Ni$ or Zn. The solid and open symbols represent T_c of Ni- and
Zn-substituted superconductors, respectively. The solid and dashed lines in (a) are linear fits
to the experimental data using the formula $T_c = T_{c0} - \alpha_M z$, where α_M is the fitting parameter. For Y123, Bi2212, Y124 and La214 superconductors, α_{Ni}/α_{Zn} equal 0.26, 0.46, 0.80 and
0.62, respectively. (b) T_c versus z for both as grown and quenched $YBa_2(Cu_{1-z}Ni_z)_3O_7$ samples, respectively. Reprinted with permission from [125]. Copyright (2002) by the American
Physical Society.

At low dopings, Γ_N is proportional to the doping concentrate z. For YBa_2
Cu_3O_7, 1% zinc concentration suppresses T_c by 1/7 of its maximal value. From
the result shown in Fig. 8.8, we estimate $\Gamma_N \approx 2T_{c0}/7$ for 1% zinc. Substituting
this value of Γ_N and $c = 0.07$ extracted from the STM experiment [109] into
Eq.(8.21), the zinc impurity potential strength V_0 is estimated to be $V_0 \sim$
1.8eV. It is smaller than but of the same order as that estimated from the
density of states. It shows that the zinc impurity is really a strong scattering
center. The suppression of T_c by Ni is only 1/4 as by Zn. The effective value of
c for nickel is also larger. Therefore, the scattering from nickel is much weaker
than zinc.

In $YBa_2Cu_4O_8$ and La_2CuO_4, the difference in the suppression of T_c induced by nickel and zinc impurities is smaller than in $YBa_2Cu_3O_7$. This is probably due to the fact that the doped zinc or nickel atoms may not all lie on the CuO_2 planes in the former two compounds.

8.6 Density of States

The density of states $\rho(\omega)$ is determined by the imaginary part of the Green's function,

$$
\begin{aligned}
\rho(\omega) &= -\frac{1}{\pi V}\mathrm{Im}\sum_k G_{11}(k,\omega) \\
&= -\frac{1}{\pi V}\mathrm{Im}\sum_k \frac{\omega - \Sigma_0(\omega)}{[\omega - \Sigma_0(\omega)]^2 - [\varepsilon_k + \Sigma_3(\omega)]^2 - \Delta_k^2} \\
&= -N_F\mathrm{Im}\underline{G}_0(\omega).
\end{aligned}
\tag{8.58}
$$

Again the summation over momentum cannot be solved exactly. Nevertheless, in the Born and unitary scattering limits, $\Sigma_3(\omega)$ can be absorbed into the chemical potential and omitted. Under this approximation, the above expression can be simplified as

$$
\begin{aligned}
\rho(\omega) &= -\frac{N_F}{\pi V}\mathrm{Im}\int\frac{d\varphi}{2\pi}\int d\varepsilon \frac{\omega - \Sigma_0(\omega)}{[\omega - \Sigma_0(\omega)]^2 - \varepsilon^2 - \Delta_\varphi^2} \\
&= \frac{2N_F}{\pi}\mathrm{Im}\bar{\omega}\int_0^1\frac{dx}{\sqrt{1-x^2}}\frac{\theta\left(\mathrm{Re}\sqrt{x^2-\bar{\omega}^2}\right)}{\sqrt{x^2-\bar{\omega}^2}}.
\end{aligned}
$$

In the limit $|\bar{\omega}|\ll 1$, the integral is contributed mainly from the region $x\ll 1$, thus $\sqrt{1-x^2}$ is approximately equal to 1 and

$$
\rho(\omega) \approx \frac{2N_F}{\pi}\mathrm{Re}\bar{\omega}\int_0^1 dx\frac{-i}{\sqrt{x^2-\bar{\omega}^2}} = \frac{2N_F}{\pi}\mathrm{Im}\left(\bar{\omega}\ln\frac{1+\sqrt{1-\bar{\omega}^2}}{\sqrt{-\bar{\omega}^2}}\right).
\tag{8.59}
$$

From this expression, we find that the impurity induced density of states is finite at the zero energy:

$$
\rho(0) \approx \frac{2N_F\Gamma_0}{\pi\Delta_0}\ln\frac{\Delta_0 + \sqrt{\Delta_0^2 + \Gamma_0^2}}{\Gamma_0},
\tag{8.60}
$$

where $\Gamma_0 = i\Sigma_0(0)$. This is different than in the ideal d-wave superconductor where the zero-energy density of states vanishes. In the Born scattering limit (see Fig. 8.10),

$$\Gamma_0 = 2\Delta_0 \exp\left(-\frac{\pi\Delta_0}{\Gamma_N}\right), \tag{8.61}$$

while in the unitary limit (see Fig. 8.11),

$$\Gamma_0 = \sqrt{\frac{\pi\Gamma_N\Delta_0}{4\ln 2}}. \tag{8.62}$$

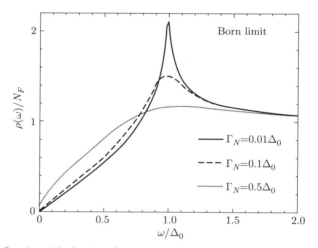

Figure 8.10   Quasiparticle density of states in the Born scattering limit.

In the gapless region, $\rho(\omega) \ll N_F$, $\rho(\omega)$ can be solved by the Taylor expansion with respect to ω. It is simple to show that the first order term in the Taylor series of ω is zero in both the Born and unitary scattering limits. Thus $\rho(\omega)$ varies as ω^2 in the low energy limit.

In the unitary scattering limit, $\rho(\omega)$ is approximately given by

$$\rho(\omega) \approx \rho(0)\left(1 - \frac{\omega^2}{4\Gamma_0^2}\right) \qquad (|\omega| \ll \Gamma_0). \tag{8.63}$$

However, in the Born scattering limit, it changes to

$$\rho(\omega) \approx \rho(0)\left(1 + \frac{\pi^2\Delta_0^2\omega^2}{2\Gamma_0^2\Gamma_N^2}\right) \qquad (|\omega| \ll \Gamma_0). \tag{8.64}$$

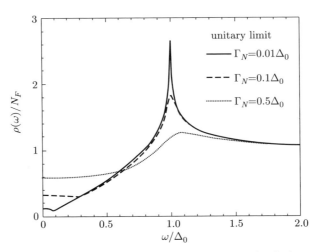

Figure 8.11   Quasiparticle density of states in the unitary scattering limit.

Clearly, $\rho(\omega)$ behaves differently in these two limits. In the Born scattering limit, $\rho(\omega)$ increases monotonically with ω. However, in the unitary scattering limit, $\rho(\omega)$ decreases with ω in the low energy limit, giving rise to a local maximum at $\omega = 0$. This local maximum results from the quasiparticle resonance states induced by the impurity scattering on the Fermi surface, but the divergence in the resonant density of states is smeared out by the disorder average.

In the intrinsic region, $\Delta \gg \omega \gg \Gamma_0$, $\rho(\omega)$ behaves similarly as in a disorder-free system. The disorder scattering only induces a small correction to $\rho(\omega)$,

$$\rho(\omega) \approx \frac{N_F \omega}{\Delta_0} \left[1 + \frac{2\Gamma_0}{\pi \omega} \left(\ln \frac{2\Delta_0}{|\omega|} - 1 \right) \right]. \tag{8.65}$$

Nonzero average of $\rho(\omega)$ on the Fermi surface is an important feature of disordered d-wave superconductors. It results from the sign change of the d-wave superconducting gap function on the Fermi surface. If the gap function only has nodes but does not change sign on the Fermi surface, for example, in an anisotropic s-wave superconductor with the gap function $\Delta_k \propto |\cos k_x - \cos k_y|$ or $\Delta_k \propto (\cos k_x - \cos k_y)^2$, the disorder scattering tends to turn the gap function more isotropic. Therefore, the correction to the gap function by the self-energy is finite, and the disorder scattering cannot yield a finite density of states at the Fermi energy. In contrast, it will eliminate the gap nodes and make the quasiparticle excitations fully gapped on the Fermi surface. Therefore, the change of impurity scattering to the superconducting properties in a d-wave superconductor differs qualitatively from that in an anisotropic s-wave superconductor. This can be used to determine whether the superconducting gap of high-T_c superconductors has the d-wave or anisotropic s-wave pairing symmetry.

8.7 Entropy and Specific Heat

Momentum is not a good quantum number in a disordered system. Nevertheless, the superconducting quasiparticles, as the solutions of the BdG equation, are still the energy eigenstates. If $\{E_n, |\phi_n\rangle\}$ is an eigen pair of the Hamiltonian for a disordered superconductor,

$$H|\phi_n\rangle = E_n|\phi_n\rangle, \tag{8.66}$$

then the entropy of this system is defined by

$$S = -k_B \sum_n \{f(E_n) \ln f(E_n) + [1 - f(E_n)] \ln [1 - f(E_n)]\}$$

$$= -k_B \int_{-\infty}^{\infty} d\omega \rho(\omega) \{f(\omega) \ln f(\omega) + [1 - f(\omega)] \ln [1 - f(\omega)]\}. \tag{8.67}$$

Taking the derivative with respect to temperature, the specific heat is found to be

$$C = T\frac{\partial S}{\partial T} = -k_B \int_{-\infty}^{\infty} d\omega \beta \omega^2 \rho(\omega) \frac{\partial f}{\partial \omega}. \tag{8.68}$$

In obtaining this expression, the temperature derivative of $\rho(\omega)$ is omitted.

In the unitary scattering limit, from the result of $\rho(\omega)$ previously obtained, we find the specific heat in the gapless region ($k_B T \ll \Gamma_0$) to be

$$C \approx \frac{\pi^2 \rho(0) k_B^2 T}{3} \left(1 - \frac{7\pi^2 k_B^2 T^2}{20\Gamma_0^2} \right). \tag{8.69}$$

The corresponding result in the Born scattering limit is

$$C \approx \frac{\pi^2 \rho(0) k_B^2 T}{3} \left(1 + \frac{7\pi^4 \Delta_0^2 k_B^2 T^2}{10\Gamma_0^2 \Gamma_N^2} \right). \tag{8.70}$$

In either limit, the specific heat of a disordered d-wave superconductor is a linear function of temperature in low temperatures. This is a direct consequence of finite density of states at the Fermi level. The temperature dependence of C/T is similar to the energy dependence of the density of states. In the resonant scattering limit, C/T should show a peak at zero temperature. This peak, however, does not exist in the Born scattering limit. Therefore, the measurement of low temperature specific heat allows us to determine in which limit the impurity scattering is. The next order correction to the specific heat by the impurity scattering is proportional to T^3, similar to the temperature dependence of the specific heat contributed by phonons.

In the intrinsic temperature region, $T \gg \Gamma_0$, the disorder scattering does not change significantly the temperature dependence of the specific heat in the d-wave superconductor. C behaves almost the same as in a pure d-wave superconductor.

Chapter 9

Superfluid Response

9.1 The Linear Response Theory

When an external magnetic field is applied to a superconductor, the magnetic field will be expelled from the interior of the superconductor by the screening effect of the supercurrent. This is the diamagnetic response of the superconductor to an external magnetic field, a fundamental property of superconducting states. A correct and microscopic description of this electromagnetic response is crucial towards a comprehensive understanding of experimental results and the determination of pairing symmetry. In high-T_c superconductors, a variety of experimental evidence for the existence of gap nodes in the pairing function are obtained through the measurement of electromagnetic response functions.

The superfluid density is an important quantity to describe the supercurrent response, and to connect microscopic mechanism of superconductivity with macroscopic electromagnetism. As revealed by the London equation, Eq. (1.1), the superfluid density is inversely proportional to the square of the penetration depth. These two quantities are intimately connected by the Meissner effect. In a superconductor, the larger is the superfluid density, the stronger is the screening effect from the supercurrent, hence the shorter is the magnetic penetration depth. The variance of the superfluid density with temperature reflects the properties of superconducting quasiparticle excitations. It is different if the pairing symmetry is different. At low temperatures, the superfluid density varies linearly with temperature in a d-wave superconductor, but exponentially in an isotropic s-wave superconductor. Thus we can acquire a great deal of valuable information of quasiparticle excitations through the measurement of the superfluid density or the magnetic penetration depth.

In the linear response theory, the response of the system to an external electromagnetic field is determined by the Kubo formula [9]

$$J_\mu(q,\omega) = -\sum_\nu K_{\mu\nu}(q,\omega)A_\nu(q,\omega), \qquad (9.1)$$

where $K_{\mu\nu}$ is the response function of the electric current J_μ to the vector field A_ν. It is determined by the electron effective mass and the current-current correlation function

$$K_{\mu\nu}(q,\omega) = \frac{e^2}{V\hbar^2}\sum_k\left\langle\frac{\partial^2\varepsilon_k}{\partial k_\mu\partial k_\nu}\right\rangle + \Pi_{\mu\nu}(q,\omega), \qquad (9.2)$$

where $\Pi_{\mu\nu}$ is the current-current correlation function defined by

$$\Pi_{\mu\nu}(q,\omega) = -\frac{i}{V\hbar^2}\int_0^\infty dt e^{i\omega t}\left\langle[J_\mu(q,t),J_\nu(-q,0)]\right\rangle, \qquad (9.3)$$

and J_μ is the electron current operator:

$$J_\mu(q) = e\sum_k\frac{\partial\varepsilon_{k+q/2}}{\partial k_\mu}c_{k+q}^\dagger c_q. \qquad (9.4)$$

In an isotropic and homogeneous system

$$\varepsilon_k = \frac{\hbar^2 k^2}{2m},$$

the first term of $K_{\mu\nu}$ is simply given by

$$\frac{e^2}{V\hbar^2}\sum_k\left\langle\frac{\partial^2\varepsilon_k}{\partial k_\mu\partial k_\nu}\right\rangle = \frac{e^2 n}{m}\delta_{\mu,\nu}.$$

It is proportional to the ratio between the electron density and the effective mass, independent of temperature. However, in real materials, this term depends on the band structure and can vary with temperature.

When an external electromagnetic field with a wavevector q and frequency ω is applied to the system, a current is generated. The response function connecting the current and the electromagnetic field is the conductivity tensor $\sigma(q,\omega)$ defined by

$$J_\mu(r,t) = \sum_\nu\sigma_{\mu\nu}(q,\omega)E_\nu(r,t). \qquad (9.5)$$

The electric field E is related to the vector potential by the equation

$$A_\mu(r,t) = -\frac{i}{\omega} E_\mu(r,t). \tag{9.6}$$

From Eq. (9.1), it is simple to show that $\sigma_{\mu\nu}$ is proportional to the response function $K_{\mu\nu}$:

$$\sigma_{\mu\nu}(q,\omega) = \frac{i}{\omega} K_{\mu\nu}(q,\omega). \tag{9.7}$$

The direct-current (DC) conductivity is the value of $\sigma_{\mu\nu}(q,\omega)$ in a uniform electric field, $q = 0$, and in the zero frequency limit, i.e. $\sigma_{\mu\nu}(0,\omega \to 0)$.

The current-current correlation function $\Pi_{\mu\nu}(q,\omega)$ can be obtained by first calculating the corresponding Matsubara Green's function in the imaginary frequency space, $\tilde{\Pi}_{\mu\nu}(q,i\omega_n)$ before taking the analytic continuation. $\tilde{\Pi}_{\mu\nu}(q,i\omega_n)$ is defined by

$$\tilde{\Pi}_{\mu\nu}(q,i\omega_n) = -\frac{1}{V\hbar^2} \int_0^\beta d\tau\, e^{i\omega_n \tau} \langle J_\mu(q,t)\, J_\nu(-q,0)\rangle. \tag{9.8}$$

Substituting the expression of $J_\mu(q,t)$ in Eq. (9.4) into Eq. (9.8) and after a simple derivation, one can express $\tilde{\Pi}_{\mu\nu}$ using the single-particle Green's function as

$$\tilde{\Pi}_{\mu\nu}(q,i\omega_n) = \frac{e^2}{\beta V\hbar^2} \sum_{k,\omega_m} \frac{\partial \varepsilon_{k+q/2}}{\partial k_\mu} \frac{\partial \varepsilon_{k+q/2}}{\partial k_\nu} \text{Tr} G(k,i\omega_m) G(k+q,i\omega_n+i\omega_m). \tag{9.9}$$

Given $\tilde{\Pi}_{\mu\nu}$, the real frequency current-current correlation function $\Pi_{\mu\nu}$ can be obtained through analytic continuation

$$\Pi_{\mu\nu}(q,\omega) = \tilde{\Pi}_{\mu\nu}(q,i\omega_n \to \omega + i0^+). \tag{9.10}$$

To perform the analytic continuation, it is usually more convenient to represent the Matsubara Green's function as an integral of the retarded Green's function $G^R(k,\omega)$ in the spectral representation:

$$G(k,i\omega_n) = -\int_{-\infty}^{\infty} \frac{d\omega}{\pi} \frac{\text{Im} G^R(k,\omega)}{i\omega_n - \omega}. \tag{9.11}$$

To substitute this expression into Eq. (9.9) and sum over the frequency ω_m, we can further express $\tilde{\Pi}_{\mu\nu}$ as

$$\tilde{\Pi}_{\mu\nu}(q, i\omega_n) = \frac{e^2}{\pi^2 V \hbar^2} \sum_k \int_{-\infty}^{\infty} d\omega_1 d\omega_2 \frac{\partial \varepsilon_{k+q/2}}{\partial k_\mu} \frac{\partial \varepsilon_{k+q/2}}{\partial k_\nu} [f(\omega_1) - f(\omega_2)]$$
$$\frac{\text{TrIm}G^R(k, \omega_1)\text{Im}G^R(k+q, \omega_2)}{i\omega_n + \omega_1 - \omega_2}. \tag{9.12}$$

The retarded current-current correlation function is then found to be

$$\Pi_{\mu\nu}(q, \omega) = \frac{e^2}{\pi^2 V \hbar^2} \sum_k \int_{-\infty}^{\infty} d\omega_1 d\omega_2 \frac{\partial \varepsilon_{k+q/2}}{\partial k_\mu} \frac{\partial \varepsilon_{k+q/2}}{\partial k_\nu} [f(\omega_1) - f(\omega_2)]$$
$$\frac{\text{TrIm}G^R(k, \omega_1)\text{Im}G^R(k+q, \omega_2)}{\omega + \omega_1 - \omega_2 + i0^+}. \tag{9.13}$$

The current-current correlation function $\Pi_{\mu\nu}$ is a complex function. Its imaginary part is proportional to the alternating-current conductivity, namely the optical conductivity. In an ideal superconductor without energy dissipation, both the resistivity and the optical conductivity vanish. In the presence of elastic or inelastic disorder scatterings, the resistivity is still zero in the superconducting state, but the optical conductivity becomes finite.

$\Pi_{\mu\nu}$ contributes mainly from the electrons around the Fermi surface. In an isotropic two-dimensional system, if we approximate the electron velocity, $v \propto [(\partial \varepsilon_k/\partial k_x)^2 + (\partial \varepsilon_k/\partial k_y)^2]^{1/2}$, by the Fermi velocity v_F, then the in-plane correlation function Π_{ab} becomes

$$\Pi_{ab}(q, \omega) = \delta_{ab} \frac{e^2 v_F^2}{2\pi^2 V} \sum_k \int_{-\infty}^{\infty} d\omega_1 d\omega_2 [f(\omega_1) - f(\omega_2)]$$
$$\frac{\text{TrIm}G^R(k, \omega_1)\text{Im}G^R(k+q, \omega_2)}{\omega + \omega_1 - \omega_2 + i0^+}. \tag{9.14}$$

Based on the discussion in Chap. 12, we find that $\Pi_{ab}(q, \omega)$ has the same expression as the spin-spin correlation function $\chi_{zz}(q, \omega)$ defined in Eq. (12.8) up to a constant factor. These two quantities are related by the equation

$$\Pi_{ab}(q, \omega) = -\frac{2e^2 v_F^2}{\gamma_e^2 \hbar^2} \chi_{zz}(q, \omega). \tag{9.15}$$

It shows that the electromagnetic response is closely related to the spin response. Nevertheless, in real experimental measurements, the ranges of momentum and frequency measured are different for different physical quantities. The optical conductivity measures the response in the long-wave length limit, which is proportional to the imaginary part of the current-current correlation function $\mathrm{Im}\Pi(q = 0, \omega)$. The real part of the zero frequency current-current correlation function at the long wave length limit, i.e. $\mathrm{Re}\Pi(q \to 0, \omega = 0)$, is the paramagnetic contribution to the superfluid density. The measurement of the spin-spin correlation function depends on experimental methods. The Knight shift of the nuclear magnetic resonance (NMR) is proportional to the real part of the spin-spin correlation function at zero momentum and zero frequency, $\mathrm{Re}\chi_{zz}(q \to 0, \omega = 0)$. It is also proportional to the paramagnetic response function of the superfluid density. As the diamagnetic part of the superfluid density is roughly temperature independent, the Knight shift and the superfluid density should exhibit similar temperature dependence. The NMR relaxation rate measures the dynamic response of spins. It measures the average of spin-spin correlation function over the entire momentum space. In comparison, the energy and momentum ranges measured by the neutron scattering are much larger than NMR, but the resolution is lower. All these independent experiment methods are complementary to each other. They are important tools for studying physical properties of d-wave superconductors.

The above electromagnetic response function is derived under the linear approximation. In principle it is valid only in the zero field limit $H \to 0$. For the s-wave superconductor, as the superconducting energy gap is finite, this formula is also valid under a finite but weak magnetic field. However, for the d-wave superconductor, the non-linear response of the superconducting state to the magnetic field is important at low temperatures due to the presence of pairing gap nodes. On the other hand, the non-local effect would also become very important at low temperatures, since the effective coherence length along the nodal direction is infinite. Both the non-linear and non-local effects can affect strongly the low temperature electromagnetic response functions.

The superfluid density is a fundamental quantity characterizing superconductivity. It is proportional to the energy scale that Cooper pairs form phase coherence. The superfluid density can be taken as an order parameter of superconductivity, because it is zero in the non-superconducting phase and finite in the superconducting phase. In the standard theory of superconductivity, the pairing energy gap is considered as the superconducting order parameter. This choice of order parameter is strictly speaking not that rigorous. It is only valid when the phase fluctuation is small. In the presence of strong phase fluctuations, electrons are often paired (namely to have a finite pairing gap) but do not exhibit long-range phase coherence. However, the long-range phase coherence is a prerequisite of superconductivity. The larger is the superfluid density, the higher is the energy cost of phase fluctuations. Thus the superfluid density is a measure of the robustness of phase coherence, which is also called the phase stiffness.

The superfluid density n_s^μ is inversely proportional to the square of the magnetic penetration depth λ. It can be obtained from the real part of $K_{\mu\mu}$. According to the London equation, Eq. (1.1), and the definition of conductivity, Eq. (9.5), it can be shown that n_s^μ is proportional to the real part of $K_{\mu\mu}$ in the long-wave length limit:

$$\mathrm{Re}K_{\mu\mu}(q_\perp \to 0, q_\| = 0, 0) = \frac{e^2 n_s^\mu}{m_\mu} = \frac{1}{\mu_0 \lambda^2}, \tag{9.16}$$

where $q_\perp$ and $q_\|$ are the components of q perpendicular and parallel to the external magnetic field, respectively. This expression shows that in the calculation of the superfluid density, one should first take the limit of $\omega \to 0$, and then the limit $q \to 0$. This sequence is opposite to the limits taken in the calculation of DC conductivity. The reason for this is not difficult to understand. The DC conductivity is the coefficient measuring the current response to an applied electric field, while the superfluid density is the coefficient measuring the supercurrent response to an applied magnetic field.

For an ideal *d*-wave superconductor, the single-particle Green's function is given by Eq. (4.3). The frequency summation in Eq. (9.9) can be evaluated rigorously and $\Pi_{\mu\nu}$ in the limit of $(\omega = 0, q \to 0)$ is obtained as

$$\Pi_{\mu\nu}(q \to 0, 0) = \frac{2e^2}{V\hbar^2} \sum_k \frac{\partial \varepsilon_k}{\partial k_\mu} \frac{\partial \varepsilon_k}{\partial k_\nu} \frac{\partial f(E_k)}{\partial E_k}. \tag{9.17}$$

Hence, the superfluid density along the μ-direction is given by

$$n_s^\mu = \frac{m_\mu}{V\hbar^2} \sum_k \left[\left\langle \frac{\partial^2 \varepsilon_k}{\partial k_\mu^2} \right\rangle + 2 \left(\frac{\partial \varepsilon_k}{\partial k_\mu} \right)^2 \frac{\partial f(E_k)}{\partial E_k} \right]. \tag{9.18}$$

Experimentally, the superfluid density, or the magnetic penetration depth λ, can be measured by the microwave attenuation on the surface of supercon-ductor, μSR, infrared spectroscopy, the AC magnetic susceptibility, etc. The microwave attenuation can accurately measure the relative values of λ at dif-ferent temperatures, but not its absolute values. μSR and the AC magnetic susceptibility can measure directly the absolute values of λ. But these exper-iments can only be performed on large samples, and the measurement errors are relatively large.

The penetration depths of high-T_c superconductors are much larger than the coherence lengths of Cooper pairs. Hence even the surface microwave ex-periments measure the bulk property of high-T_c superconductors, not just their surface properties. Furthermore, the analysis of the penetration depth measure-ment data is relatively simple. There are no background contributions from the normal excitation states, neither from phonons.

At low temperatures, corrections to the above formula arise from the non-linear and non-local effects. They change the temperature dependence as well as the magnetic field dependence of the penetration depth. In particular, they alter the temperature dependence of the penetration depth from T to T^2 at very low temperature, which is actually important for the stability of the d-wave superconductor. Otherwise, if the penetration depth remains a linear function of temperature down to zero temperature, it can be shown that the zero temperature entropy is finite and the third law of thermodynamics, i.e. the Nernst law, is violated [126, 127]. This is of course unphysical. Both the linear and local approximations are valid conditionally. It is inappropriate to extend them to the zero temperature limit.

In the *d*-wave superconductor, the non-linear and non-local effects have a common feature. They are strongly anisotropic, depending on the relative angle between the nodal direction and the superconductor surface. This dependence can be used to probe the nodal direction on the Fermi surface, which in turn can be used to determine the pairing symmetry. In real materials, the non-linear and non-local effects are intertwined. Within certain parameter regimes, or under certain boundary conditions, the anisotropy induced by one effect could be weakened by another. A comprehensive analysis of these two effects is needed in order to analyze correctly the experimental results.

9.2 The In-Plane Superfluid Density

In high-T_c superconductors, the electron energy-momentum dispersion relation is approximately given by [128, 129, 49]

$$\varepsilon_k \approx \frac{\hbar^2 (k_x^2 + k_y^2)}{2m} - \frac{\hbar^2}{m_c} \cos k_z \cos^2 (2\varphi), \tag{9.19}$$

where m and m_c are the effective masses of electrons parallel and perpendicular to the CuO_2 plane, respectively. m_c is much larger than m and φ is the azimuthal angle of the in-plane momentum (k_x, k_y). In this expression, the anisotropy of the band structure in the CuO_2 plane is neglected, and the dependence of the c-axis dispersion on the (k_x, k_y) is also slightly simplified. These simplifications do not change the physical results qualitatively, but do affect the coefficients of the temperature dependence of the superfluid density.

In the CuO_2 plane, if we neglect the effect of the m_c-term on the electron dispersion in the ab-plane, then

$$\frac{\partial \varepsilon_k}{\partial k_x} = \frac{\hbar^2 k_x}{m},$$

$$\frac{\partial \varepsilon_k}{\partial k_y} = \frac{\hbar^2 k_y}{m},$$

$$\frac{\partial^2 \varepsilon_k}{\partial k_x^2} = \frac{\partial^2 \varepsilon_k}{\partial k_x^2} = \frac{\hbar^2}{m}.$$

Substituting them into Eq. (9.18), the superfluid density along the ab plane is obtained as

$$n_s^{ab} = n + \frac{\hbar^2}{mV} \sum_k (k_x^2 + k_y^2) \frac{\partial f(E_k)}{\partial E_k}, \tag{9.20}$$

where n is the electron concentration. As the momentum summation contributes mainly from the electrons around the Fermi surface, we can take the approximation $k_x^2 + k_y^2 \approx k_F^2$ and simplify the above expression as

$$n_s^{ab} \approx n - \frac{\hbar^2 k_F^2}{m} Y(T), \tag{9.21}$$

where

$$Y(T) = -\frac{1}{V} \sum_k \frac{\partial f(E_k)}{\partial E_k} = -\int_{-\infty}^{\infty} dE \rho(E) \frac{\partial f(E)}{\partial E} \tag{9.22}$$

is the Yoshida function. The integral on the right-hand side of Eq. (9.21) cannot be solved analytically. At $T \ll \Delta$, the contribution arises mainly from the terms with $E \approx k_B T$. In this case, $\rho(E)$ is approximately given by Eq. (3.32), and

$$n_s^{ab} \approx n + \frac{2\hbar^2 k_F^2 N_F}{m\Delta_0} \int_0^{\infty} dE E \frac{\partial f(E)}{\partial E} = n \left[1 - \frac{(2\ln 2)k_B T}{\Delta_0} \right].$$

The low temperature dependence of the magnetic penetration depth in the CuO_2 plane is then found to be

$$\lambda_{ab}(T) \approx \lambda(0) \left[1 + \frac{(\ln 2)k_B T}{\Delta_0} \right].$$

Thus the penetration depth varies linearly with temperature at low temperatures. This is an important feature of the d-wave or other superconductors with gap nodes. Generally it can be shown that the exponent in the leading temperature dependence of the penetration depth is equal to the exponent in the leading energy dependence of the density of states. Hence the latter can be obtained by measuring the penetration depth.

In the isotropic s-wave superconductor, there are very few quasiparticles that can be excited above the energy gap at low temperatures. The penetration depth is thermal activated and can be approximately expressed as[130]

$$\lambda_{ab}(T) - \lambda_{ab}(0) \propto \exp(-\Delta/k_B T). \tag{9.23}$$

The difference in the low temperature dependences between the *s*- and *d*-wave superconductors is an important criterion in determining the pairing symmetry of high-T_c superconductors from the penetration depth.

In the early stage of the high-T_c study, due to the relatively poor sample quality, the experimental errors were large. Even though the high-T_c superconductivity was predicted to have the *d*-wave pairing symmetry, it was not supported by the penetration depth measurements.

In 1993, Hardy and co-workers found for the first time the linear temperature dependence of λ in the high quality YBCO twin crystals [131]. Their results are consistent with the *d*-wave pairing symmetry. Their experimental results demonstrate the importance of the penetration depth measurement in the study of high-T_c superconducting mechanism. The improvement of sample quality is crucial because the impurity scattering has a strong effect on the low temperature superfluid density. After their work, many labs made great efforts to measure accurately the penetration depth and found that the low temperature λ_{ab} indeed varies linearly with temperature:

$$\lambda_{ab}(T) \approx \lambda_{ab}(0) + \alpha_{ab}T \qquad (9.24)$$

in nearly all high quality samples, including the YBCO single crystal [132, 133], YBCO twin crystal[134, 135], YBCO twin crystal film [136], BSCCO single crystal [137, 138, 139], HgBaCaCuO and TlBaCuO powders [140] and other high-T_c compounds. It shows that even at low temperatures, there are still quasiparticle excitations, indicating the existence of gap nodes. This is a strong support to the $d_{x^2-y^2}$-wave pairing theory of high-T_c superconductivity.

The value of the zero temperature penetration depth $\lambda_{ab}(0)$ depends on the sample quality. The better is the sample quality, the stronger is the supercurrent screening and the smaller is $\lambda_{ab}(0)$. The experimental results are consistent with this expectation. The sample quality can be judged based on the measurement of normal state resistance and properties in the critical region of the superconducting phase transition. For high quality samples, the structure is more homogeneous and the critical transition range is narrower. The corresponding residual microwave resistance and the normal state resistance are also small.

Fig. 9.1 shows the temperature dependence of the superfluid density in the CuO_2 plane for a class of representative high-T_c superconductors [145]. As mentioned before, the low temperature superfluid density varies linearly with temperature within measurement errors. In addition, the slopes of the normalized superfluid density $1-\lambda_{ab}^2(0K)/\lambda_{ab}(T)$ are nearly the same for all the samples shown in the figure, independent of the doping level (e.g. underdoping, optimal doping, and overdoping). From the slope and the behavior of the low temperature normalized superfluid density

$$1 - \frac{\lambda_{ab}^2(0K)}{\lambda_{ab}^2(T)} \approx \frac{(2\ln 2)k_B T}{\Delta_0}, \qquad (9.25)$$

the amplitude of Δ_0 can be extracted. Δ_0 determined in this way scales almost linearly with T_c (Fig. 9.2)[145], consistent with the d-wave BCS mean-field result, $\Delta \approx 2.14T_c$. It should be emphasized that Δ_0 in Eq. (9.25) is only the gap amplitude around the nodal points, or more precisely, the slope of

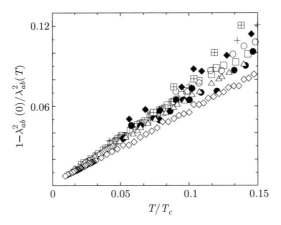

Figure 9.1   Temperature dependence of the normalized superfluid density in the CuO_2 plane for $HgBa_2Ca_2 Cu_3O_{8+\delta}$ (hollow diamond) [141], $HgBa_2CuO_{4+\delta}$ (triangle)[141], $YBa_2Cu_3O_7$ (solid circle)[142, 143], $YBa_2Cu_3O_{6.7}$ (hollow circle)[143], $YBa_2Cu_3O_{6.57}$(square) [143], $La_{2-x}Sr_xCuO_4$ [144] with x =0.2 (cross), 0.22 (solid diamond), and 0.24 (square with cross). (From Ref. [145])

the gap function at the nodal points. It is not necessary to be the maximal energy gap. In the underdoped superconductors, due to the presence of the pseudogap, the momentum dependence of the gap function is not as simple as $\Delta_k = \Delta_0 \cos(2\varphi)$, and the maximal energy gap is larger than the value predicted by the BCS theory.

The approximate linear scaling behavior between Δ_0 and T_c, as shown in Fig. 9.2, differs significantly from the non-scaling behavior between the maximal energy gap and T_c in the underdoped regime. The existence of this scaling behavior is likely an intrinsic property of high-T_c superconductors. It suggests that in the vicinity of gap nodes, the quasiparticle excitations can still be described by the BCS mean-field theory even in the underdoped regime. This

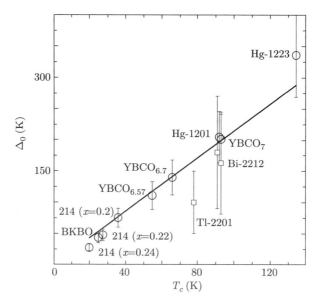

Figure 9.2   The scaling relation between the gap magnitude in the nodal region and the superconducting transition temperature T_c (from [145]). In addition to the superconductors shown in Fig. 9.1, the results for $Ba_2Sr_2CaCu_2O_{8+\delta}$ [137, 138], $Tl_2Ba_2CuO_{6+\delta}$ [146], and the s-wave superconductors $Bi_{0.6}K_{0.4}BiO_3$[140] are also included. The solid line is the theoretical result for the weak coupling d-wave superconductor, $\Delta_0 = 2.14T_c$.

is related to the Fermi surface structure of high-T_c cuprates. In underdoped materials, the ARPES measurements found that there exist arc-like Fermi surfaces, which are centered around $(\pm\pi/2, \pm\pi/2)$ and extend towards $(\pm\pi, 0)$ and $(0, \pm\pi)$[64, 65]. The pseudogap, on the other hand, begins to spread from the anti-nodal regions around $(\pm\pi, 0)$ and $(0, \pm\pi)$ toward the nodal points only at some temperature not much higher than T_c. Thus the pseudogap will not completely suppress the Fermi surface at T_c and the pairing on the remnant Fermi arcs will dominate low energy excitations in the superconducting state, leading to a linear behavior between the gap slope at the nodal point, Δ_0, and T_c.

However, it should be pointed out that theoretical explanations to the experimental results of linear penetration depth are not unique. Both the pairing phase fluctuation[16, 147] and the proximity effect can be also used to explain the linear temperature dependence of λ_{ab}. But it is difficult to use these effects to understand why the impurity scattering can change the temperature dependence of λ_{ab} from T to T^2. It should also be noted that the in-plane superfluid density depends only on the quasiparticle density of states, which is not sensitive to the phase of the gap function and the locations of the gap nodes. Thus it is inadequate to fully determine the pairing symmetry only based on the measurement of the in-plane penetration depth.

In the $YBa_2Cu_3O_{7-x}$ superconductors, there exist CuO chains and the magnetic penetration depth along the chain direction (the b-axis) differs from that along the perpendicular direction (the a-axis)[133, 148]. For the optimally doped YBCO superconductor, $\lambda_a(0) = 1600\text{Å}$ is about 1.6 times larger than $\lambda_b(0) = 1030\text{Å}$ [133, 148]. For the underdoped $YBa_2Cu_3O_{7-x}$ superconductors, the difference between $\lambda_a(0)$ and $\lambda_b(0)$ is smaller [148]. For $YBa_2Cu_4O_8$, the difference between $\lambda_a(0)$ and $\lambda_b(0)$ is larger, $\lambda_a(0) = 2000\text{Å}$ and $\lambda_b(0) = 800\text{Å}$, hence $\lambda_a(0)$ is about 2.5 times larger than $\lambda_b(0)$ [148]. Nevertheless, as shown in Fig. 9.3, no matter how large the difference between $\lambda_a(0)$ and $\lambda_b(0)$, the temperature derivatives of $\lambda_\mu(T)$ ($\mu = a, b$), after normalized by $\lambda(T = 0)$, are nearly the same along these two directions.

The difference between the lattice constants along the a- and b-axes of

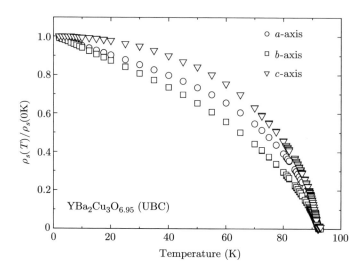

Figure 9.3   Temperature dependence of the normalized superfluid density for the
$YBa_2Cu_3O_{6.95}$ single crystal along three principle axes. (From Ref. [133])

YBCO is less than 2.5%, thus the difference between the superfluid densities
should come mainly from the contribution of CuO chains. A CuO chain can
be viewed as a quasi-1D system. It contributes to λ_b, but barely to λ_a. If
we attribute the origin of the difference between λ_a and λ_b completely to the
CuO chains, then in the optimally doped $YBa_2Cu_3O_{7-x}$ sample, about 60% of
the b-axis superfluid density comes from the contribution of CuO chains. For
$YBa_2Cu_4O_8$, the chain contribution even increases to 80%. For the underdoped
$YBa_2Cu_3O_{7-x}$, electrons on the CuO chains are easy to be localized by impurity scattering. Their contribution to the superfluid density is weakened, hence
the difference between λ_a and λ_b is suppressed. In the normal state, the CuO
chains have also significant contributions to the electric [149, 150] and thermal
conductivities [151]. For example, in the single crystal of $YBa_2Cu_3O_{6.95}$, the
electric and thermal conductivities along the b-axis are about twice larger than
those along the a-axis. The anisotropic ratios are nearly the same as for the
superfluid density.

9.3 The Superfluid Density along the c-Axis

Along the c-axis, the energy dispersion of electrons and its derivatives depend on the in-plane momentum:

$$\frac{\partial \varepsilon_k}{\partial k_z} = \frac{\hbar^2}{m_c} \sin k_z \cos^2(2\varphi),$$

$$\frac{\partial^2 \varepsilon_k}{\partial k_z^2} = \frac{\hbar^2}{m_c} \cos k_z \cos^2(2\varphi).$$

The temperature dependence of the superfluid density now becomes complicated and is determined by the following equation

$$n_s^c = \frac{1}{V} \sum_k \left[-\cos k_z \cos^2(2\varphi) \frac{\varepsilon_k}{E_k} \tanh \frac{\beta E_k}{2} + \frac{2\hbar^2}{m_c} \sin^2 k_z \cos^4(2\varphi) \frac{\partial f(E_k)}{\partial E_k} \right].$$

$$(9.26)$$

The first term is the diamagnetic contribution. It is nearly temperature-independent and equals the superfluid density at zero temperature:

$$n_s^c(0) = \frac{3\hbar^2 N_F}{16 m_c}, \tag{9.27}$$

if the band width is much larger than Δ. The second term in Eq. (9.26) is paramagnetic and temperature dependent. It is difficult to calculate this term analytically. Nevertheless, in the limit $k_B T \ll \Delta_0$, the momentum summation in Eq. (9.26) can be approximately evaluated, from which the low temperature superfluid density is obtained as [49, 50]

$$n_s^c(T) \approx n_s^c(0) \left[1 - 450\varsigma(5) \left(\frac{k_B T}{\Delta_0} \right)^5 \right], \tag{9.28}$$

where $\varsigma(5) \approx 1.04$ and the terms higher than T^5 are neglected.

The T^5-dependence of n_s^c rather than the linear dependence is caused by the $\cos 2\varphi$ term in the c-axis dispersion relation. It shows that the anisotropy of the electron structure in the CuO_2 plane has a significant effect on the c-axis superfluid density. This is a peculiar property of high-T_c cuprates. In fact, the power of the temperature dependence of n_s^c can be obtained by simple

dimensional analysis. The quasiparticle density of states in the d-wave superconductor is linear, which contributes one power of T. The additional power of
T^4 is generated by $\cos^4(2\varphi)$ because $\Delta\cos(2\varphi)$ is of the dimension of energy.

The T^5-dependence is a consequence of the coincidence of the anisotropy
of the c-axis hopping matrix element and the anisotropy of the d-wave pairing
gap function. In particular, the zeros of the former are also those of the latter.
Thus through the measurement of the c-axis superfluid density, not only can
we determine if there are nodes in the gap function, but also determine the
locations of these nodes on the Fermi surface.

Due to the quasi-two-dimensional nature of high-T_c superconductors, λ_c
is about one or two orders larger than λ_{ab} [133, 137]. As the change of λ_c
with temperature is very weak at low temperatures, it is quite difficult to determine the power of the temperature dependence of λ_c. Experimentally, the
T^5-temperature dependence of λ_c was first confirmed in $HgBa_2CuO_{4+\delta}$. This
intrinsic temperature dependence of the penetration depth along the c-axis
could be measured because the anisotropy between the c-axis and the ab-plane
penetration depth is small and the coherent interlayer hopping is relatively large
in this superconductor. Fig. 9.4 shows the experimental data obtained based on
the AC magnetic susceptibility measurement for $HgBa_2CuO_{4+\delta}$. Later on, this
T^5-behavior was also observed in the Bi2212 samples with larger anisotropy
[152]. The agreement between the theoretical and experimental results indicates that as long as the sample is clean, the c-axis hopping of electron is
predominantly coherent. In other words, the contribution from the incoherent
hopping to the c-axis transport properties is small and electrons are not dynamically confined within each CuO_2 plane. This agreement also shows that
there are gap nodes on the Fermi surface and the nodes are located along the
diagonal directions, in support of the $d_{x^2-y^2}$-wave pairing symmetry.

The anisotropy between the c-axis and the ab-plane is relatively small in
$YBa_2Cu_3O_{6+\delta}$. The coherent interlayer hopping of electrons should also be
important for the low temperature electromagnetic response functions along the
c-axis. However, due to the buckling of the CuO_2 planes and the existence of
CuO chains, the a- and b-axes are not symmetric. The interlayer hopping matrix

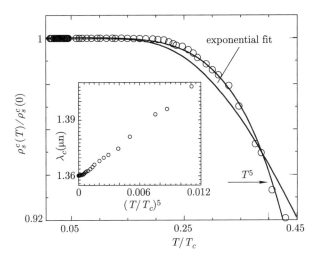

Figure 9.4   Temperature dependence of the normalized c-axis superfluid density, $\rho_s^c(T)/\rho_s^c(0K)$, for HgBa$_2$CuO$_{4+\delta}$ at low temperatures (from [50]). The hollow circles are the experiment data, and the solid lines are the theory fitting curves. The experiment data are consistent with the T^5 behavior predicted by theoretical calculation, which can be more clearly seen from the linear dependence of the penetration depth on $(T/T_c)^5$ in the inset.

element is not simply proportional to $t_c \propto (\cos k_x - \cos k_y)^2$, and its zeros do not coincide with the d-wave pairing gap nodes. Thus the temperature dependence of λ_c is changed. It does not possess the T^5-temperature dependence as in a tetragonal compound.

9.4 Impurity Correction

Impurity scattering affects strongly the low temperature behavior of superfluid density. When the impurity scattering is sufficiently strong, it can change the temperature dependence of the in-plane superfluid density from T to T^2 [153]. This property is important in the analysis of experiment data, especially in the case the sample quality is not that good. Experiments in the early days of the high-T_c study showed that the temperature dependence of the in-plane penetration depth is very weak at low temperatures, which deviates from the intrinsic

linear behavior of *d*-wave superconductors but resembles the exponential behavior of *s*-wave superconductors. This was once regarded as an evidence for
the *s*-wave symmetry of high-T_c pairing. However, after a more careful analysis
by subtracting the contribution from the paramagnetic impurities, it was found
that the low-temperature superfluid density varies quadratically instead of exponentially with temperature. This quadratic temperature dependence of the
superfluid density is a common feature of disordered *d*-wave superconductors.
It lends support to the *d*-wave pairing theory of high-T_c superconductivity.

As discussed previously, $K_{\mu\mu}$ is determined by the sum of the following two
terms,

$$K_{\mu\mu}(q \to 0, 0) = \frac{e^2}{\hbar^2 V} \sum_k \left\langle \frac{\partial^2 \varepsilon_k}{\partial k_\mu^2} \right\rangle + \Pi_{\mu\mu}(q \to 0, 0). \qquad (9.29)$$

The first term is diamagnetic, proportional to the average of the inverse effective mass of electrons in the Brillouin zone. This term is barely influenced by
the impurity scattering and nearly temperature independent. It approximately
equals the value in the clean system at zero temperature. The second term
is paramagnetic. It is affected by the impurity scattering, leading to a finite
correction to the superfluid density at zero temperature. Thus the zero temperature superfluid density is not completely the contribution from the first
term.

The second term is determined by the current-current correlation function

$$\Pi_{\mu\mu}(q \to 0, 0) = \frac{e^2}{\pi^2 \hbar^2 V} \sum_k \left(\frac{\partial \varepsilon_k}{\partial k_\mu} \right)^2 Z_k, \qquad (9.30)$$

where

$$Z_k = \int_{-\infty}^{\infty} d\omega_1 d\omega_2 \left[f(\omega_1) - f(\omega_2) \right] \frac{\text{TrIm} G^R(k, \omega_1) \text{Im} G^R(k, \omega_2)}{\omega_1 - \omega_2 + i0^+}. \qquad (9.31)$$

At low temperatures, the temperature dependence of the superfluid density
reflects the energy dependence of low energy quasiparticle density of states.
Disorder scattering changes the density of states around the nodal points, which
in turn changes the temperature dependence of the superfluid density.

The retarded Green's function $G^R(k,\omega)$ and $G^{R*}(k,\omega)$ are analytic in the upper and lower half complex plane of ω, respectively. From the residue theorem, it is simple to show that the following equations are valid:

$$\int_{-\infty}^{\infty} d\omega_2 \frac{G^R(k,\omega_2)}{\omega_1 - \omega_2 + i0^+} = -2\pi i G^R(k,\omega_1),$$

$$\int_{-\infty}^{\infty} d\omega_1 \frac{G^{R*}(k,\omega_1)}{\omega_1 - \omega + i0^+} = -2\pi i G^{R*}(k,\omega),$$

$$\int_{-\infty}^{\infty} d\omega_1 \frac{G^R(k,\omega_1)}{\omega_1 - \omega + i0^+} = \int_{-\infty}^{\infty} d\omega_2 \frac{G^{R*}(k,\omega_2)}{\omega_1 - \omega_2 + i0^+} = 0.$$

Substituting them into Eq. (9.32), Z_k can be simplified as

$$Z_k = -\pi \int_{-\infty}^{\infty} d\omega f(\omega) \, \mathrm{Im} \mathrm{Tr} G^R(k,\omega) G^R(k,\omega). \tag{9.32}$$

In the gapless regime, $G^R(k,\omega)$ is determined by Eq. (8.41). Substituting it into Eq. (9.32) gives

$$Z_k = -2\pi \mathrm{Im} \int_{-\infty}^{\infty} d\omega f(\omega) \frac{(a\omega + i\Gamma_0)^2 + E_k^2}{\left[(a\omega + i\Gamma_0)^2 - E_k^2\right]^2}. \tag{9.33}$$

The integral over ω can be performed through the Sommerfeld expansion introduced in Appendix E. The first two leading terms in the expansion are given by

$$Z_k = -2\pi \left[\frac{\Gamma_0}{a(\Gamma_0^2 + E_k^2)} - \frac{a\Gamma_0 (\Gamma_0^2 - 3E_k^2) k_B^2 T^2}{3(\Gamma_0^2 + E_k^2)^3} + o(T^4) \right]. \tag{9.34}$$

The first term is temperature independent, which is just the correction from the paramagnetic current-current correlation function to the zero temperature superfluid density. The second term is proportional to T^2. It leads to the T^2-dependence of the low temperature superfluid density, different than the linear temperature dependence in the intrinsic d-wave superconductor. The T^2-dependence results from the fact that the density of states of electrons on the Fermi surface in the disordered d-wave superconductor is finite and the Sommerfeld expansion can be used in the limit $T \ll \Gamma_0$. This is also a universal behavior of d-wave superconductors, independent of the detail of impurity scattering.

In the CuO_2 plane, the momentum summation of Z_k can be done using the standard method. In the limit $\Gamma_0 \ll \Delta_0$, the result is approximately given by

$$\frac{1}{V} \sum_k Z_k \approx -4\pi N_F \left(\frac{\Gamma_0}{a\Delta_0} \ln \frac{2\Delta_0}{\Gamma_0} + \frac{ak_B^2 T^2}{6\Gamma_0 \Delta_0} + o\left(T^4\right) \right). \tag{9.35}$$

The temperature dependence of the ab-plane superfluid density is then obtained as

$$n_s^{ab} \approx n \left(1 - \frac{2\Gamma_0}{\pi a\Delta_0} \ln \frac{2\Delta_0}{\Gamma_0} - \frac{ak_B^2 T^2}{3\pi\Gamma_0 \Delta_0} + o\left(T^4\right) \right), \tag{9.36}$$

where $n = N_F m v_F^2$ is the concentration of electrons. The corresponding magnetic penetration depth is given by [153]

$$\lambda_{ab} \approx \lambda_0 \left(1 + \frac{\Gamma_0}{\pi a\Delta_0} \ln \frac{2\Delta_0}{\Gamma_0} + \frac{ak_B^2 T^2}{6\pi\Gamma_0 \Delta_0} + o\left(T^4\right) \right), \tag{9.37}$$

where λ_0 is the penetration depth of the intrinsic d-wave superconductor at zero temperature. As already mentioned, λ_{ab} varies quadratically with temperature.

In the intrinsic regime, $\Gamma_0 \ll T$, λ is still a linear function of T, same as in an intrinsic d-wave superconductor. The correction from the disorder scattering to λ is a higher-order small quantity of Γ_0/T.

For high-T_c superconductors, besides the experimental confirmation of the intrinsic linear temperature dependence of the magnetic penetration depth, the T^2-dependence of the penetration depth in the disordered d-wave superconductors were also supported by a vast of experimental measurements. In the zinc or nickel-doped YBCO single crystals [154, 142], or other superconducting films or single crystals without doping but with relatively poor quality [136, 155, 156], it was indeed found that λ_{ab} varies as T^2 at low temperatures, consistent with the prediction of disordered d-wave superconductors.

Along the c-axis, substituting the expression of $\partial\varepsilon_k/\partial k_z$ and Eq. (9.34) into Eq. (9.30), the momentum summation of $(\partial\varepsilon_k/\partial k_z)^2 Z_k$ in the gapless regime is approximately found to be

$$\frac{1}{V} \sum_k \left(\frac{\partial\varepsilon_k}{\partial k_z} \right)^2 Z_k \approx -\frac{2\hbar^4 N_F}{m_c^2} \left[\frac{\pi\Gamma_0}{4\Delta_0 a} + \frac{\pi a\Gamma_0 k_B^2 T^2}{4\Delta_0^3} + o\left(T^4\right) \right]. \tag{9.38}$$

By further substituting this equation into the expression of the superfluid density, we obtain the c-axis superfluid density in the gapless regime as

$$n_s^c = \frac{3\hbar^2 N_F}{16m_c} \left[1 - \frac{8\Gamma_0}{3\pi\Delta_0 a} - \frac{8a\Gamma_0 k_B^2 T^2}{3\pi\Delta_0^3} + o\left(T^4\right) \right]. \qquad (9.39)$$

In obtaining this expression, the correction of the impurity scattering to the
diamagnetic term is neglected [50]. n_s^c scales as T^2 at low temperatures, showing
a stronger temperature dependence than the T^5-dependence in the intrinsic d-
wave superconductor.

The above results are derived under the assumption that the interlayer hopping is coherent, namely the momentum is conserved during the hopping. This
assumption is violated in the presence of disordered potentials. In particular, if the impurity concentration becomes significantly high, the correction to
the superfluid density from the incoherent interlayer scattering induced by the
impurities needs to be considered.

In comparison to λ_{ab}, λ_c is more strongly affected by the impurity scattering because the interlayer hopping matrix elements are much smaller than
in-plane ones. When the incoherent interlayer hopping dominates over the coherent one, the T^5-law of λ_c is no longer valid. Eq. (9.28) should be replaced
by Eq. (9.39), and λ_c scales as T^2 at low temperatures. This quadratic temperature dependence of λ_c was found in the most of high-T_c superconductors
experimentally.

9.5 Superfluid Response in Weakly Coupled Two-Band Superconductors

We have discussed the superfluid response of a single-band system. However,
in many superconductors, there exist two or even more Fermi surfaces, whose
superconducting response functions are significantly different from the single-
band case. A thorough understanding of this difference is essential to the understanding of experimental results.

Among the multi-band systems, we often meet a class of superconductors
in which the interband coupling is weak. A typical weakly coupled multi-band

superconductor is MgB_2 [157, 158], which has two electron-like Fermi surfaces and one hole-like Fermi surface. One of these three bands has stronger pairing interaction, which drives the system into the superconducting phase at 39 K. The pairing interactions in the other two bands are relatively weak. As the two main bands that are responsible for superconductivity carry different parity numbers, the coupling between them is very weak [159].

In a single-band superconductor, the temperature dependence of the super-fluid density exhibits a common feature that the curvature of the superfluid density as a function of temperature, $d^2\rho_s(T)/dT^2$, is always negative. In a two-band system with strong inter-band coupling, if the superfluid responses from the two bands are locked, then the curvature of the superfluid density remains negative. However, in a weakly coupled two-band superconductor, the curvature is modified. It becomes positive in a certain temperature range. This is an intrinsic and characteristic feature of the two-band superconductor with weak interband coupling. A simple physical picture for understanding this is illustrated in Fig. 9.5. In the absence of inter-band coupling, let us assume

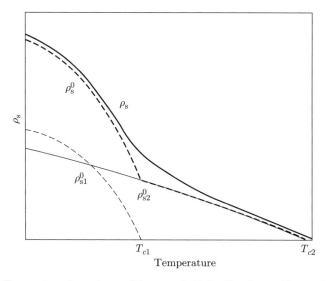

Figure 9.5   Temperature dependence of the superfluid density of a weakly coupled two-band superconductor (From [160]).

that T_{c1} and T_{c2} ($> T_{c1}$) are the transition temperatures for the two bands, respectively. The corresponding superfluid densities are denoted as ρ_{s1}^0 and ρ_{s2}^0, respectively. The curvatures of ρ_{s1}^0 and ρ_{s2}^0 are all negative as in the single-band case. The total superfluid density is just $\rho_s^0 = \rho_{s1}^0 + \rho_{s2}^0$, which exhibits a cusp at $T = T_{c1}$. Once the inter-band coupling is turned on, only the superconducting transition around T_{c2} survives. The transition at T_{c1} is now rounded off, and the corresponding temperature dependence curve of the superfluid density is also smoothed. However, the tendency towards a positive curvature around T_{c1} cannot be completely wiped out. It is replaced by a smooth curve with a positive curvature. This picture holds for all weakly coupled two-band superconductors, irrespective of the band structures and the pairing interactions.

The weakly coupled two-band superconductor can be described by a reduced two-band BCS model as

$$
\begin{aligned}
H = &\sum_{i,k,\sigma} \xi_{ik} c_{ik\sigma}^\dagger c_{ik\sigma} + \sum_{ikk'} V_{ikk'} c_{ik'\uparrow}^\dagger c_{i-k'\downarrow}^\dagger c_{i-k\uparrow} c_{ik\downarrow} \\
&+ \sum_{k,k'} \left(V_{3kk'} c_{1k'\uparrow}^\dagger c_{1-k'\downarrow}^\dagger c_{2-k\uparrow} c_{2k\downarrow} + h.c. \right),
\end{aligned}
\tag{9.40}
$$

where $c_{ik\sigma}$ ($i = 1, 2$) is the annihilation operator of electrons in band i. $V_{1kk'}$ and $V_{2kk'}$ are the pairing potentials of band 1 and 2, respectively. $V_{3kk'}$ is the inter-band pairing potential. A commonly used approximation is to assume that $V_{ikk'}$ ($i = 1, 2, 3$) are factorizable, i.e.

$$
V_{1kk'} = g_1 \gamma_{1k} \gamma_{1k'},
$$
$$
V_{2kk'} = g_2 \gamma_{2k} \gamma_{2k'},
$$
$$
V_{3kk'} = g_3 \gamma_{1k} \gamma_{2k'},
$$

where g_1, g_2, and g_3 are the coupling constants. γ_{1k} and γ_{2k} are the pairing symmetry functions of band 1 and 2, respectively. In principle, γ_{1k} and γ_{2k} can be different, but it is energetically favorable for them to share the same symmetry.

In real superconductors, the collective modes that induce the superconducting pairing of electrons could couple to both bands, giving rise to the superconducting instabilities in both bands. Nevertheless, the superconducting transition temperature is determined mainly by the band with stronger pairing energy. Another possibility is that the pairing interaction only explicitly shows up in one of the bands. The other band is not superconducting itself, but can become superconducting via the interband coupling. In this case, one of the intra-band pairing interactions, $V_{1kk'}$ and $V_{2kk'}$, equals zero.

In Eq. (9.40), the interband coupling is achieved through pairing hopping. In real superconductors, the two bands could also be coupled through the single particle hopping, or hybridization. Which coupling is more important is determined by the pairing mechanism. In some systems, the pairing hopping is stronger, while in other systems the single particle coupling becomes stronger. The role of the interband single particle hopping is to renormalize the energy dispersion, it can be absorbed into the kinetic energy terms by redefining the band dispersions as well as the pairing potentials. Hence, although the interband hybridization term is not explicitly included in Eq. (9.40), the band structure $\xi_{ik\sigma}$ and the pairing interaction $V_{ikl'}$ could be understood as that they have already contained the correction from this hybridization.

The pairing interactions in the model described by Eq. (9.40) can be decoupled following the BCS mean-field theory. Under this approximation, the quasiparticle excitation spectrum of band i is given by

$$E_{ik} = \sqrt{\xi_{ik}^2 + \Delta_i^2 \gamma_k^2}, \tag{9.41}$$

where Δ_i is the gap amplitude. They are determined by the following self-consistent equations:

$$\Delta_1 = \sum_k \gamma_k (g_1 \langle c_{1-k\downarrow} c_{1k\uparrow} \rangle + g_3 \langle c_{2-k\downarrow} c_{2k\uparrow} \rangle), \tag{9.42}$$

$$\Delta_2 = \sum_k \gamma_k (g_2 \langle c_{2-k\downarrow} c_{2k\uparrow} \rangle + g_3 \langle c_{1-k\downarrow} c_{1k\uparrow} \rangle), \tag{9.43}$$

where $\langle ... \rangle$ represents the thermodynamic average. Apparently, the two gap functions are coupled with each other and cannot vary independently.

Under the above mean-field approximation, the total superfluid density is just the sum of the superfluid densities in these two bands:

$$\rho_s = \rho_{s1} + \rho_{s2}. \qquad (9.44)$$

ρ_{s1} and ρ_{s2} are correlated through the above two gap equations. Once the temperature dependences of Δ_1 and Δ_2 are obtained, ρ_{s1} and ρ_{s2} can be derived through the formulae previously introduced for the single-band superconductor.

YBCO contains both the CuO_2 planes and the CuO chains, and is a two-band system. Besides to provide charge carriers, the CuO chains have also contribution to superconductivity. There are two kinds of interactions that can drive the CuO chains to superconduct. One is the intrinsic intrachain pairing interaction, the other is that induced by the proximity effect from the pairing interaction in the CuO_2 plane [160]. The temperature dependence of the penetration depth along the chain λ_b is very different in these two cases. In the former case, λ_b varies slowly with temperature at low temperatures, behaving just like in a single-band system. In the latter case, the contribution of CuO chains to the superfluid density becomes prominent at low temperatures due to the proximity effect. Consequently, λ_b decrease quickly with decreasing temperature, and the curvature of the superfluid density changes to negative [49]. This indicates that through the measurement of the curvature of λ_b, we in principle can determine whether pairing interaction, i.e. the direct pairing interaction or that induced by the proximity effect, is more important in the pairing of electrons on the CuO chains.

For $YBa_2Cu_3O_{7-x}$, the negative curvature of $\lambda_b(T)$ was not observed experimentally. Thus it is unlikely that the chain superconductivity is induced by the proximity effect in this material. This shows that in $YBa_2Cu_3O_{7-x}$, the CuO chains couple strongly with the CuO_2 planes and their contributions to the superfluid density are not independent. Hence this material should not be treated as a weakly coupled two-band system.

However, the situation for $YBa_2Cu_4O_8$ is different. $YBa_2Cu_4O_8$ is intrinsically underdoped. Each unit cell along the c-axis contains two layers of CuO chains. These two chains are off-set by half of the lattice constant along the chain direction (the b-axis), reducing significantly the chain-plane coupling. It leads to the strong anisotropy between the a- and b-axis penetration depths. This implies that the contributions from the CuO chains and CuO$_2$ planes are nearly independent, and this material can be considered as a weakly coupled two-band system and the superconductivity is mainly driven by the electrons in the CuO$_2$ planes. According to the discussion above, it is natural to predict that the curvature of the in-plane superfluid density as a function of temperature is positive at low temperatures. This prediction was confirmed experimentally [161]. Fig. 9.6 shows the temperature dependence of the superfluid density for

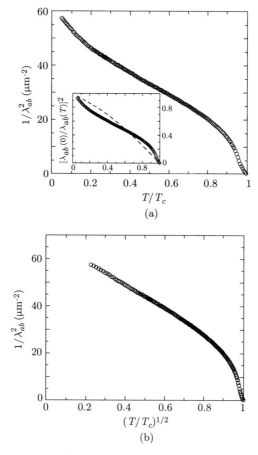

Figure 9.6   (a) The curve of $1/\lambda_{ab}^2$ v.s. the reduced temperature T/T_c for YBa$_2$Cu$_4$O$_8$; (b) the dependence of $1/\lambda_{ab}^2$ v.s. $(T/T_c)^{1/2}$ (from [161]).

YBa$_2$Cu$_4$O$_8$. Indeed, the curvature of $1/\lambda_{ab}^2$ is positive at low temperatures. A further analysis shows the low temperature $1/\lambda_{ab}^2$ scales as $T^{1/2}$. This is also consistent with the prediction made based on the proximity effect [160].

Besides YBa$_2$Cu$_4$O$_8$, the electron-doped high-T_c superconductors are also two-band systems with weak interband coupling[162]. This will be discussed in Sect. 9.6. Unlike YBa$_2$Cu$_4$O$_8$, it is not so simple to justify that the electron-doped high-T_c cuprates are weakly-coupled two-band systems. Nevertheless, this scenario is supported by the penetration depth as well as many other experimental measurements. Many seemingly contradictory phenomena in the electron-doped high-T_c cuprates can be naturally and consistently interpreted in the framework of weakly coupled two-band model.

9.6 The Electron-Doped High-T_c Superconductors

Up to now, it is commonly accepted that the hole-doped high-T_c supercon-ductors possess the $d_{x^2-y^2}$ pairing symmetry. However, for the electron-doped cuprates, it is still under debate whether the superconducting pairing has the $d_{x^2-y^2}$ or other symmetry. Most of the experiment measurements, including the ARPES[163, 164], Raman scattering[165], and phase-sensitive measurements [99, 166], supported the theory of the $d_{x^2-y^2}$-wave pairing symmetry. However, there are still discrepancies in the interpretation on other experimental results [167, 168, 169, 170, 171]. Kokales *et. al.*[168] and Prozorov *et. al.* [169, 170] measured the magnetic penetration depth and found the T^2-dependence in the low temperature superfluid density, consistent with the results of disordered d-wave superconductors. However, the temperature dependence of the super-fluid density observed by Kim *et. al.* [171] is more complicated. It does scale as T^2 in the underdoped regime, but in the overdoped regime, it shows a tem-perature dependence which is more close to the exponential behavior of s-wave superconductors. Based on this observation, they speculated that there is a phase transition from the a d-wave pairing state in the underdoped regime to the anisotropic s-wave pairing state in the overdoped regime. However, the analysis of Kim *et. al.* [171] was made based on the single-band assumption.

As already mentioned, the electron-doped high-T_c cuprates are actually weakly coupled two-band systems [162]. Analysis based on the single-band model cannot correctly describe the superfluid response of these materials.

A common feature of the electron-doped high-T_c superconductor is that the superfluid density exhibits a positive curvature around T_c [171, 172, 173, 174]. The temperature range for the positive curvature is narrow in the underdoped regime, but very broad in the overdoped regime. Apparently, this is not due to the superconducting fluctuations. The appearance of the positive curvature in the superfluid density, as explained before, is a characteristic feature of weakly coupled two-band superconductor. In contrast, the curvature of the superfluid density of the single-band system is always negative below T_c. This indicates that different from the most of hole-doped high-T_c superconductors, the electron-doped cuprates are weakly coupled two-band systems.

In addition to the superfluid density, many other experimental results also support the two-band picture for electron-doped high-T_c superconductors. The most direct one is the ARPES. For the $Nd_{2-x}Ce_xCuO_4$ (NCCO) superconductor at low doping, a small Fermi surface of electrons appears around $(\pi, 0)$ in the Brillouin zone [175]. Whereas in the hole-doped case, the Fermi surface of holes first appears around $\left(\dfrac{\pi}{2}, \dfrac{\pi}{2}\right)$ [176]. The difference in the momenta of the Fermi surfaces in these two cases is due to the sign change of the next-nearest-neighbor hopping constant t' in the corresponding effective t-J model [177], which breaks the particle-hole symmetry. With the increase of the doping level, another small Fermi surface appears around $\left(\dfrac{\pi}{2}, \dfrac{\pi}{2}\right)$ in electron-doped cuprates. The superconducting phase emerges only when this Fermi surface rises above the Fermi level, similar as in the hole-doped materials. It indicates that the emergence of superconductivity is closely related to the pairing correlation of electrons around $\left(\dfrac{\pi}{2}, \dfrac{\pi}{2}\right)$. In addition, a variety of transport measurements showed that in order to understand comprehensively the temperature dependences of the Hall coefficient, magnetoresistance and other physical quantities, one needs to assume that there are two kinds of charge carriers, i.e. electrons and holes, in electron-doped materials[178, 179, 180]. This would also imply that the electron-doped cuprates are two-band systems.

The two disconnected small Fermi surfaces in electron-doped high-T_c cuprates may hail from the upper and lower Hubbard bands, respectively [181]. They may also arise from the band folding due to the antiferromagnetic correlation [182, 183]. In either case, these two small Fermi surfaces can be described by a two-band mode with a weak inter-band coupling [183, 181]. The interband coupling is weak because these two bands are not directly coupled by the main interaction, i.e. the antiferromagnetic interaction, in this system.

Let us denote the bands around $\left(\dfrac{\pi}{2}, \dfrac{\pi}{2}\right)$ and $(\pi, 0)$ as band 1 and 2, respectively. Similar as in hole-doped superconductors, we assume that the superconducting pairing of band 1 electrons possesses the $d_{x^2-y^2}$-wave symmetry. At low temperatures, the contribution to the superfluid density from band 1 should be the same as in the single-band d-wave superconductor, exhibiting a linear temperature dependence,

$$\rho_{s,1}(T) \approx \rho_{s,1}(0)\left(1 - \frac{T}{T_c}\right). \tag{9.45}$$

If the Cooper pairing in band 2 is induced by the proximity effect from band 1, the pairing symmetry in band 2 should also be $d_{x^2-y^2}$. As the Fermi surfaces of band 2 and the gap nodal lines do not intersect, the quasiparticle excitations of band 2 are gapped even if it has the $d_{x^2-y^2}$-wave pairing symmetry. Thus the contribution to the low-temperature superfluid density from band 2, $\rho_{s,2}(T)$, is thermally activated similar as in the s-wave superconductor. It varies exponentially with temperature as

$$\rho_{s,2}(T) \sim \rho_{s,2}(0)\left(1 - ae^{-\Delta_1'/k_B T}\right), \tag{9.46}$$

where Δ_1' is the minimal gap on the Fermi surfaces of band 2, and a is a doping dependent constant.

Under the mean-field approximation, the total superfluid density equals the sum of $\rho_{s,1}$ and $\rho_{s,2}$:

$$\rho_s(T) = \rho_{s,1}(T) + \rho_{s,2}(T). \tag{9.47}$$

Thus, in the limit $T \ll T_c$, the normalized superfluid density is approximately given by

$$\frac{\rho_s(T)}{\rho_s(0)} \approx 1 - \frac{\rho_{s,1}(0)}{\rho_s(0)} \frac{T}{T_c} - \frac{\rho_{s,2}(0)}{\rho_s(0)} a e^{-\Delta_1'/k_B T}, \qquad (9.48)$$

where $\rho_s(0) = \rho_{s,1}(0) + \rho_{s,2}(0)$. At low temperatures, $\rho_s(T)/\rho_s(0)$ is predominantly determined by the linear T term. But unlike in the single-band case, the linear-T coefficient contains a correction factor $\rho_{s,1}(0)/\rho_s(0)$. At zero temperature, the superfluid density of $\rho_{s,i}(0)$ is proportional to the electron density and inversely proportional to its effective mass of band i. As doped electrons appear first at band 2, and then at band 1 at relatively high doping level, $\rho_{s,2}(0)$ is expected to be much larger than $\rho_{s,1}(0)$. This implies that $\rho_{s,1}(0)/\rho_s(0) \ll 1$ and the linear T-term in $\rho_s(T)/\rho_s(0)$ is significantly suppressed compared to the single-band system.

In real samples, the temperature dependence of $\rho_{s,1}(T)$ is further changed by the impurity scattering. At low temperatures, it is no longer a linear function of T. Instead, it varies as T^2:

$$\rho_{s,1}(T) \sim \rho_{s,1}(0) \left(1 - \frac{k_B^2 T^2}{6\pi\Gamma_0\Delta_2}\right), \qquad (9.49)$$

where Γ_0 is the impurity scattering rate. In this case, $\rho_s(T)/\rho_s(0)$ is given by

$$\frac{\rho_s(T)}{\rho_s(0)} \approx 1 - \frac{\rho_{s,1}(0)}{\rho_s(0)} \frac{k_B^2 T^2}{6\pi\Gamma_0\Delta_2} - \frac{\rho_{s,2}(0)}{\rho_s(0)} a e^{-\Delta_1'/k_B T}. \qquad (9.50)$$

It shows that the temperature dependence of $\rho_s(T)/\rho_s(0)$ can be further suppressed by impurity scatterings. Thus the power-law temperature dependence of the superfluid density in the electron-doped high-T_c superconductors is weakened at low temperatures, and the overall temperature dependence is dominated by the thermal activated behavior. This caused difficulty in identifying the pairing symmetry from the experiment data and led to the discrepancy among the interpretations from different experiment groups .

Fig. 9.7 compares the experimental data with the fitting curves obtained using Eq. (9.50). The agreement is very good. It indicates that Eq. (9.50) catches correctly the intrinsic feature of low temperature superfluid density for electron-doped high-T_c cuprates.

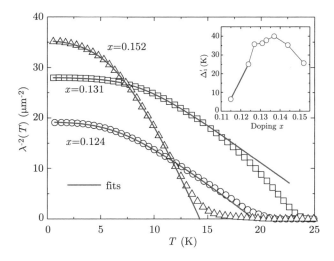

Figure 9.7　The low temperature superfluid density of the electron-doped high-T_c super-conductor PCCO. The fitting is based on Eq. (9.50). (From [162]). The doping levels of $x = 0.124, 0.131$, and 0.152 are in the underdoping, optimal doping, and overdoping regimes, respectively. The circle, square, and triangle represent experimental results. The inset shows the doping dependence of Δ'_1.

Around T_c, it is difficult to obtain an analytic expression for the superfluid density, and numerical calculations are desired. With a reasonable assumption of the energy-momentum dispersions for these two bands [181], the temperature dependence of the superfluid density can be evaluated by first solving the gap equations (9.42) and (9.43). As shown in Fig. 9.8, the theoretical results [162] agree with experimental ones [171]. Here the impurity correction to the super-fluid density is not considered in the theoretical calculations. The agreement between the theoretical and experimental results could be further improved if this correction is included.

The above analysis shows that even though the electron-doped and hole-doped cuprate superconductors behave quite differently in terms of the super-fluid density, they all have the $d_{x^2-y^2}$-wave pairing symmetry. It implies that the pairing mechanism is not fundamentally different in these two kinds of high-T_c superconductors.

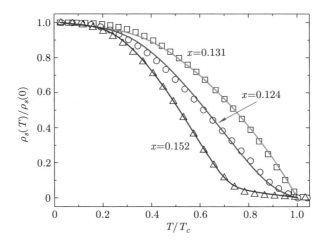

Figure 9.8   The experiment results of the normalized superfluid density of PCCO in comparison with theoretical results based on the weakly coupled two-band model. (The experiment data comes from Ref. [171])

9.7 The Non-Linear Effect

In a small but finite external magnetic field, the non-linear response of the *d*-wave superconductor becomes strong at low temperatures. An external magnetic field can change dramatically the energy dispersion of quasiparticles around the gap nodes. It can close some nodal points by opening a small gap, and at the same time broaden other nodal points into a small gapless area. These changes, particularly the latter one, changes significantly the properties of low energy excitations, giving rise to a strong nonlinear correction to the low temperature electromagnetic response function.

The Hamiltonian of superconducting quasiparticles in an external magnetic field is defined by

$$H(r,r') = \begin{pmatrix} \dfrac{1}{2m}\left(p - \dfrac{e}{c}A\right)^2 - \varepsilon_F & \Delta(r,r')e^{i\phi(R)} \\ \Delta(r,r')e^{-i\phi(R)} & -\dfrac{1}{2m}\left(p + \dfrac{e}{c}A\right)^2 + \varepsilon_F \end{pmatrix}, \qquad (9.51)$$

where $R = (r + r')/2$. $\phi(R)$ is the phase factor of the gap function. It can be gauged away by taking the following unitary transformation

$$U(\mathbf{r}, \mathbf{r}') = \begin{pmatrix} e^{i\phi(R)/2} & 0 \\ 0 & e^{-i\phi(R)/2} \end{pmatrix}, \tag{9.52}$$

resulting in an equivalent Hamiltonian given by

$$H(r, r') = i\hbar v_s \cdot \nabla + \left(-\frac{\hbar^2 \nabla^2}{2m} + \frac{m}{2} v_s^2 - \varepsilon_F \right) \sigma_3 + \Delta(r, r')\sigma_1, \tag{9.53}$$

where the gauge $\nabla \cdot v_s = 0$ is taken and

$$v_s = \frac{eA}{mc} - \frac{\hbar \nabla \phi}{2m} \tag{9.54}$$

is the superfluid velocity.

The transformation $U(r, r')$ is single-valued only if $\phi(R)$ is single-valued. If the gap function is singular at certain points in real space, for example, in the presence of the magnetic vortex line, $\phi(R)$ does not return to itself after winding around the vortex line once. Instead it changes to $\phi(R) + 2\pi$. In this case, the above transformation is singular, and special care is needed in order to handle properly the boundary condition. Here we assume that the system is in the Meissner phase. Since no vortex lines exist, the above transformation is single-valued. For simplicity we take the London gauge $\nabla \phi = 0$.

In Eq. (9.53), if we assume that the supercurrent velocity v_s is spatially independent, and the gap function Δ possesses the d-wave symmetry, then the Hamiltonian is diagonal in momentum space:

$$H = \sum_k \left[\hbar k \cdot v_s + \left(\varepsilon_k + \frac{m}{2} v_s^2 \right) \sigma_3 + \Delta_k \sigma_1 \right], \tag{9.55}$$

where $\varepsilon_k = \hbar^2 k^2 / 2m - \varepsilon_F$. The v_s^2 term can be viewed as a correction to the chemical potential and absorbed into ε_F. The above equation is then simplified as

$$H = \sum_k \left(\hbar k \cdot v_s + \varepsilon_k \sigma_3 + \Delta_k \sigma_1 \right). \tag{9.56}$$

Its eigen spectra are given by

$$W_{\pm,k} = \hbar k \cdot v_s \pm \sqrt{\varepsilon_k^2 + \Delta_k^2}. \tag{9.57}$$

Compared to the zero field dispersion, the quasiparticle spectra are shifted by $\hbar k \cdot v_s$, which is dubbed as the Doppler shift. The characteristic energy scale for the non-linear effect is $E_{nonlin} = \hbar k_F v_s$. When the temperature or other energy scales are smaller than E_{nonlin}, the non-linear effect becomes important. Otherwise, the linear approximation remains valid.

Under the Doppler shift approximation, the quasiparticle contribution to the free energy is given by

$$F_q = -\frac{1}{\beta V} \sum_k \ln \left(1 + e^{-\beta W_{+,k}}\right) \left(1 + e^{-\beta W_{-,k}}\right). \tag{9.58}$$

After taking derivatives, the current vector of quasiparticles is found to be

$$j_q = -\frac{e}{m} \frac{\partial F_q}{\partial v_s} = -\frac{e\hbar}{mV} \sum_k k \left[f\left(W_{+,k}\right) + f\left(W_{-,k}\right)\right]. \tag{9.59}$$

It is simple to verify that the quasiparticle contribution to the supercurrent is opposite to the direction of the supercurrent arising from the condensate. The total supercurrent equals the sum of these two terms:

$$j_s = -env_s - \frac{e\hbar}{mV} \sum_k k \left[f\left(W_{+,k}\right) + f\left(W_{-,k}\right)\right]. \tag{9.60}$$

It shows that if the quasiparticle contribution to the supercurrent cannot be treated by the linear approximation, the dependence of the supercurrent on the external field or the superfluid velocity is non-linear.

The above discussion shows that the Doppler shift induces corrections to the quasiparticle excitation spectra. It also induces a non-linear correction to the electromagnetic response function. This correction is negligible in the s-wave superconductors, but becomes significantly important in the d-wave superconductors due to the presence of gap nodes.

In order to describe correctly the non-linear effect, let us consider the superfluid response of the system to an external magnetic field. At zero temperature, the calculation is relatively easy and F_q can be simplified as

$$F_q = \int_{-\infty}^{0} d\omega \left[\rho_+(\omega) + \rho_-(\omega)\right] \omega, \tag{9.61}$$

where

$$\rho_+ (\omega) = \frac{1}{4} \sum_i \rho \left(\omega - \hbar k_F v_{s,i} \right) \theta \left(\omega - \hbar k_F v_{s,i} \right), \qquad (9.62)$$

$$\rho_- (\omega) = \frac{1}{4} \sum_i \rho \left(\hbar k_F v_{s,i} - \omega \right) \theta \left(\hbar k_F v_{s,i} - \omega \right), \qquad (9.63)$$

are the quasiparticle density of states corresponding to $W_\pm$. Substituting these equations into Eq. (9.61) and using the linear behavior of $\rho(\omega)$ at low frequencies, we find F_q at the low field to be

$$F_q = -\frac{N_F}{12\Delta_0} \sum_i \left(\hbar k_F v_{s,i} \right)^3 \theta \left(v_{s,i} \right) - \int_0^\infty \mathrm{d}\omega \rho \left(\omega \right) \omega. \qquad (9.64)$$

The second term on the right hand side of the equation is the quasiparticle contribution to the free energy in the absence of external magnetic field.

Taking the derivative of F_q with respect to v_s, we obtain the quasiparticle correction to the supercurrent as

$$j_q = -\frac{e}{m} \frac{\partial F_q}{\partial v_s} = \begin{cases} env_s^2/2v_c, & v_s \parallel \text{node}, \\ env_s^2/2\sqrt{2}v_c, & v_s \parallel \text{antinode}. \end{cases} \qquad (9.65)$$

Combined with the contribution from the superfluid condensate, we arrive at the supercurrent at zero temperature as

$$j_s(T = 0K, v_s) = \begin{cases} -env_s \left(1 - v_s/2v_c \right), & v_s \parallel \text{node}, \\ -env_s \left(1 - v_s/2\sqrt{2}v_c \right), & v_s \parallel \text{antinode}. \end{cases} \qquad (9.66)$$

According to the definition of the superfluid density, $j_s = -en_s v_s$, we can further obtain the superfluid density in the ab-plane at zero temperature as

$$n_s^{ab} = \begin{cases} n \left(1 - v_s/2v_c \right), & v_s \parallel \text{node}, \\ n \left(1 - v_s/2\sqrt{2}v_c \right), & v_s \parallel \text{antinode}. \end{cases} \qquad (9.67)$$

This is the result that was first obtained by Yip and Sauls [184].

Since v_s is proportional to H, Eq. (9.67) shows that $n_{ab}^s(0)$ varies linearly with H at zero temperature. This linear field dependence results from the nodal structure of the d-wave pairing gap with the linear low-energy density of states.

In the gapped *s*-wave superconductors, the field dependence of $n_{ab}^s(0)$ is zero at low fields. Hence through the measurement of the field dependence of $n_{ab}^s(0)$, one can determine whether there exist gap nodes or not. Moreover, the change of n_{ab}^s with the magnetic field varies along different directions. By measuring this direction dependence, one can also determine the positions of gap nodes on the Fermi surface.

At finite temperatures, the supercurrent vector can be decomposed into the zero temperature contribution and the finite temperature correction as

$$j_s(T, v_s) = j_s(0, v_s) + \int_0^T dT \frac{\partial j(T, v_s)}{\partial T}, \tag{9.68}$$

where $j_s = j_s \hat{v}_s$ ($\hat{v}_s$ is the unit vector along the direction of v_s). The second term on the right hand side is mainly the contribution of quasiparticles. Using the property of the Fermi distribution function,

$$\frac{\partial f(-x)}{\partial T} = -\frac{\partial f(x)}{\partial T},$$

we find

$$\frac{\partial j_s(T, v_s)}{\partial T} = -\frac{2e\hbar}{mV} \sum_k k \cdot \hat{v}_s \frac{\partial}{\partial T} f(\hbar k \cdot v_s + E_k). \tag{9.69}$$

In the case the temperature is much smaller than the characteristic energy scale of the non-linear effect $E_{nonlin} = \hbar k_F v_s$, i.e. $k_B T \ll E_{nonlin}$, the quasiparticle correction to the supercurrent contributes mainly from the nodal area. Around each nodal point, E_k can be linearized and expressed as

$$E_k = \sqrt{(\hbar v_F k_\parallel)^2 + (2\hbar v_c k_\perp)^2},$$

where $k_\parallel$ and $k_\perp$ are the momentum components parallel and perpendicular to the Fermi surface at the nodal point, respectively. The momentum summation in Eq. (9.69) can now be readily evaluated. By further integrating over T, the value of j_s, up to the leading order terms in temperature, is found to be

$$j_s(T, v_s) \approx \begin{cases} j_s(0, v_s) + \dfrac{\pi e k_B^2 T^2}{12\hbar^2 v_c}, & v_s \parallel \text{node}, \\[3mm] j_s(0, v_s) + \dfrac{\sqrt{2}\pi e k_B^2 T^2}{12\hbar^2 v_c}, & v_s \parallel \text{antinode}. \end{cases} \tag{9.70}$$

The next order correction is proportional to $T^3 v_s$. This result shows that at low temperatures, the dependences of the supercurrent on the magnetic field and temperature are nearly independent. Fixing temperature, j exhibits the same magnetic field dependence as at zero temperature. On the other hand, fixing magnetic field, j varies quadratically with temperature.

When the temperature is much larger than the nonlinear energy scale E_{nonlin} but remains much smaller than the maximal gap value Δ_0, $x = \beta \hbar k \cdot v_s$ is a small quantity. The supercurrent given in Eq. (9.60) can be expanded according to the power of this parameter as

$$j_s = -env_s - \frac{2e\hbar}{mV} \sum_{n,k} \frac{k}{n!} \frac{\partial^n f(E_k)}{\partial E_k^n} (\hbar k \cdot v_s)^n. \qquad (9.71)$$

It is easy to show that the even power terms of v_s vanish after the momentum summation, and only the odd power terms have contribution to j_s. The leading order term, which is correct up to the linear order of v_s, is just the result obtained with the linear approximation. The lowest order non-linear correction starts from the cubic order of v_s. Up to this order of correction, the supercurrent is given by

$$j_s \approx -env_s - \frac{2e\hbar}{mV} \sum_{n,k} k \left[\frac{\partial f(E_k)}{\partial E_k} \hbar k \cdot v_s + \frac{1}{3!} \frac{\partial^3 f(E_k)}{\partial E_k^3} (\hbar k \cdot v_s)^3 \right]. \qquad (9.72)$$

Hence, if $E_{nonlin} \ll k_B T \ll \Delta_0$, the non-linear correction to the supercurrent function j_s is proportional to H^3, and its correction to the superfluid density is proportional to H^2. In the s-wave superconductor, there is a finite energy gap in the quasiparticle excitations, the quasiparticle population is proportional to $\exp(-\Delta/k_B T)$ and the non-linear effect is much weaker than the d-wave case.

The Doppler shift defines an energy scale for the nonlinear effect $E_{nonlin} = \hbar k_F v_s \approx \hbar k_F e\lambda H_0/mc$, and Eq. (9.67) is valid when $T \ll E_{nonlin}$. Thus in order to observe the non-linear field dependence of the superfluid density and its spatial anisotropy, temperature should be much lower than E_{nonlin}. On the other hand, the measurement on the non-linear effect can only be performed in the Meissner phase, where the magnetic field must be below the lower critical field. Otherwise, magnetic vortex lines emerge and the above results are no longer valid. This means that E_{nonlin} is upper bounded by the lower critical field. For the YBCO superconductor, the lower critical field is at a few hundred Gauss, and E_{nonlin} is estimated around 1K (10^{-4}eV). Hence, the non-linear effect can only be observed in this material at very low temperatures.

Experimentally, the anisotropy of the magnetic penetration depth induced by the non-linear effect has not been observed yet in high-T_c superconductors [185, 186]. There are probably two reasons for this. First, the measurement temperature was still not low enough to observe the nonlinear effect. Second, the magnetic field was too high so that the system is in the mixed state rather than in the Meissner state. Due to the existence of vortex lines, the supercurrent velocity distributions are complicated. In the Meissner state the supercurrent only flows parallel to the surface. Both could reduce the anisotropy of the penetration depth along different directions. In order to probe the anisotropy induced by the non-linear effect, the measurement of the penetration depth needs to be performed in the Meissiner phase at very low temperatures.

9.8   Relationship between the Magnetic Penetration Depth and the Superfluid Density

The above discussion was done under the assumption that v_s is spatially independent. However, in realistic systems, the magnetic field decays at the surface of the superconductor and the characteristic length scale is just the penetration depth. Thus v_s cannot be a quantity that is spatially homogeneous. Nevertheless, if the spatial variation of v_s is not too sharp, the Hamiltonian Eq. (9.56) is still locally valid, and the quasiparticle spectra obtained with this Hamiltonian are just those at the corresponding spatial point. Below we discuss the spatial variation of the magnetic field under this approximation, and then determine how the magnetic penetration depth varies in superconductor.

Under the linear approximation, the penetration depth is inversely proportional to the square root of the superfluid density and is spatially independent. The decay of the magnetic field along the direction perpendicular to a semi-infinitely large superconducting plate is determined by the London equation,

Eq. (1.1). The solution is give by Eq. (1.4):

$$H^{(0)}(x) = H_0 e^{-x/\lambda}, \tag{9.73}$$

where H_0 is the magnetic field on the surface of the superconductor ($x = 0$). The superscript (0) here is to emphasize that $H^{(0)}(x)$ is the solution of the London equation without considering the non-linear correction. The corresponding vector potential is

$$A^{(0)}(x) = H^{(0)}(x)\lambda, \tag{9.74}$$

and the spatial dependence of the supercurrent velocity is found to be

$$v_s^{(0)} = \frac{e\lambda H^{(0)}(x)}{mc} \tag{9.75}$$

within the linear approximation.

Without considering the correction from the non-linear effect, v_s is proportional to H. Thus the magnetic field decay is equivalent to the decay of supercurrent velocity, and the London equation can also be expressed as the equation for the supercurrent velocity

$$\frac{\partial^2 v_s}{\partial x^2} = \frac{v_s}{\lambda^2}. \tag{9.76}$$

Considering the nonlinear effect, apparently this equation needs to be corrected. Under the local approximation, λ^{-2} should be replaced by Eq. (9.67). This leads to the nonlinear effect corrected equation for the supercurrent velocity

$$\frac{\partial^2 v_s}{\partial x^2} = \frac{v_s}{\lambda^2}(1 - \alpha v_s), \tag{9.77}$$

where α is a direction dependent quantity. $\alpha = 1/2v_c$ or $1/2\sqrt{2}v_c$ if v_s is along the nodal or anti-nodal direction.

The non-linear equation Eq. (9.77) does not have analytic solutions. In order to solve the magnetic penetration depth and compare with experiments, usually numeric solutions need to be performed. However, on the superconducting surface, if we use v_s obtained under the linear approximation to replace the v_s in the parentheses on the right hand side of Eq. (9.77), the magnetic penetration depth on the superconductor surface is approximately given by

$$\lambda_{eff}(x=0) \approx \lambda \left(1 - \frac{\alpha e \lambda H_0}{mc}\right)^{-1/2}. \tag{9.78}$$

9.9 The Non-Local Effect

The preceding discussion on the electromagnetic response function is made based on the local approximation. This approximation is valid when the coherence length $\xi_0 = v_F/\pi\Delta_0$ is much shorter than the penetration depth, $\xi_0 \ll \lambda_0$. However, in high-T_c superconductors, the electron coherence length is anisotropic. It depends on the direction of electron momentum. If the momentum is along the nodal direction, the effective coherence length of $\xi_k = v_F/\pi\Delta_k$ diverges, and the condition for the local approximation $\xi_k \ll \lambda_0$ is no longer satisfied. This means that near the gap nodes and within the energy scale

$$E_{nonloc} \approx |\Delta_k| \approx (\xi_0/\lambda_0)\Delta_0,$$

the non-local effect is important and must be considered [187, 188].

E_{nonloc} defines an energy scale around the Fermi surface within which the quasiparticle excitations have contribution to the non-local effect. It also defines a characteristic energy scale of the non-local effect. Corresponding to E_{nonloc}, one can also define a characteristic temperature to describe the non-local effect as

$$T_{nonloc} = E_{nonloc}/k_B.$$

In high-T_c superconductors, the region of the Fermi surface that contributes to the non-local effect is small compared to the entire Fermi surface as $\xi_0 \ll \lambda_0$. Thus the correction due to the non-local effect is typically small and negligible. However, when $T \ll T_{nonloc}$, the quasiparticle excitations contribute mainly by the electrons around the nodal points, and the non-local effect is no longer negligible. In real materials, this issue becomes complicated because at low temperatures the impurity scattering effect usually becomes strong, which can completely suppress the non-local effect.

The value of T_{nonloc} is determined by the three fundamental parameters of superconductors, $(\lambda_0, \xi_0, \Delta_0)$. For high-$T_c$ superconductors, the typical values of these parameters are $\lambda_0 \approx (1 \sim 3) \times 10^3 \text{Å}$, $\xi_0 \approx 15 \sim 30\text{Å}$, $\Delta_0/k_B \approx 200 \sim 300\text{K}$. The corresponding value of T_{nonloc} is estimated to be $1 \sim 3\text{K}$.

The response function Eq. (9.2) is defined in an infinitely large medium. In order to calculate the correction from the non-local effect to the magnetic penetration depth, the boundary condition of the magnetic field on the surface of superconductor must be handled carefully. In order to study a semi-infinite superconductor using the results of an infinite system, a commonly used approximation is to use a zero thickness current sheet to replace the superconducting surface. The magnetic fields such generated point to opposite directions on the two sides of the boundary, and the corresponding vector potentials are mirror symmetric with respect to the boundary layer. In this case, electrons are completely reflected on the surface. The magnetic penetration depth is determined by the following integral [130]:

$$\lambda_\alpha^{spec} = \frac{2}{\pi} \int_0^\infty \frac{dq}{\mu_0 K_{\alpha\alpha}(q\hat{n}, 0) + q^2}, \tag{9.79}$$

where $\hat{n}$ is the unit vector normal to the superconducting surface. $K_{\alpha\alpha}(q, \omega)$ is the electromagnetic response function defined in Eq. (3.58). If $K(q, 0)$ is q-independent,

$$K(q, 0) = \frac{1}{\mu_0 \lambda_L^2}, \tag{9.80}$$

then

$$\lambda_\alpha^{spec} = \lambda_L. \tag{9.81}$$

This is just the result that is obtained under the local approximation.

Another frequently used method to effectively handle the surface problem is to assume that the electron motion on the surface is diffusive. In this case, the formula for calculating the magnetic penetration depth is different from Eq. (9.79). These two methods yield qualitatively the same results although quantitatively they are different.

For a given wavevector q, the current-current correlation function at zero frequency is determined by

$$\Pi_{xx}(q, \omega = 0) = \frac{2e^2}{\beta \hbar^2 V} \sum_{k\omega_m} \left(\frac{\partial \xi_k}{\partial k_x}\right)^2 \frac{(i\omega_m)^2 + \xi_- \xi_+ + \Delta_- \Delta_+}{\left[(i\omega_m)^2 - E_+^2\right]\left[(i\omega_m)^2 - E_-^2\right]}, \quad (9.82)$$

where $\xi_\pm = \xi_{k\pm q/2}$, $\Delta_\pm = \Delta_{k\pm q/2}$, and $E_\pm = E_{k\pm q/2}$. When $|q|$ is very small, these quantities can be expanded in terms of q. Correct to the first order of q, we have

$$\xi_\pm = \xi_k \pm \nabla_k \xi_k \cdot \frac{q}{2},$$

$$\Delta_\pm = \Delta_k \pm \nabla_k \Delta_k \cdot \frac{q}{2},$$

$$E_\pm = E_k \pm \nabla_k E_k \cdot \frac{q}{2}.$$

By employing the identities

$$E_k \nabla_k E_k = \xi_k \nabla_k \xi_k + \Delta_k \nabla_k \Delta_k,$$

$$\left(\nabla_k E_k\right)^2 = \left(\nabla_k \xi_k\right)^2 + \left(\nabla_k \Delta_k\right)^2,$$

we can simplify Eq. (9.82) as

$$\Pi_{xx}(q, 0) = \frac{2e^2}{\beta \hbar^2 V} \sum_{k\omega_m} \left(\frac{\partial \xi_k}{\partial k_x}\right)^2 \frac{(i\omega_m)^2 + W_+ W_-}{\left[(i\omega_m)^2 - W_+^2\right]\left[(i\omega_m)^2 - W_-^2\right]}, \quad (9.83)$$

where

$$W_\pm = E_k \pm \frac{1}{2}\alpha, \qquad \alpha = |\nabla_k E_k \cdot q\hat{n}|.$$

Performing the frequency summation, we arrive at

$$\Pi_{xx}(q, 0) = \frac{2e^2}{\hbar^2 V} \sum_k \left(\frac{\partial \xi_k}{\partial k_x}\right)^2 \frac{f\left(E_k + \frac{1}{2}\alpha\right) - f\left(E_k - \frac{1}{2}\alpha\right)}{\alpha}$$

$$= \frac{e^2 v_F^2}{\hbar^2} \int_0^\infty d\omega \rho(\omega) \frac{f\left(\omega + \frac{1}{2}\alpha\right) - f\left(\omega - \frac{1}{2}\alpha\right)}{\alpha}. \quad (9.84)$$

In the calculation below, we assume that the superconducting surface is perpendicular to the *x*-axis, and the magnetic field is along the *z*-axis. In this case, the low energy excitations concentrate in the nodal area, and $\alpha = |q\partial_{k_x} E_{k_{node}}|$ whose values around the four nodes are approximately the same. As $\rho(\omega)$ is linear in the low energy limit, we have

$$\Pi_{xx}(q,0) \approx \frac{e^2 v_F^2 N_F}{\alpha \hbar^2 \Delta_0} \int_0^{\infty} d\omega \omega \left[f\left(\omega + \frac{1}{2}\alpha \right) - f\left(\omega - \frac{1}{2}\alpha \right) \right]. \qquad (9.85)$$

After simplification, the above expression can be further expressed as

$$\Pi_{xx}(q,0) \approx \frac{e^2 v_F^2 N_F}{\alpha \hbar^2 \Delta_0} \left[-\frac{2}{\beta^2} \int_0^{\beta\alpha/2} dx \frac{x}{e^x + 1} - \frac{\alpha^2}{8} - \frac{\alpha}{\beta} \ln\left(1 + e^{-\alpha\beta/2} \right) \right]. \qquad (9.86)$$

When $k_B T \gg \alpha$, the non-local effect is negligible. Eq. (9.86) recovers the result that is obtained under the local approximation. However, when $k_B T \ll \alpha$, the non-local effect is strong, and the upper limit of the integral on the right hand side of Eq. (9.86) can be set to ∞, then

$$\Pi_{xx}(q,0) \approx -\frac{e^2 v_F^2 N_F}{\alpha \hbar^2 \Delta_0} \left(\frac{\alpha^2}{8} + \frac{\pi^2 k_B^2 T^2}{6} \right). \qquad (9.87)$$

The first term on the right hand side is temperature independent, which is a correction to the zero temperature superfluid density, or zero temperature magnetic penetration depth λ_0. Compared with the results under the local approximation, Π_{xx} approximately varies as T^2. Using Eq. (3.58) and Eq. (9.79), it can also be shown that λ_x^{spec} varies as T^2 at low temperatures. Thus similar to the non-linear effect, the non-local effect can also change the low temperature behavior of λ [188]. This change, as mentioned before, is very important to maintain the stability of the d-wave superconducting state without violating the third law of thermodynamics.

Chapter 10

Optical and Thermal Conductivities

10.1 Optical Conductivity

In an ideal superconductor, there is no energy dissipation, both the resistivity and the light absorption vanish. In real materials, however, there are always elastic and inelastic scatterings. Due to the existence of superfluid in a superconductor, the direct-current resistivity remains zero, but the light absorption is no longer zero. The optical and thermal conductivities are two fundamental quantities in characterizing the transport of electrons in a superconductor. They measure the responses of superconducting quasiparticles to an applied electromagnetic field and a temperature gradient, respectively. The experimental measurement and theoretical analysis of these quantities has played an important role in the study of high-T_c superconductivity. Not only can it be used to probe the pairing symmetry, but also to provide vital information on the interaction between superconducting quasiparticles and other low-lying excitations.

Both Cooper pairs and quasiparticles can absorb light. Through the light absorption, a Cooper pair can be depaired and a quasiparticle can transit from a low energy state to a high energy one. Due to the existence of pairing gap, a condensed Cooper pair becomes two normal electrons and contribute to the optical conductivity only when the frequency of the absorbed light exceeds the pairing bound energy, namely twice of the single particle energy gap.

The light absorption rate, or the optical conductivity, is closely related to the mean-free-path l of normal electrons and the coherence length ξ of Cooper pairs. There are two limits that deserve special attention. One is the dirty limit with $l \ll \xi$. In this limit, the paired electrons, after being excited by a light above the energy gap, may experience many times of impurity scatterings and lose their initial momentum correlations within a characteristic time scale ξ/v_f. The other is the clean limit with $l \gg \xi$. In this limit, there is almost no impurity scattering within the time scale ξ/v_f and the electron momenta are essentially conserved.

In the most of metal-based superconductors, the coherence length is much larger than the mean free path, $\xi \gg l$. Hence these superconductors are in the dirty limit. Indeed, the measured infra-red absorption spectra in these superconductors agree well with the theoretical calculation obtained in the dirty limit. In contrast, in high-T_c superconductors, the coherence length $\xi \approx 20 \sim 30\text{Å}$, is less than the mean free path and is generally considered to be in the clean limit. However, this judgement of clean limit may not be always correct due to the strong anisotropy of d-wave pairing function whose coherence length diverges along the nodal direction, much longer than the mean free path. Thus the quasiparticles around the gap nodes are always in the dirty limit. This suggests that in the low frequency or temperature regime where the contribution of nodal quasiparticle excitations dominates, the optical conductivity of high-T_c superconductors is in the dirty limit. On the contrary, if the contribution from high energy quasiparticles dominates, the system is in the clean limit. In this limit, the light absorption depends on the scattering processes of electrons. In addition to the elastic scattering of impurities, the characteristic energy scale of inelastic scattering from spin excitations is large and comparable to $k_B T_c$, which can also affect the infrared absorption.

In the dirty limit, the momentum correlations between excited electrons are completely destructed by impurity scattering. The detail and the mechanism of scattering processes are therefore not important. In this case, the light absorption with the corresponding optical conductivity at finite frequencies can be accurately calculated. In the non-local electromagnetic response theory, the conductivity depends on the initial and final coordinates of electron, r and r'. The correlation length between these two coordiantes ξ_c is determined by the mean free path l and the coherence length ξ in the absence of disorder:

$$\frac{1}{\xi_c} = \frac{1}{l} + \frac{1}{\xi}.$$

In the limit $l \ll \xi$, $\xi_c \approx l$ is the characteristic length scale of electromagnetic response function. It is also the coherence length of excited electrons. Within this length scale, the optical absorption is approximately dissipationless. As the characteristic length scale of dissipationless systems is ξ which is significantly larger than l, the optical conductivity σ in the dirty limit is approximately proportional to the corresponding conductivity $\sigma^{(0)}$ in a dissipationless system in the limit $(r' - r) \to 0$ [189]:

$$\sigma_\mu(\omega) \propto \sigma_\mu^{(0)}(r' - r \to 0, \omega) = \frac{1}{V} \sum_q \sigma_\mu^{(0)}(q, \omega). \tag{10.1}$$

Hence σ is given by the average of $\sigma^0(\omega, q)$ over all the momenta q. It reflects the uncertainty of the momenta of excited electron pairs after scattering in the limit $l \ll \xi$.

In the clean limit, there is no universal theory to describe the optical conductivity of superconductors. In particular, different scattering centers or processes affect the optical absorption differently, and these effects need to be studied independently. Generally speaking, our understanding on the conducting behavior of superconducting quasiparticles is incomplete and there is not a commonly accepted microscopic theory. Both the optical absorption theory of Mattis and Bardeen in the dirty limit [189] and all semiphenomenological theories of optical conductivity proposed in the clean limit have their own limitations. They should not be blindly used before a comprehensive understanding of physical properties of the system is achieved.

10.2 The Optical Sum Rule

Compared with the normal state, the optical absorption spectral weight at finite frequencies is reduced in the superconducting state. The reduced weight is transferred to the zero frequency and becomes the superfluid density. Nevertheless, the total spectral weight, including the zero frequency part, is conserved. This is just the statement of optical sum rule for the superconducting

state, which is also called the FGT (Ferrell, Glover, Tinkham) optical sum rule
[190, 191].

The FGT sum rule, or the general conductivity sum rule represented by
Eq. (10.6) given below, results from the charge conservation. The proof of this
sum rule is straightforward. To do this, let us start from Eq. (9.8). By inserting
a complete set of eigen bases of the Hamiltonian, we can express the current-
current correlation function as

$$\tilde{\Pi}_{\mu\nu}(q, i\omega_m)$$
$$= -\frac{1}{V\hbar^2 Z} \sum_{nm} \int_0^\beta d\tau e^{i\omega_m \tau} e^{(E_n - E_m)\tau} e^{-\beta E_n} \langle n| J_\mu(q) |m\rangle \langle m| J_\nu(-q) |n\rangle$$
$$= -\frac{1}{V\hbar^2 Z} \sum_{nm} \frac{e^{-E_m\beta} - e^{-\beta E_n}}{i\omega_m + E_n - E_m} \langle n| J_\mu(q) |m\rangle \langle m| J_\nu(-q) |n\rangle .$$

To convert the imaginary frequency to the real frequency by the analytic continuation, we then obtain the retarded current-current correlation function:

$$\Pi_{\mu\nu}(q, \omega) = -\frac{1}{V\hbar^2 Z} \sum_{nm} \frac{e^{-E_m\beta} - e^{-\beta E_n}}{\omega + E_n - E_m + i0^+} \langle n| J_\mu(q) |m\rangle \langle m| J_\nu(-q) |n\rangle .$$

$$(10.2)$$

Using the identity
$$\frac{1}{x + i0^+} = \frac{1}{x} - i\pi\delta(x),$$
we obtain the following equation

$$\int_{-\infty}^\infty d\omega \frac{\mathrm{Im}\Pi_{\mu\nu}(q, \omega)}{\omega} = \pi \mathrm{Re}\Pi_{\mu\nu}(q, 0). \qquad (10.3)$$

From Eq. (9.7), the complex conductivity is found to be

$$\sigma_{\mu\nu}(q, \omega) = \frac{i}{\omega + i0^+} \left[\frac{c^2}{V\hbar^2} \sum_k \left\langle \frac{\partial^2 \varepsilon_k}{\partial k_\mu \partial k_\nu} \right\rangle + \Pi_{\mu\nu}(q, \omega) \right]. \qquad (10.4)$$

Its real part is given by

$$\mathrm{Re}\sigma_{\mu\nu}(q, \omega) = \left[\frac{\pi e^2}{V\hbar^2} \sum_k \left\langle \frac{\partial^2 \varepsilon_k}{\partial k_\mu \partial k_\nu} \right\rangle + \pi \mathrm{Re}\Pi_{\mu\nu}(q, \omega) \right] \delta(\omega) - \frac{\mathrm{Im}\Pi_{\mu\nu}(q, \omega)}{\omega}.$$

$$(10.5)$$

Using Eq. (10.3) and performing the integral over ω, we have

$$\int_{-\infty}^{\infty} d\omega \operatorname{Re}\sigma_{\mu\nu}(q,\omega) = \frac{\pi e^2}{V\hbar^2} \sum_k \left\langle \frac{\partial^2 \varepsilon_k}{\partial k_\mu \partial k_\nu} \right\rangle. \tag{10.6}$$

This is just the generalized sum rule of electric conductivity, which is valid for any momentum q.

The optical conductivity is the diagonal component of $\sigma_{\mu\nu}(q,\omega)$ in the long-wave-length limit $(q \to 0)$. Because $\operatorname{Re}\sigma_{\mu\mu}(\omega) = \operatorname{Re}\sigma_{\mu\mu}(0,\omega)$ is an even function of ω, Eq. (10.6) is simplified as

$$\int_0^{\infty} d\omega \operatorname{Re}\sigma_{\mu\mu}(\omega) = \frac{\pi e^2}{2V\hbar^2} \sum_k \left\langle \frac{\partial^2 \varepsilon_k}{\partial k_\mu^2} \right\rangle. \tag{10.7}$$

In the superconducting state, the conductivity contains two terms, contributed by the superconducting electrons and the normal quasiparticles, respectively:

$$\operatorname{Re}\sigma_{\mu\mu}(\omega) = \frac{\pi e^2 n_s^\mu}{m_\mu}\delta(\omega) + \operatorname{Re}\sigma_\mu(\omega). \tag{10.8}$$

The first term is the contribution of superfluid, and n_s^μ is the superfluid density. $\sigma_\mu(\omega)$ is the contribution of normal electrons and is regular at $\omega = 0$. From the above equations, we then obtain the following FGT sum rule:

$$\int_{0+}^{\infty} d\omega \operatorname{Re}\sigma_\mu(\omega) + \frac{\pi e^2 n_s^\mu}{2m_\mu} = \frac{\pi e^2}{2V\hbar^2} \sum_k \left\langle \frac{\partial^2 \varepsilon_k}{\partial k_\mu^2} \right\rangle. \tag{10.9}$$

If the conductivity $\operatorname{Re}\sigma_\mu$ is measured at two different temperatures, below $(T < T_c)$ and above $(T > T_c)$ the superconducting transition temperature, respectively. Then the difference of $\operatorname{Re}\sigma_\mu$ between these two temperatures, $\delta\sigma_\mu$, satisfies the equation

$$\int_{0+}^{\infty} d\omega\, \delta\sigma_\mu(\omega) = \frac{\pi e^2 n_s^\mu}{2m_\mu} + \frac{\pi e^2}{2V\hbar^2} \sum_k \left(\left\langle \frac{\partial^2 \varepsilon_k}{\partial k_\mu^2} \right\rangle_n - \left\langle \frac{\partial^2 \varepsilon_k}{\partial k_\mu^2} \right\rangle_s \right). \tag{10.10}$$

Dividing both sides by the first term on the right-hand side gives

$$R(\omega \to +\infty) = 1 + \frac{m_\mu}{V\hbar^2 n_s^\mu} \sum_k \left(\left\langle \frac{\partial^2 \varepsilon_k}{\partial k_\mu^2} \right\rangle_n - \left\langle \frac{\partial^2 \varepsilon_k}{\partial k_\mu^2} \right\rangle_s \right), \tag{10.11}$$

where

$$R(\omega) = \frac{2m_\mu}{\pi e^2 n_s^\mu} \int_{0+}^{\omega} d\omega\, \delta\sigma_\mu(\omega). \tag{10.12}$$

In an isotropic free electron system, the effective mass $m_\mu = m$ is momentum independent and

$$\varepsilon_k = \frac{\hbar^2 k^2}{2m}. \tag{10.13}$$

Eq. (10.9) becomes

$$\int_{0+}^{\infty} d\omega \mathrm{Re}\sigma_\mu(\omega) + \frac{\pi e^2 n_s^\mu}{2m} = \frac{\pi e^2 n}{2m}. \tag{10.14}$$

This is the expression of optical sum rule most frequently used in literature [130]. But this equation is valid only when the energy dispersion of electron is defined by Eq. (10.13). In this case, Eq. (10.11) simply becomes

$$R(\omega \to +\infty) = 1. \tag{10.15}$$

Equation (10.15) indicates that the integral of $\delta\sigma_\mu$ times the factor $(2m_\mu)/(\pi e^2 n_s^\mu)$ equals 1 in the limit $\omega \to \infty$ if the energy dispersion is given by Eq. (10.13). In real superconductors, only low energy electrons participate in the Cooper pairing and $\delta\sigma_\mu$ is very small when ω is much larger than the gap value Δ_0. Generelly $R(\omega)$ is close to 1 when $\omega \sim 6\Delta_0$ [192]. Thus the FGT sum rule can be tested through the measurement of low frequency conductivity.

For the optimally doped high-T_c superconductor, it was found from experimental measurements that $R(\omega)$ indeed reaches the saturated value at a relatively low frequency ($\approx 800\mathrm{cm}^{-1}$), in good agreement with the theoretical prediction [193, 192]. However, for the underdoped high-T_c superconductors, the infrared spectroscopy measurements show that the integral of the c-axis optical conductivity $R(\omega)$ remains much less than the theoretical prediction even when the integral upper limit is taken much larger than $6\Delta_0$. It does not even show any tendency of saturation [193]. This indicates that the superfluid density is not purely the contribution of low frequency spectra. High energy spectra have also contribution to the superconducting condensation [194]. Thus high-T_c superconductivity is a phenomenon of multiple energy scales even though the superconducting condensation occurs at low temperatures. In conventional quantum field theory that is based on the idea of renormalization group, an effective low energy model (e.g. the t-J model) is investigated by integrating out all high energy excitation states. This kind of approaches seems to be incomplete or incorrect in the study of high-T_c superconductivity.

The violation of the optical sum rule, or the missing of low frequency spectral weight, in the underdoped high-T_c cuprates, is closely related to the loss of low energy entropy or the pseudogap effect introduced in Sect. 3.3. Currently there is not a satisfactory explanation to this phenomenon. For high-T_c cuprates, the energy-momentum dispersion relation deviates strongly from the free electron dispersion defined in Eq. (10.13), thus the correction from the second term of Eq. (10.11) to Eq. (10.15) is non-negligible. This may be part of the reason for the violation of the optical sum rule in the CuO_2 plane [194] . However, along the c-axis, $R(\omega)$ of the underdoped superconductor can only reach 50% of the expectation even if ω reaches the infrared regime. It is definitely not sufficient to only consider the correction from the band structure to Eq. (10.15) [195].

10.3 Light Absorption in the Dirty Limit

In the dirty limit, the frequency dependence of the optical conductivity is determined by Eq. (10.1). In order to determine its value, we need to first calculate the conductivity $\sigma_\mu^{(0)}(q, \omega)$ in an ideal BCS superconductor.

Substituting the free Green's function Eq. (4.3) into Eq. (9.9), we obtain the following current-current correlation function:

$$\tilde{\Pi}_{\mu v}(q, i\omega_n) = \frac{e^2}{V\hbar^2} \sum_k \frac{\partial \varepsilon_{k+\frac{q}{2}}}{\partial k_\mu} \frac{\partial \varepsilon_{k+\frac{q}{2}}}{\partial k_v} A(k, q, i\omega_n), \qquad (10.16)$$

where

$$A(k, q, i\omega_n) = \frac{1}{\beta} \sum_{\omega_m} \text{Tr} G^{(0)}(k, i\omega_m) G^{(0)}(k + q, i\omega_m + i\omega_n)$$

$$= \frac{E_k(E_k + i\omega_n) + \varepsilon_k \varepsilon_{k+q} + \Delta_k \Delta_{k+q}}{E_k\left[(E_k + i\omega_n)^2 - E_{k+q}^2\right]} f(E_k)$$

$$-\frac{E_k \left(E_k - i\omega_n\right) + \varepsilon_k \varepsilon_{k+q} + \Delta_k \Delta_{k+q}}{E_k \left[\left(E_k - i\omega_n\right)^2 - E_{k+q}^2\right]} \left[1 - f\left(E_k\right)\right]$$

$$+\frac{E_{k+q} \left(E_{k+q} - i\omega_n\right) + \varepsilon_k \varepsilon_{k+q} + \Delta_k \Delta_{k+q}}{E_{k+q} \left[\left(E_{k+q} - i\omega_n\right)^2 - E_k^2\right]} f\left(E_{k+q}\right)$$

$$-\frac{E_{k+q} \left(E_{k+q} + i\omega_n\right) + \varepsilon_k \varepsilon_{k+q} + \Delta_k \Delta_{k+q}}{E_{k+q} \left[\left(E_{k+q} + i\omega_n\right)^2 - E_k^2\right]} \left[1 - f\left(E_{k+q}\right)\right].$$

At zero temperature, $f\left(E_k\right) = 0$, the above expression can be simplified as

$$A\left(k, q, i\omega_n\right) = -\frac{E_k \left(E_k - i\omega_n\right) + \varepsilon_k \varepsilon_{k+q} + \Delta_k \Delta_{k+q}}{E_k \left[\left(E_k - i\omega_n\right)^2 - E_{k+q}^2\right]}$$

$$-\frac{E_{k+q} \left(E_{k+q} + i\omega_n\right) + \varepsilon_k \varepsilon_{k+q} + \Delta_k \Delta_{k+q}}{E_{k+q} \left[\left(E_{k+q} + i\omega_n\right)^2 - E_k^2\right]}.$$

Substituting this equation into the expression of $\tilde{\Pi}_{\mu\nu}$ and performing the analytic continuation, we obtain the retarded current-current correlation function as

$$\Pi_{\mu\nu}\left(q, \omega\right) = -\frac{e^2}{V\hbar^2} \sum_k \frac{\partial \varepsilon_{k+\frac{q}{2}}}{\partial k_\mu} \frac{\partial \varepsilon_{k+\frac{q}{2}}}{\partial k_\nu} \left[\frac{E_k \left(E_k - \omega\right) + \varepsilon_k \varepsilon_{k+q} + \Delta_k \Delta_{k+q}}{E_k \left[\left(E_k - \omega - i\delta\right)^2 - E_{k+q}^2\right]}\right.$$

$$\left.+\frac{E_{k+q} \left(E_{k+q} + \omega\right) + \varepsilon_k \varepsilon_{k+q} + \Delta_k \Delta_{k+q}}{E_{k+q} \left[\left(E_{k+q} + \omega + i\delta\right)^2 - E_k^2\right]}\right]. \tag{10.17}$$

Its imaginary part is given by

$$\mathrm{Im}\Pi_{\mu\nu}\left(q, \omega\right) = \frac{\pi e^2}{2V\hbar^2} \sum_k \frac{\partial \varepsilon_{k+\frac{q}{2}}}{\partial k_\mu} \frac{\partial \varepsilon_{k+\frac{q}{2}}}{\partial k_\nu} \frac{E_k E_{k+q} - \varepsilon_k \varepsilon_{k+q} - \Delta_k \Delta_{k+q}}{E_k E_{k+q}}$$

$$\left[\delta\left(\omega + E_k + E_{k+q}\right) + \delta\left(\omega - E_k - E_{k+q}\right)\right]. \tag{10.18}$$

The real part of the electric conductivity $\sigma_\mu^{(0)}\left(q, \omega\right)$ is proportional to $\mathrm{Im}\Pi_{\mu\mu}$ $\left(q, \omega\right)$. From Eq. (9.7), we have

$$\mathrm{Re}\sigma_\mu^{(0)}\left(q, \omega\right) = \frac{\pi e^2}{2\hbar^2\omega} \sum_k \left(\frac{\partial \varepsilon_{k+\frac{q}{2}}}{\partial k_\mu}\right)^2 \frac{E_k E_{k+q} - \varepsilon_k \varepsilon_{k+q} - \Delta_k \Delta_{k+q}}{E_k E_{k+q}}$$

$$\delta\left(\omega - E_k - E_{k+q}\right). \tag{10.19}$$

Substituting this equation into Eq. (10.1), we obtain the zero temperature optical conductivity in the dirty limit:

$$\mathrm{Re}\sigma_\mu\left(\omega\right) = \frac{\pi e^2 v_F^2}{2d\omega} \sum_{k,q} \left(1 - \frac{\varepsilon_k \varepsilon_q + \Delta_k \Delta_q}{E_k E_q}\right) \delta\left(\omega - E_k - E_q\right), \qquad (10.20)$$

where d is the spatial dimension. The term proportional to $\varepsilon_k \varepsilon_{k+q}$ is an odd function of both ε_k and ε_q, and is zero after momentum summation. Eq. (10.20) can be simplified as

$$\mathrm{Re}\sigma_\mu\left(\omega\right) = \frac{\pi e^2 v_F^2}{2d\omega} \sum_{k,q} \left(1 - \frac{\Delta_k \Delta_q}{E_k E_q}\right) \delta\left(\omega - E_k - E_q\right). \qquad (10.21)$$

In an isotropic s-wave superconductor, $\Delta_k = \Delta$, the above equation becomes

$$\mathrm{Re}\sigma\left(\omega\right) = \frac{\pi e^2 v_F^2 N_F^2}{2d\omega} \int_{-\infty}^{\infty} d\varepsilon_2 d\varepsilon_1 \frac{E_1 E_2 - \Delta^2}{E_1 E_2} \delta\left(\omega - E_1 - E_2\right), \qquad (10.22)$$

where $E_i = \sqrt{\varepsilon_i^2 + \Delta^2}$ and N_F is the density of states of electrons at the Fermi level in the normal state. Using the identity

$$\int_{-\infty}^{\infty} d\varepsilon_i = 2 \int_{\Delta}^{\infty} dE_i \frac{E_i}{\sqrt{E_i^2 - \Delta^2}},$$

we can reexpress $\mathrm{Re}\sigma(\omega)$ as

$$\mathrm{Re}\sigma\left(\omega\right) = \frac{2\pi e^2 v_F^2 N_F^2}{d\omega} \theta\left(\omega - 2\Delta\right) \int_{\Delta}^{\omega - \Delta} dE \frac{E\left(\omega - E\right) - \Delta^2}{\sqrt{E^2 - \Delta^2}\sqrt{\left(\omega - E\right)^2 - \Delta^2}}. \qquad (10.23)$$

Eq. (10.23) shows that $\mathrm{Re}\sigma(\omega)$ has a threshold or an absorption edge. $\mathrm{Re}\sigma(\omega)$ is finite only when $\omega > 2\Delta$. This is an important property of s-wave superconductors. It is also an important criterion for testing and measuring the energy gap of an s-wave superconductor. Physically, this is because there are no quasiparticle excitations at zero temperature, light can be absorbed by electrons only when the light frequency exceeds the binding energy of the Cooper pair.

To define

$$E = \frac{\omega + x\left(\omega - 2\Delta\right)}{2},$$

$\text{Re}\sigma(w)$ is rewritten as

$$\text{Re}\sigma(w) = \frac{2\pi e^2 v_F^2 N_F^2 (w - 2\Delta)}{\hbar^2 dw} \theta(w - 2\Delta) F(\alpha), \tag{10.24}$$

where

$$F(\alpha) = \int_0^1 dx \frac{1 - \alpha x^2}{\sqrt{1 - x^2}\sqrt{1 - \alpha^2 x^2}}, \qquad \alpha = \frac{w - 2\Delta}{w + 2\Delta}.$$

In the normal state, $\Delta = 0$ and $\alpha = 1$, hence $F(1) = 1$. Thus in the dirty limit, the normal state optical conductivity $\text{Re}\sigma(w)$ is w-independent. The ratio between the optical conductivity in the superconducting state, $\text{Re}\sigma^s$, and that in the normal state, $\text{Re}\sigma^n$, is given by

$$\frac{\text{Re}\sigma^s(w)}{\text{Re}\sigma^n(w)} = \frac{w - 2\Delta}{w} F(\alpha) \theta(w - 2\Delta). \tag{10.25}$$

This is a frequently used formula in the analysis of optical conductivity of isotropic s-wave superconductors. It is valid in the dirty limit. This formula was first derived by Mattis and Bardeen [189]. It shows that $\text{Re}\sigma^s(w)/\text{Re}\sigma^n(w)$ is a universal function of w/Δ, independent on the details of scattering processes.

In a d-wave superconductor, the gap function Δ_k changes sign when the momentum k is rotated by $\pi/2$. In Eq. (10.21), the momentum summation of the $\Delta_k \Delta_q$ term also becomes zero. The optical conductivity now becomes

$$\text{Re}\sigma(w) = \frac{\pi e^2 v_F^2 N_F^2}{4w} \int_{-\infty}^{\infty} d\varepsilon_1 d\varepsilon_2 \int_0^{2\pi} \frac{d\varphi_1}{2\pi} \frac{d\varphi_2}{2\pi} \delta(w - E_1 - E_2), \tag{10.26}$$

where

$$E_i = \sqrt{\varepsilon_i^2 + \Delta^2 \cos^2 2\varphi_i} \qquad (i = 1, 2).$$

Using the expression of the quasiparticle density of states in the d-wave super-conductor,

$$\rho(E) = N_F \int_{-\infty}^{\infty} d\varepsilon \int_0^{2\pi} \frac{d\varphi_1}{2\pi} \delta\left(E - \sqrt{\varepsilon^2 + \Delta^2 \cos^2 2\varphi}\right),$$

$\text{Re}\sigma(w)$ can be further expressed as

$$\text{Re}\sigma(w) = \frac{\pi e^2 v_F^2}{4w} \int_0^{\infty} dE_1 \int_0^{\infty} dE_2 \delta(w - E_1 - E_2) \rho(E_1) \rho(E_2)$$

$$= \frac{\pi e^2 v_F^2}{4w} \int_0^w dE \rho(E) \rho(w - E). \tag{10.27}$$

Hence $\mathrm{Re}\sigma(\omega)$ is determined by the convolution of the density of states of *d*-wave quasiparticles. It equals the sum of all the probabilities for exciting two quasiparticles to the energies at E and $\omega - E$, respectively. Unlike in an *s*-wave superconductor, $\mathrm{Re}\sigma(\omega)$ does not have an absorption edge due to the presence of the nodal points in the *d*-wave gap function.

The integral in Eq. (10.27) cannot be solved analytically. But in the low frequency limit $E < \omega \ll \Delta$, $\rho(E)$ is approximately a linear function of E,

$$\rho(E) \approx a_0 E.$$

In this case, the optical conductivity is approximately given by

$$\mathrm{Re}\sigma_\mu(\omega) \approx \frac{\pi e^2 v_F^2 a_0^2}{4\omega} \int_0^\omega dE \left(\omega E - E^2\right) = \frac{\pi e^2 v_F^2 a_0^2 \omega^2}{24}, \qquad (10.28)$$

which varies quadratically with ω.

In the normal state, $\Delta = 0$, ρ equals approximately the density of states at the Fermi level, independent on ω. $\mathrm{Re}\sigma$ is also ω-independent. The ratio between the optical conductivity in the superconducting state and that in the normal state is given by

$$\frac{\mathrm{Re}\sigma^s(\omega)}{\mathrm{Re}\sigma^n(\omega)} = \frac{1}{N_F^2 \omega} \int_0^\omega dE \rho(E) \rho(\omega - E). \qquad (10.29)$$

Fig. 10.1 compares the frequency dependence of the optical conductivity for the *s* and *d*-wave superconductors in the dirty limit. The difference lies mainly in the low frequency regime. There is an absorption edge in the *s*-wave superconductor but not in the *d*-wave one. In the *d*-wave superconductor, $\mathrm{Re}\sigma$ drops almost linearly with decreasing ω as $\Delta < \omega < 2\Delta$, and changes gradually to the ω^2-dependence as $\omega < \Delta$. When ω is slightly larger than 2Δ, the optical conductivity of the *d*-wave superconductor is already approaching the value in the normal state, much larger than the value in the *s*-wave case.

In high-T_c superconductors, the system responses are predominantly determined by high energy quasiparticles when the measured temperature or frequency is comparable to the gap value Δ_0. As the coherence lengths of these quasiparticles are shorter than the mean-free-length, the system lies in the clean

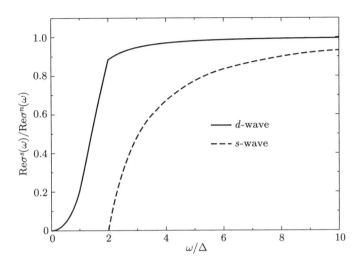

Figure 10.1   The zero temperature infrared conductivity as a function of frequency for the s- and d-wave superconductors in the dirty limit.

limit. We cannot directly use the above result to quantitatively understand or interpret the experimental data of high-T_c cuprates. On the other hand, if the measurement frequency is much lower than Δ_0, the absorption is determined by the low-lying excitations around the nodal points. In this case, the effective coherence length of nodal quasiparticles is longer than the mean free path, and the above result is applicable.

The imaginary part of the optical conductivity can be obtained through the Kramers-Kronig relation:

$$\mathrm{Im}\sigma(\omega) = \frac{2\omega}{\pi} \int_0^\infty d\omega' \frac{\mathrm{Re}\sigma(\omega')}{\omega'^2 - \omega^2}. \tag{10.30}$$

Fig. 10.2 shows the frequency dependence of the imaginary optical conductivity of the d-wave superconductor in the dirty limit. $\mathrm{Im}\sigma(\omega)$ varies linearly with ω when $\omega < \Delta$ and exhibits a peak between $\omega = \Delta$ and 2Δ. The curvature of $\mathrm{Im}\sigma(\omega)$ changes at $\omega = 2\Delta$.

From the real and imaginary parts of the optical conductivity obtained above, we can calculate the dielectric function and the reflection index using

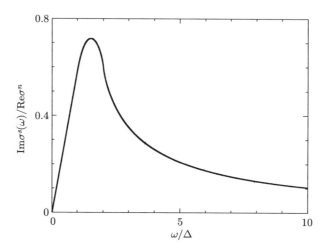

Figure 10.2   The imaginary part of the optical conductivity as a function of frequency for the *d*-wave superconductor in the dirty limit at zero temperature.

the formula

$$\epsilon = \epsilon_\infty + \frac{4\pi i}{\omega}(\text{Re}\sigma + i\text{Im}\sigma) = \epsilon_1 + i\epsilon_2, \tag{10.31}$$

$$n = \frac{1}{\sqrt{2}}\sqrt{\sqrt{\epsilon_1^2 + \epsilon_2^2} + \epsilon_1}, \tag{10.32}$$

$$k = \frac{1}{\sqrt{2}}\sqrt{\sqrt{\epsilon_1^2 + \epsilon_2^2} - \epsilon_1}. \tag{10.33}$$

The reflectivity $R(\omega)$ of the *d*-wave superconductor in the dirty limit is then given by

$$R(\omega) = \left|\frac{n + ik - 1}{n + ik + 1}\right|^2. \tag{10.34}$$

$R(\omega)$ can be measured directly from the infrared spectroscopy experiments. Fig. 10.3 shows $R(\omega)$ as a function of frequency for the *d*-wave superconductor. Below the gap energy, $\omega < \Delta$, $R(\omega)$ drops linearly with ω, resulting from the linear density of states of the *d*-wave superconductor. The curvature of R is negative when $\Delta < \omega < 2\Delta$, and becomes positive when $\omega > 2\Delta$.

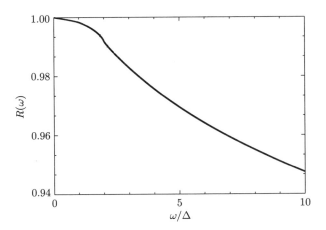

Figure 10.3   Typical frequency dependence of the reflectivity of the d-wave superconductor in the dirty limit at zero temperature.

10.4 Effect of Elastic Impurity Scattering

The quasiparticle conductivity is determined by the imaginary part of the current-current correlation function. It can be expressed using the single-particle spectral function as

$$\mathrm{Re}\sigma_\mu(\omega) = \frac{e^2}{\pi\hbar^2 V}\sum_k \left(\frac{\partial\varepsilon_k}{\partial k_\mu}\right)^2 \int d\omega_1 \frac{f(\omega_1)-f(\omega_1+\omega)}{\omega}$$
$$\mathrm{TrIm}G^R(k,\omega_1)\mathrm{Im}G^R(k,\omega_1+\omega). \tag{10.35}$$

At low temperatures or low frequencies, the elastic impurity scattering is the major scattering channel, and the contribution from inelastic scattering is relatively small and can be neglected [196, 197].

At zero temperature, the conductivity is proportional to the density of states, and inversely proportional to the electron scattering rate. In normal metals, the electron density of states is energy independent and the conductivity is mainly determined by the scattering rate. However, in a d-wave superconductor, the quasiparticle density of states on the Fermi surface is proportional to the scattering rate, and these two effects cancel each other. As a consequence, the quasiparticle conductance at zero temperature is a universal constant, independent on the impurity scattering potential. Lee *et. al.* first found this universal behavior of the *d*-wave superconductor. He also derived the formula of the universal conductance [198]. However, as will be discussed below, this universal property is only valid in the weak scattering limit. In the strong scattering limit, the conductance is no longer universal.

In the limit of zero frequency at zero temperature, Eq. (10.35) can be simplified as

$$\mathrm{Re}\sigma_\mu(0) = \frac{e^2}{\pi\hbar^2 V} \sum_k \left(\frac{\partial\varepsilon_k}{\partial k_\mu}\right)^2 \mathrm{TrIm}G^R(k,0)\mathrm{Im}G^R(k,0). \qquad (10.36)$$

The imaginary part of the zero frequency Green's function is given by

$$\mathrm{Im}G^R(k,0) = -\frac{\Gamma_0}{\Gamma_0^2 + \varepsilon_k^2 + \Delta_k^2}.$$

Substituting it into Eq. (10.36) gives the expression of the quasi-two-dimensional conductivity:

$$
\begin{aligned}
\mathrm{Re}\sigma_{ab}(0) &= \frac{e^2 v_F^2}{\pi V} \sum_k \frac{\Gamma_0^2}{(\Gamma_0^2 + \varepsilon_k^2 + \Delta_k^2)^2} \\
&= \frac{4e^2 v_F^2 N_F}{\pi^2} \int_0^{\pi/4} d\varphi \int_{-\infty}^{\infty} d\varepsilon \frac{\Gamma_0^2}{(\Gamma_0^2 + \varepsilon^2 + \Delta_0^2 \cos^2 2\varphi)^2} \\
&= \frac{e^2 v_F^2 N_F}{\pi\Delta_0} \int_0^{\Delta_0/\Gamma_0} \frac{dx}{\sqrt{1 - \left(\frac{\Gamma_0 x}{\Delta_0}\right)^2}} \frac{1}{(1+x^2)^{3/2}}.
\end{aligned}
$$

For most of superconductors, Δ_0 is much larger than Γ_0. The above integral can be expanded in terms of Γ_0/Δ_0. Up to the order of $(\Gamma_0^2/\Delta_0^2)\ln(2\Delta_0/\Gamma_0)$, the result is given by

$$\mathrm{Re}\sigma_{ab}(0) = \sigma_0 \left[1 + \frac{\Gamma_0^2}{\Delta_0^2}\ln\frac{2\Delta_0}{\Gamma_0} + o\left(\frac{\Gamma_0^2}{\Delta_0^2}\right)\right], \qquad (10.37)$$

where

$$\sigma_0 = \frac{e^2 v_F^2 N_F}{\pi\Delta_0} \qquad (10.38)$$

is the universal conductance. It depends only on the intrinsic parameters of *d*-wave superconductors, but not on the impurity potential.

The above formula shows that the zero temperature conductivity $\mathrm{Re}\sigma_{ab}$ equals a universal value, in the limit $\Gamma_0 \ll \Delta_0$. The condition $\Gamma_0 \ll \Delta_0$ is valid in most of high-T_c superconductors. The second term in Eq. (10.37) is not neglectable when the sample is not very clean and the gap is not very big. In this case, $\mathrm{Re}\sigma_{ab}$ depends on the scattering rate and is no longer universal.

In high-T_c superconductors, the electron velocity along the c-axis is given by $v_c \approx v_\perp \cos^2(2\varphi)$. At zero temperature, to the leading order of Γ_0/Δ_0, the c-axis conductivity in the limit $\Gamma_0 \ll \Delta_0$ is determined by the formula

$$\mathrm{Re}\sigma_c(0) = \frac{e^2 v_\perp^2}{\pi V} \sum_k \frac{\Gamma_0^2 \cos^4 2\varphi}{(\Gamma_0^2 + \varepsilon_k^2 + \Delta_k^2)^2} \approx \frac{e^2 v_\perp^2 N_F \Gamma_0^2}{\pi \Delta_0^3}. \tag{10.39}$$

$\mathrm{Re}\sigma_c$ depends on the scattering rate. The reason $\mathrm{Re}\sigma_c$ is not universal because the c-axis velocity of electrons v_c depends strongly on the angle of the in-plane momentum. It changes the weight of conductivity contributed by the quasiparticles on different parts of the Fermi surface. In particular, the contribution of quasiparticles around the nodal points is suppressed. The cancellation between the density of states and the scattering rate does not occur again.

At low temperatures, the in-plane conductivity of quasiparticles is determined by the equation

$$\mathrm{Re}\sigma_{ab}(T) = -\frac{e^2 v_F^2}{2\pi V} \sum_k \int d\omega \frac{\partial f}{\partial \omega} \mathrm{TrIm}G^R(k, \omega) \mathrm{Im}G^R(k, \omega). \tag{10.40}$$

In the gapless regime, $T \ll \Gamma_0$, $\partial f/\partial \omega$ decays exponentially with ω and the integral in Eq. (10.40) can be calculated using the Sommerfeld expansion. The first two leading terms are given by

$$\mathrm{Re}\sigma_{ab}(T) \approx \frac{e^2 v_F^2}{2\pi V} \sum_k \left[g_k(0) + \frac{\pi^2 T^2}{6} g_k''(0) \right], \tag{10.41}$$

with

$$g_k(\omega) = \mathrm{TrIm}G^R(k, \omega) \mathrm{Im}G^R(k, \omega). \tag{10.42}$$

The first term in Eq. (10.41) gives the zero temperature conductivity, and the second term is the correction resulting from the finite temperature thermal fluctuation.

At low frequencies, the second order derivative of $g_k(\omega)$ with respect to ω can be evaluated using the expression of the retarded Green's function given in Eq. (8.41). In the limit $\Gamma_0 \ll \Delta_0$, the momentum summation of $g_k''(0)$ is approximately given by

$$\frac{1}{V} \sum_k g_k''(0) \approx \frac{4N_F a^2}{\Delta_0 \Gamma_0^2}. \tag{10.43}$$

The corresponding in-plane conductivity is

$$\mathrm{Re}\sigma_{ab} \approx \sigma_0 \left(1 + \frac{\pi^2 a^2 T^2}{3\Gamma_0^2}\right). \tag{10.44}$$

$\mathrm{Re}\sigma_{ab}$ at low temperatures scales as T^2. This is a consequence of the Sommerfeld expansion at $T \ll \Gamma_0$, independent on the value of Γ_0. This T^2-dependence of the conductivity is also a universal property of d-wave superconductors.

In the intrinsic regime, $\Gamma(\omega) \ll \omega \ll \Delta_0$, using Eq. (8.44) and the identity

$$\mathrm{Im}G^R(k,\omega)\mathrm{Im}G^R(k,\omega) = \frac{1}{2}\mathrm{Re}G^R(k,\omega)\left[G^{R*}(k,\omega) - G^R(\mathrm{k},\omega)\right], \tag{10.45}$$

it can be shown that

$$\frac{1}{V} \sum_k \mathrm{TrIm}G^R(k,\omega)\mathrm{Im}G^R(k,\omega) \approx \frac{\pi \omega N_F}{\Delta_0 \Gamma(\omega)}. \tag{10.46}$$

Thus the in-plane conductivity is proportional to the average of $|\omega|/\Gamma(\omega)$ on the Fermi surface,

$$\mathrm{Re}\sigma_{ab}(T) = -\frac{e^2 v_F^2 N_F}{2\Delta_0} \int d\omega \frac{\omega}{\Gamma(\omega)}\frac{\partial f}{\partial \omega}, \tag{10.47}$$

consistent with the result obtained by the Boltzmann transport theory of quasiparticles.

In the Born scattering limit, substituting the value of Γ in Eq.(8.45) to Eq. (10.47), we find the in-plane conductivity to be

$$\mathrm{Re}\sigma_{ab}(T) \approx \frac{e^2 v_F^2 N_F}{\Gamma_N}. \tag{10.48}$$

It is temperature independent and equals the conductivity in the normal state. This is a special property of d-wave superconductors in the Born scattering limit. In real materials, the intrinsic region is generally very narrow and this temperature independent conductivity is valid only in a very narrow temperature range, which is difficult to be detected.

In the unitary limit, $\Gamma(\omega)$ is given by Eq. (8.46). Substituting it into Eq. (10.47) and keeping the leading order term, we have

$$\mathrm{Re}\sigma_{ab}(T) \approx -\frac{4e^2 v_F^2 N_F}{\pi^2 \Delta_0^2 \Gamma_N} \int d\omega \omega^2 \ln^2 \frac{2\Delta_0}{|\omega|} \frac{\partial f}{\partial \omega}$$

$$\approx \frac{4e^2 v_F^2 N_F k_B^2 T^2}{3\Delta_0^2 \Gamma_N} \ln^2 \frac{2\Delta_0}{k_B T}. \tag{10.49}$$

$\mathrm{Re}\sigma_{ab}$ is now a function of $T^2 \ln T$, different than in the Born limit. If we neglect the weak logarithmic correction, it approximately scales as T^2, close to the temperature dependence in the gapless regime. Nevertheless, the coefficients of the T^2 terms are different in these two regions. The T^2 behavior of the conductivity in the intrinsic region is not a continuation of that in the gapless region.

At zero temperature and low frequency, the optical conductivity is given by

$$\mathrm{Re}\sigma_\mu(\omega) = \frac{e^2 v_F^2}{4\pi\omega V} \sum_k A(k,\omega),$$

where

$$A(k,\omega) = \int_{-\omega}^{0} d\omega_1 \mathrm{Re}\,\mathrm{Tr}\left[G^{R*}(k,\omega_1) - G^R(k,\omega_1) \right] G^R(k,\omega_1+\omega). \tag{10.50}$$

In the limit of low frequency, the quasiparticle Green's function is approximately given by Eq. (8.41). Substituting it into Eq. (10.50) and integrating over ω_1, we obtain

$$A(k,\omega) = \mathrm{Re}\frac{2}{2ia\Gamma_0 + a^2\omega} \ln \frac{E_k^2 + \Gamma_0^2}{E_k^2 + (\Gamma_0 - ia\omega)^2}$$

$$-\frac{1}{a^2\omega} \ln \frac{\left[E_k^2 + (\Gamma_0 + ia\omega)^2 \right]\left[E_k^2 + (\Gamma_0 - ia\omega)^2 \right]}{(E_k^2 + \Gamma_0^2)^2}.$$

This expression of $A(k, \omega)$ is rather complicated and difficult to handle. Nevertheless, at low frequencies, $A(k, \omega)$ can be expanded in terms of ω. The leading two terms are

$$A(k, \omega) = \frac{4\Gamma_0^2 \omega}{(E_k^2 + \Gamma_0^2)^2} + \frac{2E_k^2 - \Gamma_0^2}{(E_k^2 + \Gamma_0^2)^4} \frac{8a^2\Gamma_0^2\omega^3}{3} + o(\omega^5).$$

The first term leads to the universal conductance at zero temperature, and the second term gives the finite frequency correction to the conductivity[196]. In the limit $\Gamma_0 \ll \Delta_0$, the optical conductivity can be expressed as

$$\mathrm{Re}\sigma_{ab}(\omega) \approx \sigma_0 \left(1 - \frac{a^2\Gamma_0^2\omega^2}{2\Delta_0^4} + o(\omega^4) \right) \qquad (\omega \ll \Gamma_0 \ll \Delta_0). \tag{10.51}$$

The finite frequency correction to the conductivity is negative. Comparing with Eq. (10.44), we find that there are both similarity and difference in the temperature and frequency dependencies of $\mathrm{Re}\sigma_{ab}$. The similarity is that $\mathrm{Re}\sigma_{ab}(T)$ depends on T^2 while $\mathrm{Re}\sigma(\omega)$ depends on ω^2. The difference is that $\mathrm{Re}\sigma_{ab}(T)$ increases with increasing temperature, while $\mathrm{Re}\sigma(\omega)$ decreases with increasing frequency.

10.5 Microwave Conductivity of High-T_c Superconductors

The low frequency conductivity of high-T_c superconductors can be determined through the measurement of surface resistance R_s. For the YBCO and BSCCO single crystals, it was found experimentally that the surface resistance drops nearly four orders of magnitude below T_c within a temperature window narrower than $T_c/5$ [138, 199]. This quick decay of R_s implies that the quasiparticle lifetime increases very rapidly below T_c, different than in a metal or alloy-based superconductor. After this quick decrease, R_s begins to grow gradually and drops again linearly with temperature after reaching a maximum at about $T_c/3$ [199].

In the superconducting state, R_s is proportional linearly to the microwave conductivity and cubically to the magnetic penetration depth, i.e. $R_s \propto \sigma\lambda^3$. Thus from the measurement of R_s and λ, we can find the temperature dependence of σ.

Fig. 10.4 shows the typical temperature dependence of the microwave conductivity along the direction parallel to the CuO_2 plane as well as that along the c-axis. In the CuO_2 plane, $Re\sigma_{ab}$ exhibits a non-monotonic temperature dependence in accordance with the non-monotonic behavior of the temperature dependence of R_s. Below T_c, $Re\sigma_{ab}$ first shows a fast increase with decreasing temperature, and then drops linearly with temperature after reaching a maximum at about $T_c/3$.

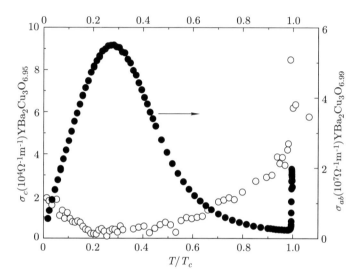

Figure 10.4   Temperature dependence of the microwave conductivities of the $YBa_2Cu_3O_x$ superconductor along both the ab-plane and the c-axis. The measured frequency is 1.14 GHz for σ_{ab} and 22 GHz for σ_c. This figure is plotted based on the data published in Ref. [199] and on the homepage of the UBC experimental group (http://www.physics.ubc.ca/supercon/supercon.html).

The non-monotonic temperature dependence of $Re\sigma_{ab}$ results from the competition between the increase of quasiparticle lifetime and the decrease of quasiparticle density of states in the superconducting state. The results shown in Fig. 10.4 can be understood using the following generalized Drude formula

$$Re\sigma_{ab} = \frac{e^2 n_e \tau}{m^*}, \tag{10.52}$$

where n_e is the density of normal electrons (or quasiparticles) and τ is the quasiparticle lifetime. In the superconducting state, n_e decreases with decreasing temperature. Meanwhile, the quasiparticle scattering is weakened, hence the scattering lifetime is increased. Near T_c, the superfluid density is very small and the change of n_e is small, but τ increases quickly. This leads to the fast increase of $\mathrm{Re}\sigma_{ab}$ just below T_c. In low temperatures, the quasiparticle lifetime does not change too much. It increases slowly with lowering temperature, but the quasiparticle density drops quickly. This leads to the drop of $\mathrm{Re}\sigma_{ab}$ in the low temperature regime.

Eq. (10.52) is a simple formula that can be used to qualitatively understand the temperature dependence of $\mathrm{Re}\sigma_{ab}$. But it is difficult to use it to describe quantitatively the behavior of $\mathrm{Re}\sigma_{ab}$. In particular, it is difficult to explain the linear temperature dependence of $\mathrm{Re}\sigma$ in low temperatures.

In fully gapped superconductors, $\mathrm{Re}\sigma$ always decays exponentially in low temperatures. Apparently, the linear conductivity cannot be explained by the *s*-wave pairing. In a *d*-wave superconductor, $\mathrm{Re}\sigma$ contributed by the elastic impurity scattering scales as T^2. Thus it is also difficult to use the *d*-wave pairing to explain this linear temperature dependence. Whether this difficulty can be resolved by further considering inelastic scatterings or electron-electron interaction needs further investigation.

The *c*-axis conductivity of high-T_c superconductors shows completely different temperature dependence than the in-plane conductivity. Just below T_c, as shown in Fig. 10.4, the *c*-axis conductivity drops with lowering temperature just below T_c, different from the in-plane conductivity. This is a common feature of high-T_c superconductors. By further lowering temperature, the *c*-axis conductivity of the $\mathrm{YBa_2Cu_3O_x}$ superconductor begins to increases (see Fig. 10.4). This non-monotonic behavior of the *c*-axis conductivity is only observed in the YBCO superconductor, not in other high-T_c superconductors. It is unknown why the *c*-axis conductivity of YBCO varies non-monotonically with temperature. The increase of the *c*-axis conductivity in low temperatures may result from the scattering of magnetic impurities in YBCO.

In high-T_c superconductors, the conductivities along the *c*-axis and the *ab*-

plane exhibit completely different temperature dependence. This difference is difficult to understand within the conventional framework of transport theory. However, considering the anisotropy of the hopping matrix elements along the c-axis,

$$t_c = -t_\perp \cos^2 2\theta, \tag{10.53}$$

it is not difficult to understand qualitatively this difference. Under appropriate approximations, one can even make quantitative predictions for the temperature dependence of conductivity [200].

The conductivity is determined by the imaginary part of the current-current correlation function. If we only consider the contribution from the coherent quasiparticles by neglecting the vertex correction, the c-axis conductivity is determined by the formula

$$\mathrm{Re}\sigma_c = -\frac{\alpha_c}{\pi}\int_{-\infty}^{\infty}d\omega\frac{\partial f(\omega)}{\partial\omega}\int_{0}^{2\pi}\frac{d\theta}{2\pi}\cos^4(2\theta)M(\theta), \tag{10.54}$$

where

$$M(\theta) = \frac{\pi}{\Gamma_\theta}\mathrm{Re}\frac{(\omega+i\Gamma_\theta)^3 - \omega\Delta_0^2\cos^2 2\theta}{\left[(\omega+i\Gamma_\theta)^2 - \Delta_0^2\cos^2 2\theta\right]^{3/2}},$$

and $\alpha_c = e^2 t_\perp^2 N_F/4$. In deriving the above formula, the single-particle Green's function is assumed to be

$$G_{ret}(k,\omega) = \frac{1}{\omega - \xi_k\tau_3 - \Delta_\theta\tau_1 + i\Gamma_\theta}, \tag{10.55}$$

where $\xi_k = \varepsilon_{ab}(k) - t_\perp \cos k_z \cos^2(\theta)$ is the energy dispersion of electrons. Γ_θ is the quasiparticle scattering rate which is a function of the azimuthal angle θ on the Fermi surface.

In the superconducting state, the integral in Eq. (10.54) cannot be evaluated analytically. Nevertheless, when the temperature is less than T_c but much larger than the scattering rate Γ_θ, $T_c > T \gg \Gamma_\theta$, we can expand the integral on the right-hand side of Eq. (10.54) in terms of Γ_θ. Up to the leading order of Γ_θ, $\mathrm{Re}\sigma_c$ can be expressed as

$$\mathrm{Re}\sigma_c \approx -\alpha_c\int_{-\infty}^{\infty}d\omega\frac{\partial f(\omega)}{\partial\omega}\int_{0}^{2\pi}\frac{d\theta}{2\pi}\frac{\cos^4(2\theta)}{\Gamma_\theta}\mathrm{Re}\frac{|\omega|}{\sqrt{\omega^2 - \Delta_\theta^2}}. \tag{10.56}$$

In the superconducting state, the average scattering rate of quasiparticles on the Fermi surface $\tau_0^{-1} = \langle \Gamma_\theta \rangle_{FS}$ can be estimated from the measurement data of the microwave conductivity and the superfluid density in the CuO_2 plane, using the generalized Drude formula Eq. (10.52). For the optimally doped YBCO[199], τ_0^{-1} is less than 1K at low temperatures. Γ_0 increases with increasing temperature. At $T = 60K$, τ_0^{-1} is approximately equal to 6K. This estimation shows that the condition $T_c > T \gg \Gamma_\theta$ is satisfied and the leading order approximation in Γ_θ is valid at least when the temperature is not too low.

In high-T_c superconductors, Γ_θ was found to be very anisotropic [201, 202]. The scattering rate of electrons along the gap nodal directions is much smaller than that along the anti-nodal directions. Ioffe and Mills[203] proposed an phenomenological "cold spot" model to describe this anisotropy. In this model, Γ_θ is assumed to have the form

$$\Gamma_\theta = \Gamma_0 \cos^2 2\theta + \tau^{-1}(T), \tag{10.57}$$

where Γ_0 is a temperature independent parameter, proportional to the anisotropy in the scattering rate. τ^{-1} is angular independent but temperature dependent. The cold spot form of Γ_θ is proposed based on the analysis of experimental data. How to derive this formula theoretically remains an open question.

Substituting Eq. (10.57) into Eq. (10.56) and performing the integral over θ, we find that σ_c is approximately given by

$$\mathrm{Re}\sigma_c \approx \frac{9\alpha_c \zeta[3] T^3}{2\Gamma_0 \Delta_0^3} - \frac{(2\ln 2) T \alpha_c}{\tau \Gamma_0^2 \Delta_0} + \frac{\alpha_c \sigma_a}{\alpha_a \tau^2 \Gamma_0^2}, \tag{10.58}$$

where

$$\sigma_a \approx -\frac{T\tau \alpha_a}{\Delta_0} \int_{-\infty}^{\infty} dx \frac{\partial f(xT)}{\partial x} \frac{|x|}{\sqrt{1 + T^2 \Gamma_0 \tau x^2 / \Delta_0^2}},$$

$\alpha_a = e^2 v_F^2 N_F / 4$ and $\zeta(3) = 1.202$.

In the high temperature limit, when the condition $\Gamma_0 \tau T^2 / \Delta_0^2 \gg 1$ is satisfied, $\mathrm{Re}\sigma_c$ is mainly determined by the first term on the right-hand side of Eq.

(10.58). In this case, $\mathrm{Re}\sigma_c$ varies approximately with the cube of temperature,

$$\mathrm{Re}\sigma_c \approx \frac{9\alpha_c\zeta(3)T^3}{2\Gamma_0\Delta_0^3}. \tag{10.59}$$

This is a remarkable result. It shows that different from $\mathrm{Re}\sigma_{ab}$, $\mathrm{Re}\sigma_c$ drops with decreasing temperature just below T_c, consistent with experimental measurements. Furthermore, the $\mathrm{Re}\sigma_c$ does not depend on the scattering rate τ^{-1}. Thus the cubic temperature dependence of $\mathrm{Re}\sigma_c$ is universal.

The universal cubic temperature dependence of $\mathrm{Re}\sigma_c$ results from the interplay between the anisotropic c-axis hopping integral and the anisotropic scattering rate of electrons. For the YBCO superconductor, as $\tau_0^{-1} > \tau^{-1}$ at $T/\Delta_0 \gg 1/\sqrt{\Gamma_0\tau_0}$, the condition that $\Gamma_0\tau T^2/\Delta_0^2 \gg 1$ is satisfied. Based on the experimental data of the normal state resistivity, Ioffe and Mills estimated Γ_0 to be around $0.6eV$ [203]. If we take $\tau_0^{-1} \sim 6K$ for YBCO at $T = 60K$, then $1/\sqrt{\Gamma_0\tau_0}$ is estimated to be around 0.03. Hence for the optimally doped YBCO superconductor, $\Gamma_0\tau T^2/\Delta_0^2 \gg 1$ is valid at least when $T/\Delta_0 \gg 0.03$, i.e. $T/T_c \gg 0.06$.

Fig. 10.5 replots the experimental data of $\mathrm{Re}\sigma_c$ shown in Fig. 10.4 as a

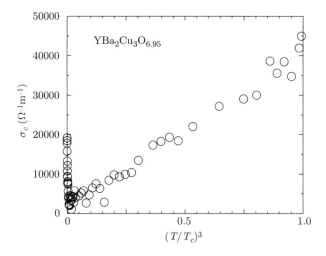

Figure 10.5   The cubic temperature dependence of the c-axis conductivity data previously shown in Fig. 10.4 for YBa$_2$Cu$_3$O$_{6.95}$.

function of $(T/T_c)^3$ for $YBa_2Cu_3O_{6.95}$ [200]. Within the measurement errors, $\text{Re}\sigma_c$ agrees with the T^3-law theoretically predicted from 30K to T_c. This universal T^3-dependence is also consistent with the low frequency measurement data of conductivity for $Bi_2Sr_2CaCu_2O_{8+x}$ and other superconductors [204]. The agreement between the theoretical analysis and the experimental result shows again that it is crucial to include the anisotropy of the c-axis hopping matrix elements in the analysis of the c-axis transport properties. It also lends support to the "cold spot" model.

At low temperatures, the impurity scattering becomes dominant, which ruins the universal behavior of the c-axis conductivity. The low-temperature conductivity along the c-axis is no longer universal. Different materials exhibit different temperature dependencies.

10.6 The Heat Current Density Operator

In the superconducting state, the system is dissipationless due to existence of the superfluid, and both the direct-current resistivity and the zero-frequency conductivity vanish. However, the thermal conductivity does not vanish. The thermal conductivity measures the response of the system to a temperature gradient. It reveals the property of quasiparticle excitations and can be used to extract information on the superconducting state.

In a d-wave superconductor, the thermal conductivity possesses certain peculiar properties. For example, in the zero temperature limit, the thermal conductivity is a universal quantity, which depends only on the Fermi velocity and the derivatives of the gap function at the nodal points, but not on the impurity scattering and the strength of the Coulomb interaction. In the above discussion, we have shown that the low frequency electric conductivity of d-wave superconductors is universal at zero temperature. However, the universality of the electric conductivity is not as robust as the thermal conductivity. Vertex corrections from interactions and impurity scatterings can destroy the universality of the electric conductivity, but they have much weaker effects on the universality of thermal conductivity. The universal behavior of thermal conductivity can be used not only to determine the pairing symmetry, but also to determine the fundamental parameters of d-wave superconductors, for example, the ratio between the Fermi velocity and the gap slope at the nodal points.

Similar to the electric conductivity, the contribution of electrons to the thermal conductivity is determined by the correlation function of heat current. The heat current represents the energy transfer in the presence of a temperature gradient. Different from the electric current density, the energy flow density depends on not only the kinetic energy, but also the interaction potential. Hence for different physical systems, the definition of the heat current density is different. Nevertheless, no matter what the system is, if the energy is measured with respect to the Fermi energy, the energy density and the corresponding heat current density must satisfy the following continuation equation:

$$\dot{h}(r) + \nabla \cdot j_E(r) = 0, \qquad (10.60)$$

where $h(r)$ is the energy density, and j_E is the heat current density. This is the energy flow continuity equation in analogy to the electric charge continuity equation. The expression of the heat current operator can be obtained based on this equation.

In momentum space, the energy density and the heat current density operators are defined by

$$h(q) = \int dr e^{-iq \cdot r} h(r),$$

$$j_E(q) = \int dr e^{-iq \cdot r} j_E(r).$$

Substituting these definitions into Eq. (10.60), we obtain the expression of the heat current continuity equation in the momentum space

$$\dot{h}(q) + iq \cdot j_E(q) = 0. \qquad (10.61)$$

Below we drive the expression of the heat current density operator for the d-wave superconductor under the mean-field approximation. There are two approaches that can be used. The first is to use the BCS mean-field Hamiltonian for the d-wave superconductor, and derive the expression of j_E based on the equations of motion and the continuity equation of electron operators. The second is to start from the BCS Hamiltonian and derive the heat current operator using the equation of motion of electrons before taking the mean-field approximation. The mean-field approximation is then taken to decouple the heat current operator into a quadratic form. These two methods yield the same result. In literature, the second approach is commonly used [205, 206]. But the first one is simpler to implement. Below we derive the expression of the heat current density j_E just using this approach.

Let us consider the following BCS mean-field Hamiltonian,

$$H = \sum_{k,\sigma} \xi_k c_{k\sigma}^\dagger c_{k\sigma} + \sum_k \Delta_k \left(c_{k\uparrow}^\dagger c_{-k\downarrow}^\dagger + c_{-k\downarrow} c_{k\uparrow} \right). \tag{10.62}$$

For simplicity, we assume that the gap function is invariant under the spatial inversion, i.e. $\Delta_{-k} = \Delta_k$. From the Fourier transformations of fermion operators

$$c_{r\sigma} = \frac{1}{\sqrt{V}} \sum_k e^{ik \cdot r} c_{k\sigma},$$

$$c_{k\sigma} = \frac{1}{\sqrt{V}} \int dr e^{-ik \cdot r} c_{r\sigma},$$

it can be shown that the energy density corresponding to H is defined by

$$h(r) = \frac{1}{V} \sum_{k,k'} e^{-i(k-k') r} \left(\sum_\sigma \frac{\xi_k + \xi_{k'}}{2} c_{k\sigma}^\dagger c_{k'\sigma} + \frac{\Delta_k + \Delta_{k'}}{2} c_{-k'\uparrow}^\dagger c_{k\downarrow}^\dagger \right.$$
$$\left. + \frac{\Delta_k + \Delta_{k'}}{2} c_{-k\downarrow} c_{k'\uparrow} \right).$$

$h(r) = h^\dagger(r)$ is Hermitian. In momentum space, the corresponding energy density is given by

$$h(q) = \sum_k \left(\sum_\sigma \frac{\xi_k + \xi_{k+q}}{2} c_{k\sigma}^\dagger c_{k+q\sigma} + \frac{\Delta_{-k} + \Delta_{-k-q}}{2} c_{-k-q\uparrow}^\dagger c_{k\downarrow}^\dagger \right.$$
$$\left. + \frac{\Delta_k + \Delta_{k+q}}{2} c_{-k\downarrow} c_{k+q\uparrow} \right). \tag{10.63}$$

Taking the time derivative of $h(q)$ and using the equations of motion of electrons

$$i\dot{c}_{k\sigma} = [c_{k\sigma}, H] = \xi_k c_{k\sigma} + \delta_{\sigma\uparrow}\Delta_k c^\dagger_{-k\downarrow} - \delta_{\sigma\downarrow}\Delta_{-k} c^\dagger_{-k\uparrow}, \quad (10.64)$$

$$i\dot{c}^\dagger_{k\sigma} = \left[c^\dagger_{k\sigma}, H\right] = -\xi_k c^\dagger_{k\sigma} - \delta_{\sigma\uparrow}\Delta_k c_{-k\downarrow} + \delta_{\sigma\downarrow}\Delta_{-k} c_{-k\uparrow}, \quad (10.65)$$

we find that

$$\dot{h}(q) = \sum_{k,\sigma} \frac{\xi_{k+q} - \xi_k}{2} \left(c^\dagger_{k\sigma}\dot{c}_{k+q\sigma} - \dot{c}^\dagger_{k\sigma}c_{k+q\sigma}\right)$$

$$+ \sum_{k} \frac{\Delta_{k+q} - \Delta_k}{2} \left(\dot{c}^\dagger_{-k-q\uparrow}c^\dagger_{k\downarrow} - c^\dagger_{-k-q\uparrow}\dot{c}^\dagger_{k\downarrow} + c_{-k\downarrow}\dot{c}_{k+q\uparrow} - \dot{c}_{-k\downarrow}c_{k+q\uparrow}\right).$$

$$(10.66)$$

In the limit of $q \to 0$, we have

$$\xi_{k+q} - \xi_k = q \cdot \partial_k \xi_k,$$

$$\Delta_{k+q} - \Delta_k = q \cdot \partial_k \Delta_k.$$

Combining this with the continuity equation, we find the heat current density operator in the long-wave length limit to be

$$j_E(q) = \frac{i}{2} \sum_{k,\sigma} \partial_k \xi_k \left(c^\dagger_{k\sigma}\dot{c}_{k+q\sigma} - \dot{c}^\dagger_{k\sigma}c_{k+q\sigma}\right)$$

$$+ \frac{i}{2} \sum_{k} \partial_k \Delta_k \left(\dot{c}^\dagger_{-k-q\uparrow}c^\dagger_{k\downarrow} - c^\dagger_{-k-q\uparrow}\dot{c}^\dagger_{k\downarrow} + c_{-k\downarrow}\dot{c}_{k+q\uparrow} - \dot{c}_{-k\downarrow}c_{k+q\uparrow}\right).$$

$$(10.67)$$

Taking the Fourier transformation with respect to time,

$$c_{k\sigma}(t) = \frac{1}{\sqrt{2\pi}} \int d\omega e^{-i\omega t} c_{k\sigma}(\omega),$$

$$c^\dagger_{k\sigma}(t) = \frac{1}{\sqrt{2\pi}} \int d\omega e^{i\omega t} c^\dagger_{k\sigma}(\omega),$$

we then obtain the expression of the heat current density operator in the frequency space as

$$j_E (q, \Omega) = \int dt e^{i\Omega t} j_E (q, t)$$

$$= \sum_k \int d\omega \left(\omega + \frac{\Omega}{2} \right) \left\{ (\partial_k \xi_k) \sum_\sigma c_{k\sigma}^\dagger (\omega) c_{k+q\sigma} (\omega + \Omega) \right.$$

$$\left. - (\partial_k \Delta_k) \left[c_{-k-q\uparrow}^\dagger (\omega) c_{k\downarrow}^\dagger (-\omega - \Omega) + c_{-k\downarrow} (\omega + \Omega) c_{k+q\uparrow} (-\omega) \right] \right\}.$$

$$(10.68)$$

This result can also be obtained directly from the BCS Hamiltonian without the mean-field approximation. The mean-field approximation is imposed only after the expression of the time-derivative of the heat current density operator is obtained. Such approach is used to derive j_E in Ref. [206]. These two kinds of approaches are equivalent. The difference lies that in the first approach the mean-field approximation is applied to the Hamiltonian, while in the second approach this approximation is applied to the heat current operator. In Ref. [206], the free electron dispersion for the kinetic energy is used. Here we have used a general electron dispersion relation ξ_k, hence Eq. (10.68) holds more generally.

10.7 The Universal Thermal Conductivity

According to the linear response theory, the thermal conductivity in the absence of electric current is determined by the equation [9]

$$\kappa = \frac{1}{k_B \hbar T^2} \text{Re} \Pi_E (\omega \to 0), \qquad (10.69)$$

where Π_E is the thermal polarization function which is determined by the correlation of heat current densities:

$$\Pi_E (i\omega_n) = \frac{i}{i\omega_n \beta} \int_0^\beta d\tau e^{i\omega_n \tau} \langle \text{Tr}\, j_E^\mu (\tau)\, j_E^\mu (0) \rangle, \qquad (10.70)$$

where $j_E = j_E(q = 0)$. In the imaginary frequency representation,

$$c_{k\sigma}(i\omega_m) = \int_0^\beta d\tau e^{i\omega_m\tau} c_{k\sigma}(\tau),$$

$$c_{k\sigma}(\tau) = \frac{1}{\beta}\sum_{\omega_m} e^{-i\omega_m\tau} c_{k\sigma}(i\omega_m),$$

the heat current density operator Eq. (10.67) becomes

$$j_E(\tau) = \frac{1}{2\beta^2}\sum_{k,\omega_m',\omega_m}(\omega_m+\omega_m')e^{i(\omega_m-\omega_m')\tau}\left\{\sum_\sigma(\partial_k\xi_k)c_{k\sigma}^\dagger(i\omega_m)c_{k\sigma}(i\omega_m')\right.$$

$$\left.+(\partial_k\Delta_k)\left[c_{k\uparrow}^\dagger(i\omega_m)c_{-k\downarrow}^\dagger(-i\omega_m')-c_{-k\downarrow}(-i\omega_m')c_{k\uparrow}(i\omega_m)\right]\right\}. \tag{10.71}$$

Substituting it into Eq. (10.70), we obtain

$$\Pi_E(i\omega_n) = -\frac{i}{4i\omega_n\beta^2}\sum_{k\omega_m}(2\omega_m+\omega_n)^2$$

$$\left[\left(\partial_{k_\mu}\xi_k\right)^2\operatorname{Tr}G(k,i\omega_m)\sigma_3 G(k,i\omega_m+i\omega_n)\sigma_3\right.$$

$$\left.+\left(\partial_{k_\mu}\Delta_k\right)^2\operatorname{Tr}G(k,i\omega_m)\sigma_1 G(k,i\omega_m+i\omega_n)\sigma_1\right], \tag{10.72}$$

where the Green's function $G(k,i\omega_n)$ is defined by

$$G(k,i\omega_n) = -\frac{1}{\beta}\left\langle:\begin{pmatrix}c_{k\uparrow}(i\omega_n)\\c_{-k\downarrow}^\dagger(-i\omega_n)\end{pmatrix}\left(c_{k\uparrow}^\dagger(i\omega_n)\quad c_{-k\downarrow}(-i\omega_n)\right):\right\rangle, \tag{10.73}$$

and $::$ represents the normal ordering of operators.

Using the spectral representation of the Green's function

$$G(k,i\omega_m) = -\int_{-\infty}^\infty \frac{d\omega}{\pi}\frac{\operatorname{Im}G^R(k,\omega)}{i\omega_m-\omega},$$

Eq. (10.72) can be represented as

$$\Pi_E(i\omega_n) = \frac{i}{i\omega_n 4\pi^2\beta}\sum_k\int_{-\infty}^\infty d\omega d\omega'\frac{(2\omega+i\omega_n)^2 f(\omega)-(2\omega'-i\omega_n)^2 f(\omega')}{\omega+i\omega_n-\omega'}$$

$$\left[\left(\partial_{k_\mu}\xi_k\right)^2\operatorname{Tr}\operatorname{Im}G^R(k,\omega)\sigma_3\operatorname{Im}G^R(k,\omega')\sigma_3\right.$$

$$\left.+\left(\partial_{k_\mu}\Delta_k\right)^2\operatorname{Tr}\operatorname{Im}G^R(k,\omega)\sigma_1\operatorname{Im}G^R(k,\omega')\sigma_1\right].$$

To perform the analytic continuation $i\omega_n \to \omega + i\delta$ and take the limit $\omega \to 0$, this expression can be simplified. The real part is given by

$$\mathrm{Re}\Pi_E(\omega \to 0) = -\frac{1}{\pi\beta} \sum_k \int_{-\infty}^{\infty} d\omega \omega^2 \frac{\partial f(\omega)}{\partial \omega}$$

$$\left[\left(\partial_{k_\mu}\xi_k\right)^2 \mathrm{Tr}\, \mathrm{Im}G^R(k,\omega)\,\sigma_3 \mathrm{Im}G^R(k,\omega)\,\sigma_3 \right.$$

$$\left. + \left(\partial_{k_\mu}\Delta_k\right)^2 \mathrm{Tr} \mathrm{Im}G^R(k,\omega)\,\sigma_1 \mathrm{Im}G^R(k,\omega)\,\sigma_1 \right]. \quad (10.74)$$

In the limit of zero temperature, $-\partial f(\omega)/\partial\omega \to \delta(\omega)$, one can set the frequency ω in the Green's function in Eq. (10.74) to zero. Furthermore, using the integral formula

$$\int_{-\infty}^{\infty} d\omega \omega^2 \frac{\partial f(\omega)}{\partial \omega} = -\frac{\pi^2}{3\beta^2},$$

$\mathrm{Re}\Pi_E$ can be simplified as

$$\mathrm{Re}\Pi_E(\omega \to 0) = \frac{\pi}{3\beta^3} \sum_k \left[\left(\partial_{k_\mu}\xi_k\right)^2 \mathrm{Tr}\, \mathrm{Im}G^R(k,0)\,\sigma_3 \mathrm{Im}G^R(k,0)\,\sigma_3 \right.$$

$$\left. + \left(\partial_{k_\mu}\Delta_k\right)^2 \mathrm{Tr}\, \mathrm{Im}G^R(k,0)\,\sigma_1 \mathrm{Im}G^R(k,0)\,\sigma_1 \right]. \quad (10.75)$$

The zero frequency retarded Green's function is given by

$$G^R(k,0) = \frac{1}{-\xi_k\sigma_3 - \Delta_k\sigma_1 + i\Gamma}, \quad (10.76)$$

where $\Gamma = \Gamma(\omega = 0)$ is the quasiparticle scattering rate. Substituting the above result into Eq. (10.75), we obtain

$$\mathrm{Re}\Pi_E(\omega \to 0) = \frac{\pi}{3\beta^3} \sum_{k_\mu} \frac{\Gamma^2}{(\Gamma^2 + E_k^2)^2} \left[\left(\partial_{k_\mu}\xi_k\right)^2 + \left(\partial_{k_\mu}\Delta_k\right)^2 \right]. \quad (10.77)$$

When Γ is small, the momentum summation in Eq. (10.77) contributes mainly from the nodal region. In this case,

$$\sum_\mu \left(\partial_{k_\mu}\xi_k\right)^2 \approx \hbar^2 v_F^2,$$

$$\sum_\mu \left(\partial_{k_\mu}\Delta_k\right)^2 \approx \hbar^2 v_2^2.$$

The momentum summation can be performed independently around the four gap nodes. The gradients of ξ_k and Δ_k are perpdendicular to each other at each nodal point, this gives

$$\sum_k \approx 4 \int \frac{dk_1 dk_2}{4\pi^2} = \frac{1}{\pi^2} \int \frac{d\xi d\Delta}{\hbar^2 v_F v_2},$$

where k_1 and k_2 are the wavevectors along the tangential direction of ξ_k and Δ_k, respectively. Thus $\mathrm{Re}\Pi_E(\omega \to 0)$ can be further expressed as

$$\mathrm{Re}\Pi_E(\omega \to 0) \approx \frac{1}{3\pi\beta^3} \frac{v_F^2 + v_2^2}{v_F v_2} \int d\xi d\Delta \frac{\Gamma^2}{(\Gamma^2 + \xi^2 + \Delta^2)^2}. \tag{10.78}$$

Generally speaking, the integral in the above equation is a function of Γ. On the other hand, this integral is dimensionless based on the dimensional analysis. Thus we expect that this integral is Γ-independent. An explicit calculation confirms this expectation:

$$\int d\xi d\Delta \frac{\Gamma^2}{(\Gamma^2 + \xi^2 + \Delta^2)^2} = 2\pi \int_0^\infty E dE \frac{\Gamma^2}{(\Gamma^2 + E^2)^2} = \pi. \tag{10.79}$$

Substituting this result into Eq. (10.69), we finally obtain the universal formula for the thermal conductivity in the limit of zero temperature:

$$\frac{\kappa}{T} \approx \frac{k_B^2}{3\hbar} \frac{v_F^2 + v_2^2}{v_F v_2}. \tag{10.80}$$

It shows that κ only depends on two fundamental parameters v_F and v_2, but not on the quasiparticle scattering rate Γ. This result was first derived by Durst and Lee [206]. They also showed that this result is robust against the vertex corrections from the impurity scattering as well as the Coulomb interaction to the leading order approximation. It suggests that the thermal conductivity is more appropriate than the electric conductivity to probe the universal transport properties of d-wave superconducting quasiparticles.

Fig. 10.6 shows the thermal conductivity coefficient κ_e/T in the zero temperature limit as a function of the quasiparticle scattering rate Γ_ρ for the $YBa_2Cu_3O_{6.9}$ superconductors without or with partial substitution of copper atoms by zinc atoms [207]. Γ_ρ is determined from the measurement data of the

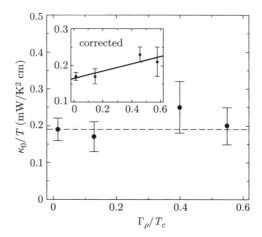

Figure 10.6   The thermal conductivity coefficient κ_e/T versus the scattering rate Γ_ρ/T_c for the YBa$_2$(Cu$_{1-x}$Zn$_x$)$_3$O$_{6.9}$ superconductor in the limit of zero temperature. The zinc concentrations of the four samples are $x = 0, 0.006, 0.02, 0.03$. The larger x is, the larger Γ_ρ would be. The inset shows the results after considering the correction from the geometric effect. Reprinted with permission from [207]. Copyright (1997) by the American Physical Society.

residual resistance. The higher the zinc impurity concentration is, the larger Γ_ρ would be. The 3% concentration of zinc impurity suppresses T_c by 20%. Within experimental errors, the coefficient of thermal conductivity κ/T is almost a constant, independent on Γ_ρ. This indicates that, as expected, the thermal conductivity is universal in the limit of zero temperature. The similar universal behavior was observed in the Bi$_2$Sr$_2$CaCu$_2$O$_8$ superconductors [208]. However, it should be pointed out that the four samples shown in Fig. 10.6 have different superconducting transition temperatures. Δ_0/v_F and v_2/v_F are also changed with the change of the zinc concentration. Thus the universality does not imply that κ/T is completely independent of doping.

The universal formula given by Eq. (10.80) indicates that v_2/v_F can be determined through the measurement of thermal conductivity. Then using the value of v_F determined by the ARPES experiments, we can further determine the value of v_2. Based on this idea, Sutherland *et. al.* [209] measured the doping dependence of the thermal conductivities for YBCO and LSCO superconductors. They found that κ/T decreases with decreasing doping level in the limit of zero temperature, but remains finite even at very low doping. This shows unambiguously that the gap nodes exist at all doping. From the measurment data, they calculated the doping dependence of Δ_0, and found that Δ_0 is not proportional to T_c, unlike what expected from the BCS theory. Instead, they found that Δ_0 increases as the doping level or T_c is decreased. The low temperature κ/T measures the quasiparticle excitations around the gap nodes, the value of Δ_0 obtained from the universal relation Eq. (10.80) is actually the amplitude or the slope of the gap function around the gap nodes. Furthermore, they found that the doping dependence of Δ_0 agrees qualitatively with that of the maximal pseudogap obtained from the ARPES or other experimental measurements. It suggests that the superconducting gap is inherently connected to the pseudogap, and they may result from the same physical origin.

Chapter 11

Raman Spectroscopy

11.1 Raman Response Function

Raman scattering occurs when electromagnetic radiation or light, usually from a laser, impinges on a crystal and interacts with electrons, phonons or other elementary excitations. The Raman spectroscopy measures the cross section of inelastic scattering of light by electrons or other quasiparticles. The spectrum of the scattered light is termed the Raman spectrum. It shows the intensity of the scattered light as a function of its frequency difference to the incident light. Through the analysis of Raman spectroscopy, one can acquire useful information on the electronic structures, the lattice and magnetic excitations of solids. The Raman spectroscopy has served as a powerful experimental tool in the study of condensed matter physics. In the superconducting phase, the superconducting pairing affects strongly the photoelectric interaction. The Raman spectroscopy can be used to explore the pairing symmetry and the pairing mechanism. It has played an important role in the study of both conventional and high-T_c superconductors.

The interaction between photons and electrons is described by the Hamiltonian

$$H_I = H_1 + H_2, \tag{11.1}$$

$$H_1 = -\frac{e}{\hbar c} \sum_q J_{\alpha,q} A_{\alpha,-q}, \tag{11.2}$$

$$H_2 = \frac{e^2}{2\hbar^2 c^2} \sum_{q_1, q_2} \tau_{q_1+q_2}^{\alpha\beta} A_{\alpha,-q_1} A_{\beta,-q_2}, \tag{11.3}$$

where $J_{\alpha,q}$ is the current operator of electrons

$$J_{\alpha,q} = \sum_{k} \frac{\partial \varepsilon_k}{\partial k_\alpha} c^{\dagger}_{k+\frac{q}{2}\sigma} c_{k-\frac{q}{2}\sigma},$$

and $\tau^{\alpha\beta}_{q_1+q_2}$ is a second-order tensor defined by

$$\tau^{\alpha\beta}_{q} = \sum_{k} \frac{\partial^2 \varepsilon_k}{\partial k_\alpha \partial k_\beta} c^{\dagger}_{k+\frac{q}{2}\sigma} c_{k-\frac{q}{2}\sigma}. \tag{11.4}$$

In the above equations, summation is implied over the repeated greek indices. If $\varepsilon_k = k^2/2m$, then $\tau^{\alpha\beta}_{q}$ is proportional to the density matrix. In order to study microscopically the light scattering process, we need to consider a quantized electromagnetic field $A_{\alpha,q}$. To define a_{qe_α} and $a^{\dagger}_{qe_\alpha}$ as the creation and annihilation operators of a photon polarized along the direction e_α, it can be shown by utilizing the Maxwell equations that

$$A_{\alpha,q} = g_q e_\alpha \left(a_{qe_\alpha} + a^{\dagger}_{-qe_\alpha} \right), \tag{11.5}$$

where

$$g_q = \sqrt{\frac{hc^2}{\omega_q V}},$$

$\omega_q = c|q|$ is the photon energy, and V is the volume of the system.

A Raman scattering involves the incident and scattered photons, and the scattered electrons. If we denote the energy, momentum, and polarization of the incident photon as $\left(\omega_i, k_i, e^I\right)$ and the initial electron state as $|i\rangle$, then the initial wavefunction of the system can be represented as

$$|I\rangle = a^{\dagger}_{k_i e^I} |i\rangle.$$

Similarly, the final state of the system can be expressed as

$$|F\rangle = a^{\dagger}_{k_f e^S} |f\rangle,$$

where $\left(\omega_f, k_f, e^S\right)$ are the energy, momentum, and polarization of the scattered photon, respectively. $|f\rangle$ is the wavefunction of the scattered electron.

The Raman scattering is a two-photon process, one photon in the initial state and another in the final state. Thus the Raman scattering measures the second order response of electromagnetic field A_k. It can be generated by the first order perturbation of H_2 and the second order perturbation of H_1. The latter involves an intermediate process during which a virtual photon is emitted or absorbed. When the energy of the emitted or absorbed virtual photon matches the energy difference between the initial and the intermediate states of electrons, a resonant transition happens. Thus this process is essentially resonant. On the contrary, the first order perturbation process of H_2 is non-resonant. It does not need a virtual intermediate state. As long as the momentum and energy are conserved, there are no special requirements for the initial and final electron states.

An important quantity in describing the Raman scattering is the transition probability $R_{I,F}$ from the initial state $|I\rangle$ to the final state $|F\rangle$. According to the Fermi golden rule, this transition probability equals

$$R_{I,F} = \frac{2\pi}{\hbar} |\langle F|M|I\rangle|^2 \, \delta \left(E_I - E_F\right), \tag{11.6}$$

where M is the effective scattering operator. We use $\langle F|M_N|I\rangle$ and $\langle F|M_R|I\rangle$ to represent the contributions from the non-resonance and resonance scattering processes, respectively. The total scattering amplitude $\langle F|M|I\rangle$ is simply a sum of these two terms

$$\langle F|M|I\rangle = \langle F|M_N|I\rangle + \langle F|M_R|I\rangle.$$

According to the perturbation theory, it can be shown that the non-resonance transition amplitude is given by

$$\langle F|M_N|I\rangle = \langle F|H_2|I\rangle = \frac{e^2 g_{k_i} g_{k_f} e_\alpha^I e_\beta^S}{\hbar^2 c^2} \langle f|\tau_q^{\alpha\beta}|i\rangle. \tag{11.7}$$

Similarly, it can be shown that the resonance transition amplitude is given by

$$\langle F|M_R|I\rangle = -\sum_J \langle F|H_1|J\rangle \frac{1}{E_J - E_I} \langle J|H_1|I\rangle$$

$$= -\frac{e^2 g_{k_i} g_{k_f} e_\alpha^I e_\beta^S}{\hbar^2 c^2}$$

$$\sum_l \left[\frac{\langle f|J_{\beta,k_f}|l\rangle\langle l|J_{\alpha,-k_i}^\alpha|i\rangle}{\varepsilon_l - \varepsilon_i - \omega_i} + \frac{\langle f|J_{\alpha,-k_i}|l\rangle\langle l|J_{\beta,k_f}|i\rangle}{\varepsilon_l - \varepsilon_i + \omega_f} \right]. \tag{11.8}$$

By substituting the above results into Eq. (11.6), and considering the thermal equilibrium distribution for the initial electron states, we find that the scattering probability is

$$\tilde{R}(q,\omega) = \frac{2\pi e^4}{\hbar^5 c^4} \sum_{i,f} \frac{e^{-\beta\varepsilon_i}}{Z} \left| g_{k_i} g_{k_f} e_\alpha^I e_\beta^S \langle f|T_q^{\alpha\beta}|i\rangle \right|^2 \delta\left(\varepsilon_f - \varepsilon_i - \omega\right), \quad (11.9)$$

where $q = k_i - k_f$, $\omega = \omega_i - \omega_f$, Z is the partition function, and

$$\langle f|T_q^{\alpha\beta}|i\rangle = \langle f|\tau_q^{\alpha\beta}|i\rangle - \sum_l \left[\frac{\langle f|J_{\beta,k_f}|l\rangle\langle l|J_{\alpha,-k_i}^\alpha|i\rangle}{\varepsilon_l - \varepsilon_i - \omega_i} + \frac{\langle f|J_{\alpha,-k_i}|l\rangle\langle l|J_{\beta,k_f}|i\rangle}{\varepsilon_l - \varepsilon_i + \omega_f} \right].$$
$$(11.10)$$

Using the fluctuation-dissipation theorem,

$$\sum_{ij} \frac{e^{-\beta\varepsilon_i}}{Z} |\langle j|\rho_\gamma(q)|i\rangle|^2 \delta\left(\varepsilon_f - \varepsilon_i - \omega\right) = -\frac{1}{\pi\left(1 - e^{-\beta\omega}\right)} \mathrm{Im} R\left(q,\omega\right), \quad (11.11)$$

we can reexpress Eq. (11.9) as

$$\tilde{R}(q,\omega) = -\frac{2e^4 g_{k_i}^2 g_{k_f}^2}{\hbar^5 c^4 \left(1 - e^{-\beta\omega}\right)} \mathrm{Im} R\left(q,\omega\right). \quad (11.12)$$

$R(q,\omega)$ is the Fourier transform of the retarded Raman response function defined by

$$R(q,t) = -i\theta(t)\langle[\rho_\gamma(q,t), \rho_\gamma(-q,0)]\rangle, \quad (11.13)$$

where $\rho_\gamma(q)$ is the Raman density operator containing both the resonance and non-resonance parts. After neglecting the contribution of surface reflection, it can be shown that $\tilde{R}(q,\omega)$ is proportional to the differential cross section of the Raman scattering

$$\frac{\partial^2\sigma}{\partial\Omega\partial\omega} \propto \tilde{R}(q,\omega). \quad (11.14)$$

Thus what measured by the Raman scattering is just the Raman response function $R(q,\omega)$.

In solids, because the light velocity is much larger than the Fermi velocity of electrons, the energy transfer ω of photons after scattering could be large, but the corresponding momentum transfer q is small. According to the uncertainty principle, the momentum transfer q is roughly of the order of the inverse

magnetic penetration depth, which can be approximately taken as zero. In low
temperatures, when the momentum transfer is very small and the frequency of
the incident light is much smaller than the band gap, there is no resonant scattering. In this case, only the non-resonant scattering needs to be considered,
and the effective mass approximation can be used for $\rho_\gamma(q)$ [210]:

$$\rho_\gamma(q) = \sum_k \gamma_k c^\dagger_{k+q/2} c_{k-q/2}, \tag{11.15}$$

$$\gamma_k = \sum_{\alpha\beta} e^S_\beta \frac{\partial^2 \varepsilon_k}{\partial k_\alpha \partial k_\beta} e^I_\alpha. \tag{11.16}$$

If $\gamma = 1$, ρ_γ is just the electron density operator defined by

$$\rho_1(q) = \sum_k c^\dagger_{k+q/2} c_{k-q/2}. \tag{11.17}$$

The intensity of the Raman scattering depends on the polarization of the
incident light e^I and that of the scattered light e^S. By adjusting the polarization
directions of the incident and scattered lights, the Raman spectroscopy can
measure the scattering of quasiparticles on different parts of the Fermi surface.
This is useful to the analysis of the momentum dependence of the gap function
over the Fermi surface. For high-T_c superconductors, the polarization vectors
of the incident and scattered lights are usually chosen along the directions of
(100), (010), (110), and (1$\bar{1}$0). The values of the Raman response functions
for different symmetry modes can be measured and derived based on different
combinations of these polarization vectors.

In the study of high-T_c superconductivity, the three symmetry modes, A_{1g},
B_{1g}, and B_{2g}, are usually measured. Table 11.1 shows the relation between
these symmetry modes and the polarization vectors of incident and scattered
lights. Under the effective mass approximation, the Raman vertex functions for
these three symmetry modes are

$$\gamma_k = \begin{cases} \dfrac{1}{2}\left(\dfrac{\partial^2 \varepsilon_k}{\partial k_x^2} + \dfrac{\partial^2 \varepsilon_k}{\partial k_y^2}\right), & A_{1g}, \\[3mm] \dfrac{1}{2}\left(\dfrac{\partial^2 \varepsilon_k}{\partial k_x^2} - \dfrac{\partial^2 \varepsilon_k}{\partial k_y^2}\right), & B_{1g}, \\[3mm] \dfrac{\partial^2 \varepsilon_k}{\partial k_x \partial k_y}, & B_{2g}. \end{cases} \tag{11.18}$$

Table 11.1   The polarization directions of incident and scattered lights and the corresponding symmetry modes commonly used in the Raman scattering experiments

	e^I	e^S
$A_{1g} + B_{2g}$	$\dfrac{1}{\sqrt{2}}(\hat{x} + \hat{y})$	$\dfrac{1}{\sqrt{2}}(\hat{x} + \hat{y})$
$A_{1g} + B_{1g}$	$\hat{y}$	$\hat{y}$
B_{1g}	$\dfrac{1}{\sqrt{2}}(\hat{x} + \hat{y})$	$\dfrac{1}{\sqrt{2}}(\hat{x} - \hat{y})$
B_{2g}	$\hat{x}$	$\hat{y}$

Clearly, the vertex function γ_k depends on the band structure of electrons. Near the Fermi surface, γ_k can be expanded using the harmonic modes of crystal [211]. To the leading order approximation, γ_k can be expressed as

$$\gamma_k = \begin{cases} \gamma_0 + \gamma(A_{1g})\cos(4\phi), & A_{1g}, \\ \gamma(B_{1g})\cos(2\phi), & B_{1g}, \\ \gamma(B_{2g})\sin(2\phi), & B_{2g}, \end{cases} \tag{11.19}$$

where $\phi = \arctan(k_y/k_x)$. These expressions of γ_k are more convenient to use in the analytical calculation.

From Eq. (11.19), it is clear that the A_{1g}-mode measures the average of the scattering cross section over the whole Fermi surface. The B_{1g}-mode has the same symmetry as the $d_{x^2-y^2}$-wave superconductor. It measures the Raman spectrum contributed mainly by the quasiparticles along the anti-nodal directions. On the contrary, the vertex function of the B_{2g}-mode has the largest absolute value along the nodal directions. It measures mainly the contribution of quasiparticles around the gap nodes. Therefore, through the measurement of these Raman modes, especially the B_{1g}- and B_{2g}-modes, we can extract useful information on the gap function.

11.2 Vertex Correction by the Coulomb Interaction

Light radiation induces a charge fluctuation in a metal through photoelectric interaction. There is a strong screen effect resulting from the long-range Coulomb interaction between the fluctuating charges. It changes the intensity of the Raman scattering and modifies the vertex function of the Raman modes. In the long wavelength limit, the Coulomb interaction only couples to the high symmetry A_{1g}-mode and changes the corresponding Raman spectrum. But it does not affect the Raman spectra for the B_{1g}, B_{2g} and other low symmetry modes.

The Coulomb interaction of fluctuating charges is described by the Hamiltonian

$$H_c = \frac{1}{2} \sum_q \rho_1(q) V_q \rho_1(-q), \tag{11.20}$$

$$V_q = 4\pi e^2/q^2.$$

After considering the correction of this Coulomb interaction to the vertex function, whose corresponding Feymann diagram is shown in Fig. 11.1, the Raman response function becomes [212, 213]

$$
\begin{aligned}
R(q,\omega) &= R^0_{\gamma,\gamma}(q,\omega) + R^0_{\gamma,1}(q,\omega) \left[V_q + V_q R^0_{1,1}(q,\omega) V_q + ... \right] R^0_{1,\gamma}(q,\omega) \\
&= R^0_{\gamma,\gamma}(q,\omega) + \frac{R^0_{\gamma,1}(q,\omega) V_q R^0_{1,\gamma}(q,\omega)}{1 - V_q R^0_{1,1}(q,\omega)}, \tag{11.21}
\end{aligned}
$$

where $R^0_{a,b}(q,\omega)$ is the response function without considering the Coulomb interaction, the subscripts a and b represent the vertex functions. The second term of the above equation can be reorganized so that the response function can be expressed as

$$R(q,\omega) = R^0_{\gamma,\gamma}(q,\omega) - \frac{R^0_{\gamma,1}(q,\omega) R^0_{1,\gamma}(q,\omega)}{R^0_{1,1}(q,\omega)} + \frac{R^0_{\gamma,1}(q,\omega) R^0_{1,\gamma}(q,\omega)}{R^0_{1,1}(q,\omega) \left[1 - V_q R^0_{1,1}(q,\omega) \right]}. \tag{11.22}$$

The third term in Eq. (11.22) is proportional to the density-density response function of electrons:

$$\chi(q,\omega) = \frac{R^0_{1,1}(q,\omega)}{1 - V_q R^0_{1,1}(q,\omega)}. \tag{11.23}$$

$$R(q,\omega) \quad = \quad R_{\gamma\gamma}^0 \quad + \quad R_{\gamma1}^0 \; V_q \; R_{1\gamma}^0 \quad + \quad R_{\gamma1}^0 \; V_q \; R_{11}^0 \; V_q \; R_{1\gamma}^0 \quad + \quad \cdots\cdots$$

Figure 11.1   The correction to the Raman response functions due to the Coulomb interaction generated by the charge fluctuation.

It can be shown that $\chi(q,\omega)$ vanishes in the limit of $q \to 0$ [213]. Thus in the long wavelength limit, the Raman response function is only determined by the first two terms in Eq. (11.22):

$$R(q,\omega) = R_{\gamma,\gamma}^0(q,\omega) - \frac{R_{\gamma,1}^0(q,\omega)R_{1,\gamma}^0(q,\omega)}{R_{1,1}^0(q,\omega)}. \tag{11.24}$$

In Eq. (11.22), $R_{\gamma,1}^0(q,\omega)$ and $R_{1,\gamma}^0(q,\omega)$ involve the average of the vertex function γ over the Fermi surface. Thus only the mode whose average of the vertex function on the whole Fermi surface has contribution to the Raman response function. For the B_{1g}, B_{2g} or other low symmetry modes, the average of the vertex function on the whole Fermi surface is zero and the correction from the charge fluctuation to the corresponding Raman response function is also zero. Thus for low symmetry modes, only the first term on the right-hand side of Eq. (11.24) is finite.

The vertex correction to the A_{1g}-mode due to the charge fluctuation is finite. This will completely screen the ϕ-independent part in the vertex function of the A_{1g}-mode. Only the $\cos 4\phi$ or higher order harmonic components contribute to the A_{1g}-spectra. For the A_{1g}-mode, if we rewrite its vertex function as

$$\gamma_k = \gamma_0 + \delta\gamma_k,$$

then it is straightforward to show that the momentum independent term, i.e. the γ_0-term, has no contribution to $R(q,\omega)$ and

$$R(q,\omega) = R_{\delta\gamma,\delta\gamma}^0(q,\omega) - \frac{R_{\delta\gamma,1}^0(q,\omega)R_{1,\delta\gamma}^0(q,\omega)}{R_{1,1}^0(q,\omega)}. \tag{11.25}$$

11.3 The Raman Response Function of *d*-Wave Supercon-ductors

To obtain the Raman response function, one can first evaluate the Matsubara Green's function corresponding to $R^0_{\gamma_1\gamma_2}(q,\omega)$:

$$R^0_{\gamma_1\gamma_2}(q,\tau) = -\langle T_\tau \rho_{\gamma_1}(q,\tau)\rho_{\gamma_2}(-q,0)\rangle, \qquad (11.26)$$

and then use Eq. (11.24) to obtain the value of $R_{\gamma_1\gamma_2}(q,\omega)$ by performing the analytic continuation from the imaginary to the real frequency axis. Using the definition of $\rho_\gamma(q)$, $R^0_{\gamma_1\gamma_2}(q,\tau)$ can be expressed as

$$R^0_{\gamma_1\gamma_2}(q,\tau) = \sum_k \gamma_{1k}\gamma_{2k}\,\mathrm{Tr}\,G\left(k+\frac{q}{2},-\tau\right)\sigma_3 G\left(k-\frac{q}{2},\tau\right)\sigma_3. \qquad (11.27)$$

Taking the Fourier transformation with respect to the imaginary time τ, we have

$$
\begin{aligned}
R^0_{\gamma_1\gamma_2}(q,i\omega_n) &= \int_0^\beta d\tau\, e^{i\omega_n\tau} R(q,\tau)\\
&= \frac{1}{\beta}\sum_{k\omega_m}\gamma_{1k}\gamma_{2k}\,\mathrm{Tr}\,G\left(k+\frac{q}{2},i\omega_m\right)\sigma_3 G\left(k-\frac{q}{2},i\omega_m+i\omega_n\right)\sigma_3.
\end{aligned}
$$
$$(11.28)$$

To substitute the BCS mean-field expression of the single-particle Green's function into the above equation, $R^0_{\gamma_1\gamma_2}(q,i\omega_n)$ can be obtained in the weak coupling limit. In the long wavelength limit, $q \to 0$, the result is

$$
R^0_{\gamma_1\gamma_2}(0,i\omega_n) = \sum_k \gamma_{1k}\gamma_{2k}\frac{2\Delta_k^2}{i\omega_n E_k}\left(\frac{1}{i\omega_n+2E_k}+\frac{1}{i\omega_n-2E_k}\right)
$$
$$
\tanh\frac{\beta E_k}{2}. \qquad (11.29)
$$

After the analytic continuation, the corresponding retarded Green function $R^0(0,\omega)$ is found to be

$$
R^0_{\gamma_1\gamma_2}(0,\omega) = \sum_k \gamma_{1k}\gamma_{2k}\frac{2\Delta_k^2}{(\omega+i\delta)E_k}\left(\frac{1}{\omega+i\delta+2E_k}+\frac{1}{\omega+i\delta-2E_k}\right)
$$
$$
\tanh\frac{\beta E_k}{2}.
$$

Its imaginary part is given by

$$\mathrm{Im}R^0_{\gamma_1\gamma_2}(0,\omega) = -\frac{4\pi}{\omega^2}\tanh\frac{\beta\omega}{4}\sum_k \gamma_{1k}\gamma_{2k}\Delta_k^2\left[\delta\left(\omega - 2E_k\right) + \delta\left(\omega + 2E_k\right)\right].$$

(11.30)

In the above expression, it is assumed that γ_k does not include any harmonic component that is ϕ-independent, hence the average of γ_k over the Fermi surface is zero.

In the low energy limit, the momentum summation can be approximately expressed as an integral over the energy ξ_k and the average of other variables over the Fermi surface:

$$\sum_k A(\xi_k, \Delta_k) = \left\langle\frac{N_F}{2}\int \mathrm{d}\xi A(\xi, \Delta_k)\right\rangle_{FS}.$$

From the integral formula of the δ-function

$$\int \mathrm{d}\xi \delta\left(\omega - 2E_k\right) = \mathrm{Re}\frac{|\omega|}{2\sqrt{\omega^2 - 4\Delta_k^2}},$$

and using Eq. (11.24), the imaginary part of the Raman response function $\mathrm{Im}R(0,\omega)$ can be simplified as

$$-\mathrm{Im}R(0,\omega) = \frac{\pi N_F}{\omega}\mathrm{Re}\left\langle\frac{\gamma_k^2\Delta_k^2}{\sqrt{\omega^2 - 4\Delta_k^2}}\right\rangle_{FS}\tanh\frac{\beta\omega}{4}.$$

(11.31)

To obtain the real part of $R^0_{\gamma_1\gamma_2}(0,\omega)$, we first integrate out ξ_k. At zero temperature, the result is

$$\mathrm{Re}R^0_{\gamma_1\gamma_2}(0,\omega) = \begin{cases} -\left\langle\dfrac{4\gamma_{1k}\gamma_{2k}N_F\Delta_k^2}{\omega\sqrt{4\Delta_k^2 - \omega^2}}\arctan\dfrac{\omega}{\sqrt{4\Delta_k^2 - \omega^2}}\right\rangle_{FS}, & \omega^2 < 4\Delta_k^2, \\[3ex] -\left\langle\dfrac{2\gamma_{1k}\gamma_{2k}N_F\Delta_k^2}{\omega\sqrt{\omega^2 - 4\Delta_k^2}}\ln\dfrac{\omega - \sqrt{\omega^2 - 4\Delta_k^2}}{\omega + \sqrt{\omega^2 - 4\Delta_k^2}}\right\rangle_{FS}, & \omega^2 > 4\Delta_k^2. \end{cases}$$

(11.32)

For the isotropic s-wave superconductor, as $\Delta_k = \Delta$ and the average of γ_k over the Fermi surface is zero, it is simple to show that $R^0_{\gamma,1}(0,\omega) = 0$. In this case, the Raman response function is completely given by the first term on the right-hand side of Eq. (11.24), we then have

$$-\mathrm{Im}R(0,\omega) = \theta(|\omega| - 2\Delta)\frac{\pi N_F\Delta^2\langle\gamma_k^2\rangle_{FS}}{\omega\sqrt{\omega^2 - 4\Delta^2}}\tanh\frac{\beta\omega}{4}.$$

(11.33)

$\mathrm{Im}R(0,\omega)$ vanishes at low frequency $|\omega| < 2\Delta$. But it exhibits a square-root divergence when ω is above and approaching 2Δ.

For the d-wave superconductor, $\Delta_k = \Delta_0 \cos 2\phi$, there is not a threshold for ω. However, as the quasiparticle density of states diverges at $\omega = \Delta_0$, $\mathrm{Im}R(0,\omega)$ of the B_{1g}-model diverges at $\omega = 2\Delta_0$. This is a logarithmic divergence, weaker than the square root divergence in the s-wave superconductor. The vertex function of the B_{2g}-mode vanishes at the anti-nodal direction. It cancels the divergence of the density of states. Therefore, there is no divergence in the Raman spectrum of the B_{2g}-mode at $\omega = 2\Delta_0$. The peak position of $\mathrm{Im}R(0,\omega)$ for this mode is determined by the average of the vertex function γ_k^2 and the density of states over the Fermi surface. It is not located at $\omega = 2\Delta_0$, instead at a lower frequency around $\omega \approx 1.6\Delta_0$ as shown in Fig. 11.2.

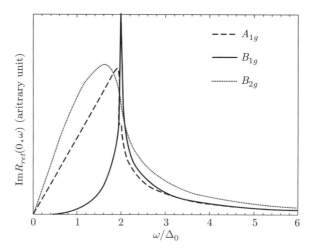

Figure 11.2   The Raman scattering spectra in the ideal d-wave superconductor at zero temperature. The vertex function is given in Eq. (11.19).

The Coulomb interaction affects strongly the Raman scattering cross section in the long wave length limit. It completely screens the long wavelength part of the effect mass, or the ϕ-angle independent part of the vertex function γ_k, under the effective mass approximation. Without considering the effect of Coulomb screening, the Roman spectrum of the A_{1g}-mode would also exhibit a logarithmic divergence at $\omega = 2\Delta_0$. However, this divergence is removed by the Coulomb screening. For the vertex function of the A_{1g}-model given in Eq. (11.19), i.e. $\gamma_k = \gamma_0 + \gamma(A_{1g})\cos 4\phi$, the spectral peak is still located at $\omega = 2\Delta_0$ (Fig. 11.2). On the other hand, if the vertex function of the A_{1g}-mode contains $\cos 8\phi$ or even higher order harmonic components, the spectral peak of the A_{1g}-mode would shift towards lower frequency. Thus, in real materials, it is expected that the spectral peak of the A_{1g}-mode is always located at a frequency lower than $2\Delta_0$, i.e. $\omega < 2\Delta_0$.

　　The Raman spectrum of the A_{1g}-mode is not directly measured experimently. It is obtained usually from the subtraction of the spectrum measured with $e^I = e^S = \hat{y}$ or $e^I = e^S = (\hat{x} + \hat{y})/\sqrt{2}$ and that of the B_{1g} or B_{2g} mode. The subtraction introduces errors in addition to the measurement error. The response function of the A_{1g}-mode is sensitive to the coefficients of higher order harmonic components. Thus it is difficult to predict quantitatively the line shape of the A_{1g}-Raman spectrum.

　　In low temperatures, the ω-dependence of the Raman spectrum of the B_{1g}-mode in the d-wave superconductor is different from those of the A_{1g} and B_{2g}-modes when $\omega \ll 2\Delta_0$. Since the vertex function of the B_{1g}-mode, $\gamma_k = \gamma(B_{1g})\cos 2\phi$, has the same symmetry as the $d_{x^2-y^2}$-wave energy gap, the contribution to the Raman spectrum of this mode from the nodal quasiparticles is suppressed. As a result of this, the low energy Raman spectra of the B_{1g}-mode are much weaker than those of the B_{2g} and A_{1g}-modes. For the latter two modes, the low energy spectra vary linearly with ω, resulting from the linear quasiparticle density of states. However, for the B_{1g}-mode, we find that the Raman spectrum scales as ω^3 in low frequencies simply based on the dimensional analysis:

$$-\mathrm{Im}R(0,\omega) = \begin{cases} \dfrac{\pi N_F \gamma^2(B_{2g})}{16\Delta_0}\omega\tanh\dfrac{\beta\omega}{4}, & B_{2g}, \\[3mm] \dfrac{3\pi N_F \gamma^2(B_{1g})}{128\Delta_0}\omega^3\tanh\dfrac{\beta\omega}{4}, & B_{1g}. \end{cases} \quad (11.34)$$

　　The Raman scattering cross section of the B_{1g}-mode exhibits a weaker frequency dependence than the A_{1g} and B_{2g}-modes. This is due to the coincidence of the zeros of the B_{1g} vertex function and the nodes of the d-wave gap function. In general, if the zeros of the vertex function of a Raman mode coincide with the gap nodes, then the ω-dependence of the corresponding Raman response function is weaker. This property can be used to determine the nodal directions of the gap function.

11.4 Effect of Non-Magnetic Impurity Scattering

The disorder scattering has a strong effect on the Raman spectra. Neglecting the vertex correction, the Raman response function can be expressed using the single-particle spectral function as

$$R(0,\omega) = \sum_k \int \frac{d\omega_1 d\omega_2}{\pi^2} \gamma_k^2 \mathrm{Tr} \left[\mathrm{Im}G^R(k,\omega_1)\sigma_3 \mathrm{Im}G^R(k,\omega_2)\sigma_3 \right]$$
$$\frac{f(\omega_1) - f(\omega_2)}{\omega + \omega_1 - \omega_2 + i0^+}. \tag{11.35}$$

At zero temperature, its imaginary part is given by

$$-\mathrm{Im}R(0,\omega) = \frac{1}{\pi} \sum_k \gamma_k^2 \int_0^\omega d\omega_1 \mathrm{TrIm}G^R(k,\omega_1-\omega)\tau_3 \mathrm{Im}G^R(k,\omega_1)\tau_3. \tag{11.36}$$

One effect of non-magnetic impurity scattering is to smear out the divergence of the B_{1g}-mode at $\omega = 2\Delta_0$. In addition to this, the correction of the impurity scattering to the Raman response function occurs mainly in the low energy part. In the gapless regime, if the energy is smaller than the quasiparticle scattering rate, i.e. $\omega \ll \Gamma_0$, the ω-dependence of the Raman response functions can be obtained through the series expansion. The zero temperature result is given by

$$-\mathrm{Im}R(0,\omega) \approx \begin{cases} \dfrac{2N(0)\gamma^2(B_{2g})}{\pi\Delta_0}\omega, & B_{2g}, \\ \dfrac{2N(0)\gamma^2(B_{1g})}{\pi\Delta_0} \left(\dfrac{\Gamma_0}{\Delta_0}\right)^2 \left(\ln\dfrac{2\Delta_0}{\Gamma_0}\right)\omega, & B_{1g}. \end{cases} \tag{11.37}$$

Compared with the results of the intrinsic d-wave superconductor, the Raman spectrum $\mathrm{Im}R(0,\omega)$ of the B_{2g}-mode is still a linear function of ω. But the linear coefficient changes. For the B_{1g}-mode, $\mathrm{Im}R(0,\omega)$ changes completely. It also varies linearly with ω and decays more slowly than in the intrinsic d-wave superconductor. In the weak scattering limit, the correction to the Raman spectrum is proportional to Γ_0/Δ_0 and $\mathrm{Im}R(0,\omega)$ is not much different from that of the intrinsic d-wave superconductor.

11.5 Experimental Results of High-T_c Superconductors

The above result indicates that the Raman scattering is a powerful tool for exploring the anisotropy of superconducting gap function. By varying the polarizations of the incident and scattered lights, the Raman spectroscopy can probe the momentum dependence of the gap function over the Fermi surface. For the isotropic s-wave superconductor, the Raman scattering is insensitive to the polarization directions. The peak positions of the Raman spectra for all symmetry modes, including A_{1g}, B_{1g} and B_{2g}, are located at $\omega \approx 2\Delta_0$. However, for the d-wave superconductor, the Raman spectral peaks are located at different frequencies for different symmetry modes. The contribution to the B_{1g}-mode comes mainly from the anti-nodal region on the Fermi surface with the maximal energy gaps. Similar as in the s-wave superconductor, the spectral peak is located at $\omega \approx 2\Delta_0$. On the other hand, the contribution to the B_{2g}-mode is mainly from the quasiparticle excitations around the nodal region of the $d_{x^2-y^2}$-wave energy gap, while the contribution from the anti-nodal region is completely suppressed. The spectral peak is located at $\omega \approx 1.6\Delta_0$. The Raman spectrum of the A_{1g}-mode is simply an algebraic average of the contribution of quasiparticles over the entire Fermi surface if ignoring the charge fluctuation induced by the incident light. But the Coulomb screening of electrons modifies significantly the spectrum. The measured spectrum of the A_{1g}-mode depends strongly on the energy dispersion of electrons and is system dependent. Its peak can appear around $\omega \approx 2\Delta_0$ or at a frequency much lower than $2\Delta_0$.

The low frequency Raman scattering cross section varies exponentially with ω in the s-wave superconductor, but exhibits a power-law dependence of ω in the d-wave superconductor. The non-magnetic scattering does not change the low frequency exponential behavior in the s-wave superconductor, but does change the low frequency behavior in the d-wave case. The impurity scattering broadens the peaks of Raman spectra but usually does not change the peak positions. In the intrinsic d-wave superconductor, the low frequency Raman response functions varies linearly with ω for the A_{1g} and B_{2g}-modes, but cubically with ω for the B_{1g}-mode. However, in a disordered d-wave superconductor, the Raman response functions of the B_{1g}-mode also scales linearly with ω in low frequencies, similar as for the B_{2g}-mode.

In Fig. 11.3, the measurement data of the Raman spectra of the A_{1g}, B_{1g}, and B_{2g}-modes for $Bi_2Sr_2CaCu_2O_8$ are shown and compared with the theoretical calculation [214]. The spectra of these Raman modes show qualitatively the same behavior in other hole-doped high-T_c superconductors [215, 216, 217, 218]. As revealed by the experimental data, the low frequency Raman scattering cross section decays linearly with ω for the A_{1g} and B_{2g}-modes. For the B_{1g}-mode, the low frequency Raman spectrum decays faster and scales approximately as ω^3 within the experimental errors. In some superconductors, the low frequency cross section of the B_{1g}-mode is also a linear function of ω. This may result from the impurity scattering. All these properties are consistent with the expected scaling behaviors of low frequency Raman spectra of d-wave superconductors.

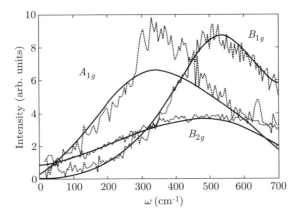

Figure 11.3 The Raman spectra of $Bi_2Sr_2CaCu_2O_8$ ($T_c = 90K$) at $T = 20K$. The solid curves are obtained by theoretical calculation. Reprinted with permission from [214]. Copyright (1996) by the American Physical Society.

In the optimally doped as well as some overdoped high-T_c superconductors, the peak energy ω_p of the B_{1g} Raman mode is located around $(6 \sim 10)T_c$, which is about 1/3 higher than those of the A_{1g} and B_{2g}-modes. The difference between the peak energies of the A_{1g} and B_{2g}-modes is very small. The peak energy of the B_{2g}-mode is slightly higher. These differences in the peak energies of the A_{1g}, B_{1g} and B_{2g}-modes agree qualitatively with the theoretical calculations by considering the vertex correction induced by the Coulomb interaction for d-wave superconductors. If we assume that the peak energy ω_p of the B_{1g} Raman mode is equal to 2Δ, then the value of $2\Delta/T_c$ such obtained is much larger than the weak-coupling BCS s or d-wave superconductors. In the underdoped high-T_c superconductors, the value of $2\Delta/T_c$ determined by the B_{1g} peak increases with decreasing doping and is generally around or larger than 10.

The fact that $2\Delta/T_c$ is larger than the theoretical prediction based on the weak coupling theory is closely related to the pseudogap effect observed in high-T_c cuprates. One possibility is that electrons are already paired at a temperature much higher than T_c, but the phase coherence among Cooper pairs is not establibled at the same temperature due to strong phase fluctuations. In the weak-coupling BCS theory, the phase fluctuation is ignored and the superconducting condensation happens immediately after electrons form Cooper pairs. Thus the superconducting transition temperature predicted in this theory is much larger than the measured value.

The Raman peaks of high-T_c superconductors are usually broad, not as sharp as those observed in the s-wave superconductors. This is likely due to the fact that the Van Hove divergence of the density of states is just logarithmic at $\omega = \Delta$ in the d-wave superconductor, much weaker than the square root divergence in the s-wave case. Elastic and inelastic scatterings both can broaden the Raman scattering peaks. The broadening due to the inelastic scattering is sensitive to temperature. It is more prominent in high temperatures. One can determine the contribution of inelastic scattering through the measurement of the variance in the temperature dependence of Raman scattering peaks.

Compare to hole-doped cuprate superconductors, the Raman spectra of electron-doped ones behave very differently. Fig. 11.4 shows the Raman spectra for electron-doped high-T_c cuprates [165, 219, 220]. The main difference than the hole-doped case is that the peak of the B_{2g}-mode appears at a higher energy than that of the B_{1g}-mode. If we still use the single-band model to analyze the Raman spectra in the electron-doped high-T_c cuprates, then the gap function would vary non-monotonically with the momentum on the Fermi surface, which is inconsistent with the $d_{x^2-y^2}$-wave pairing symmetry.

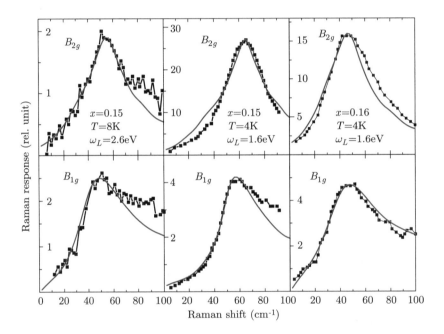

Figure 11.4   The Raman spectra of the electron-doped high-T_c superconductor $Nd_{2-x}Ce_xCuO_4$ and the comparison with the calculation based on the two-band theory. The experiment results in the first column is from Ref. [165], and those from the second and third column are from Ref. [219, 220]. Reprinted with permission from [221]. Copyright (2006) by the American Physical Society.

However, as pointed out in Sect. 9.6, the electron-doped cuprate is not a single-band system. There exist two disconnected Fermi surfaces. One of them is electron-like, located at $(\pi, 0)$ and its equivalent points. The other is hole-like, located around the nodal directions of the $d_{x^2-y^2}$-wave gap function. These two non-equivalent Fermi surfaces can be effectively described by a two-band model [162]. The weight of contribution of these two bands to a specific Raman mode is different. The vertex function of the B_{1g}-mode is zero along the nodal direction of the $d_{x^2-y^2}$-gap function and reaches the maximum at $(\pi, 0)$. The Raman spectrum of this mode contributes mainly by the quasiparticle excitations around $(\pi, 0)$ on the Fermi surfaces. On the contrary, the vertex function of the B_{2g}-model reaches the maximum along the nodal direction and becomes zero at $(\pi, 0)$. It contributes mainly by the quasiparticle excitations along the nodal direction. This suggests that unlike in the hole-doped case, we should not use simply the single-band model to understand the experimental results of Raman spectroscopy for electron-doped high-T_c superconductors.

Indeed, if the weak coupling two-band model introduced in Sect. 9.6 is adopted to analyze the Raman spectra, we find that there is no contradictory between the experimental results shown in Fig. 11.4 and the $d_{x^2-y^2}$-wave pairing symmetry. The Raman spectra of the B_{1g} and B_{2g}-modes are mainly the contributions of the hole and electron bands, respectively. Therefore, the peak energies of these two modes are determined by the energy gaps on two different bands. The peak energy of the B_{2g} mode is higher than that of the B_{1g} mode because the gap amplitude of the hole band is larger than that of the electron band. The theoretical curves shown in Fig. 11.4 are obtained from numerical calculation by considering the contribution of non-resonant Raman scattering based on the two-band model introduced in Sect. 9.6. The agreement between the calculation and the experimental result lends strong support to the assumption that the superconducting pairing in electron-doped high-T_c cuprate has the $d_{x^2-y^2}$-wave symmetry.

Chapter 12

Nuclear Magnetic Resonance

12.1 Spin Correlation Function

Effect of magnetic correlations plays an important role in the study of d-wave superconductivity. In particular, the spin fluctuation is very strong in high-T_c cuprates and its coupling with electrons is widely believed to be the driving force to glue electrons into Cooper pairs. A thorough investigation of magnetic fluctuations is crucial to the understanding of high-T_c pairing mechanism.

Experimentally, properties of magnetic fluctuations are investigated by measuring dynamic spin correlation functions of electrons using neutron scattering and nuclear magnetic resonance (NMR) techniques. These two kinds of experimental probes are complementary to each other, providing a comprehensive understanding of the spin dynamics of d-wave superconductors.

NMR spectroscopy relies on the measurement of a resonant phenomenon in which nuclei in a magnetic field absorb and re-emit electromagnetic radiation. It measures the Knight shift, the nuclear spin-lattice relaxation, and other physical quantities of electrons [222] and is one of the principal techniques for studying magnetic correlations in solids. The nuclear spin resonance frequency equals the Zeeman splitting energy of a nuclear spin and is proportional to the applied magnetic field. This resonance frequency is typically of the order of 10^{-7}eV, much lower than other energy scales of electrons in solids. Hence the energy window measured by NMR is very narrow. The resonance frequencies are different for different nuclei. This can be used to resolve the contribution of different nuclei. Thus the spatial resolution of NMR is very high. On the contrary, the momentum resolution of NMR is very poor according to the uncertainty principle.

In a solid, the spin magnetization of electron is defined by

$$M_e(r) = -\gamma_e\hbar \sum_i S_i\delta(r - r_i),$$ (12.1)

where $\gamma_e = e/mc$ is the gyromagnetic ratio of electron. $S_i = c_i^\dagger(\sigma/2)c_i$ is the electron spin operator at site i and $c_i^\dagger = (c_{i\uparrow}^\dagger, c_{i\downarrow}^\dagger)$. The Zeeman energy corresponding to $M_e(r)$ is given by

$$H_M = -\int dr M_e \cdot H(r,t) = \gamma_e\hbar \sum_q S(-q) \cdot H(q,t),$$ (12.2)

where $S(q)$ and $H(q)$ are the Fourier components of the electron spin and the applied magnetic field, respectively

$$S(q) = \sum_i S_i e^{iq\cdot r_i} = \sum_k c_{k+q}^\dagger \frac{\sigma}{2} c_k,$$

$$H(q) = \frac{1}{V}\sum_i e^{iq\cdot r_i} H(r_i).$$

In an external magnetic field, the magnetic response of electrons is determined by the spin susceptibility, χ, defined by

$$\langle M_\mu(q,t)\rangle = \chi_{\mu\nu}(q,t)H_\nu(q,t).$$ (12.3)

Under the linear approximation, $\chi_{\mu\nu}$ is determined by the dynamical spin-spin correlation function as defined in the Kubo formula

$$\chi_{\mu\nu}(q,t) = i\gamma_e^2\hbar^2 \langle[S_\mu(q,t), S_\nu(-q,0)]\rangle\,\theta(t).$$ (12.4)

The corresponding Matsubara function is defined by

$$\chi_{\mu\nu}(q,\tau) = \gamma_e^2\hbar^2 \langle T_\tau S_\mu(q,\tau) S_\nu(-q,0)\rangle.$$ (12.5)

In an isotropic system, if the vertex correction is negligible and $G(-k,\tau) = G(k,\tau)$, then the diagonal component of $\chi_{\mu\nu}$ can be expressed using the single-particle Green's function as

$$\chi_{\mu\mu}(q,\tau) = \chi_{zz}(q,\tau) = -\frac{\gamma_e^2\hbar^2}{4}\sum_k \mathrm{Tr}\, G(k,\tau) G(q+k,0).$$ (12.6)

After the Fourier transformation, χ_{zz} becomes

$$\chi_{zz}(q, i\omega_n) = -\frac{\gamma_e^2 \hbar^2}{4\beta V} \sum_{k, p_m} \text{Tr}\, G(k, ip_m) G(q+k, i\omega_n + ip_m). \qquad (12.7)$$

It can be further expressed using the single-particle spectral function as

$$\chi_{zz}(q, i\omega_n) = -\frac{\gamma_e^2 \hbar^2}{4\pi^2 V} \sum_k \int d\omega_1 d\omega_2 \frac{f(\omega_1) - f(\omega_2)}{i\omega_n + \omega_1 - \omega_2}$$

$$\text{TrIm}G^R(k, \omega_1)\, \text{Im}G^R(q+k, \omega_2). \qquad (12.8)$$

In an ideal d-wave superconductor, the single particle Green's function is determined by Eq. (4.3). In this case, the summation over p_m in Eq. (12.7) can be readily done. Taking analytic continuation by setting $i\omega_n \to \omega + i0^+$, we arrive at

$$\chi_{zz}(q, \omega) = -\frac{\gamma_e^2 \hbar^2}{4V} \sum_k [2f(E_k) - 1]\, [g_{k,q}(\omega) + g_{k,q}(-\omega)], \qquad (12.9)$$

where

$$g_{k,q}(\omega) = \frac{E_k(\omega + E_k) + \xi_k \xi_{k+q} + \Delta_k \Delta_{k+q}}{E_k\left[(\omega + E_k)^2 - E_{k+q}^2\right]}.$$

In the long wavelength limit, the static magnetic susceptibility is found to be

$$\lim_{q \to 0} \chi_{zz}(q, 0) = -\frac{\gamma_e^2 \hbar^2}{2V} \sum_k \frac{\partial f(E_k)}{\partial E_k} = \frac{\gamma_e^2 \hbar^2}{4} Y(T). \qquad (12.10)$$

Clearly, the static susceptibility of electrons is temperature dependent. It is determined purely by the quasiparticle density of states, and scales linearly with T in low temperatures:

$$\chi_{zz}(0, 0) \approx \frac{(\ln 2)\gamma_e^2 \hbar^2 N_F k_B T}{2\Delta_0}, \qquad T \ll T_c. \qquad (12.11)$$

12.2 Hyperfine Interaction

The energy level of nuclear spin is split under an external magnetic field. This split is determined by the Zeeman energy of nuclear spin

$$H_{Zeeman} = -\gamma_N \hbar I \cdot H. \qquad (12.12)$$

The gap between two adjacent energy levels is

$$\hbar\omega_0 = -\gamma_N \hbar H, \tag{12.13}$$

where γ_N is the gyromagnetic ratio of nuclear magnetic moment. I is the nuclear spin. In this nuclear spin system, if we further apply an alternating electromagnetic field to generate the transition between two different energy levels, resonant absorption by nuclear spins occurs when the frequency of the apply electromagnetic field equals ω_0. Such magnetic resonance frequency typically corresponds to the radio frequency range of the electromagnetic spectrum for magnetic fields up to roughly 20T. It is this magnetic resonant absorption which is detected in NMR.

In solid systems, in addition to the applied magnetic field, nuclear spin also experience the effective magnetic field generated by electrons around the nuclei. Due to the interaction between the nuclear spin and the magnetic moment of electron, there is a shift in the resonance frequency. The magnetic moment of electron is contributed by both the orbital and spin angular momenta. The shift induced by the orbital magnetic moment of electron is called the chemical shift. It is also called the Van Vleck shift. The shift induced by the spin moment of electron, on the other hand, is called the Knight shift, which is more pronounced in a metallic system.

The chemical shift is determined by the Hamiltonian

$$H_{chem} = -\frac{\gamma_N \hbar}{c} I \cdot \int dr \frac{r \times j(r)}{r^3}, \tag{12.14}$$

where j is the gauge invariant current density operator of electrons defined by

$$j = \frac{-ie\hbar}{2m} (\psi^* \nabla \psi - \psi \nabla \psi^*) - \frac{e^2}{mc} A \psi^* \psi. \tag{12.15}$$

As the electric current surrounding a nucleus is strongly screened by electromagnetic interactions, the chemical shift is also strongly affected by this effect. The electromagnetic screening, on the other hand, is affected by the chemical environment. Thus a nuclear spin can have different chemical shifts under different chemical environments. The orbital angular momentum of electron on a *s*-orbital is zero. The contribution of this electron to the chemical shift comes mainly from the A term in the current operator j. This contribution is diamagnetic, which is to lower the resonance frequency. The chemical shift induced by an electron with a finite orbital angular momentum results from the contribution of both the paramagnetic and diamagnetic current terms. It can be either positive or negative, depending on the relative contribution of these two kinds of currents. Usually, the chemical shift is positive, namely dominated by the paramagnetic contribution. But its value is generally very small. The chemical shift is not sensitive to the change of temperature, neither to the superconducting transition. Hence limited information can be extracted from the measurement of chemical shift on the magnetic and superconducting correlations of electrons.

The Knight shift is much larger than the chemical shift in conductors. The Knight shift results from the hyperfine interaction between a *s*-orbital electron and a nuclear spin, which is also called the Fermi contact interaction. It is determined by the wavefunction of electrons and governed by the Hamiltonian

$$H_{hyper} = \frac{8\pi}{3} \gamma_e \gamma_N \hbar^2 \sum_i S_i \cdot I_i |u(r_i)|^2, \qquad (12.16)$$

where $|u(r_i)|^2$ is the probability of electron at the nuclear site.

There exists a magnetic dipolar interaction between an electron with finite orbital angular momentum (i.e. non-*s*-orbital) and a nuclear spin. But the energy scale of this dipolar interaction is generally very small in comparison with the hyperfine interaction. Nevertheless, non-*s*-orbital electrons can affect the distribution of electrons at inner shell *s*-orbitals, which in turn induces indirectly an effective contact interaction between these non-*s*-orbital electrons and nuclear spins. Furthermore, the spin moment of a magnetic ion can be transferred to a *s*-orbital on one of its neighboring sites via the orbital hybridization. It interacts indirectly with a neighboring nuclear spin through the Fermi-contact interaction, generating a transferred hyperfine interaction. This transferred interaction has significant contribution to NMR in magnetic materials.

In high-T_c superconductors, the local moment of a Cu^{2+} cation interacts strongly with its neighboring nuclear spins. This interaction is even stronger than the direct hyperfine interaction on Cu. Thus in the analysis of NMR experiments of high-T_c cuprates, one should consider carefully the transferred hyperfine interactions induced by the hybridization of Cu^{2+} $3d_{x^2-y^2}$ or $4s$ orbitals with other atoms.

Assuming a magnetic field is applied along the c-axis, Mila and Rice proposed a phenomenological model to describe the transferred hyperfine interaction between copper and oxygen atoms in the CuO_2 plane of high-T_c cuprates [223]. The model Hamiltonian is defined by

$$H_{Hyper} = \sum_i \left[A^{63}I_i \cdot S_i + \sum_{\delta=\pm\hat{x},\pm\hat{y}} \left(B^{63}I_{i+\delta} \cdot S_i + C^{17}I_{i+\delta/2} \cdot S_i \right) \right],$$
(12.17)

where ^{63}I and ^{17}I are the nuclear spins of ^{63}Cu and ^{17}O, respectively. (A, B, C) are the coefficients of the hyperfine interactions. The hyperfine interaction of ^{63}Cu contains two terms. One is the interaction between the electron spin and the nuclear spin of Cu on the same site. The other is the transferred hyperfine interaction between electron and nuclear spins on two neighboring Cu sites. This transferred hyperfine interaction does not exist in conventional metals, but it is finite in high-T_c cuprates due to the strong antiferromagnetic exchange interaction between Cu^{2+} spins. The transferred hyperfine interaction between the nuclear spin of oxygen and the magnetic moment of Cu^{2+} on one of its neighboring sites is also strong. But the hyperfine interactions within an O^{2-} anion itself is small and negligible. The effective hyperfine interactions for other atoms in high-T_c superconductors, for example, the yttrium atom in YBCO, can be similarly constructed.

In the momentum space, Eq. (12.17) can be expressed as

$$H_{hyper} = \sum_q \left[F_{Cu}(q)\,^{63}I + F_O(q)\,^{17}I \right] \cdot S(q),$$
(12.18)

where $F_{Cu}(q)$ and $F_O(q)$ are the structural factors of the hyperfine interactions for copper and oxygen unclear spins, respectively:

$$F_{Cu}(q) = A + 2B(\cos q_x + \cos q_y), \qquad (12.19)$$

$$F_O(q) = 2C\cos(q_x/2). \qquad (12.20)$$

A more general expression of the hyperfine interaction has the form

$$H_{hyper} = \sum_{q,\mu} F(q)I \cdot S(q), \qquad (12.21)$$

where $F(q)$ is the structural factor.

12.3 Knight Shift

In an external magnetic field H_0, both the electron and nuclear spins are polarized. These polarized spins of electrons introduce an "extra" effective field at the nuclear site. As the resonance frequency is determined by the difference between the energy levels of nuclear spin in the whole magnetic field, the field induced by the electron spins leads to a shift in the resonance frequency. The relative shift in the resonance frequency is referred as the Knight shift, which is defined by the relative change of the resonance frequency for atoms in a metal compared with the same atoms in a nonmetallic environment. This shift was first observed in a paramagnetic substance by Walter D. Knight in 1949.

The Knight shift is determined by the magnetic susceptibility of electrons and the hyperfine interaction. For the hyperfine interaction defined in Eq. (12.21), the effective magnetic field generated by the polarized electron spins on the nuclear sites equals

$$\delta H_\mu = -\frac{F(q)\langle S_\mu(q)\rangle \delta_{q,0}}{\gamma_N \hbar} = \frac{F(0)\chi_{\mu\mu}(0,0)H_0}{\gamma_N \gamma_e \hbar^2}, \qquad (12.22)$$

where μ is the direction of the applied magnetic field. It induces a shift in the resonance frequency as

$$\delta\omega = \gamma_N \delta H_\mu. \qquad (12.23)$$

The Knight shift K_μ is defined by the ratio between $\delta\omega$ and the resonance frequency purely induced by the applied magnetic field:

$$K_\mu = \frac{\delta\omega}{\gamma_N H_0} = \frac{F(0)\chi_{\mu\mu}(0,0)}{\gamma_N \gamma_e \hbar^2}. \qquad (12.24)$$

Clearly, the Knight shift is direction-dependent. It is proportional to the electron static magnetic susceptibility.

In conventional metals, without considering the contribution of electron-electron interactions, $\chi(0,0)$ simply equals the Pauli paramagnetic susceptibility, proportional to the electron density of states at the Fermi level:

$$\chi_{zz} = \frac{\gamma_e^2 \hbar^2 N_F}{2}. \tag{12.25}$$

The corresponding Knight shift is given by

$$K = \frac{F(0)\gamma_e N_F}{2\gamma_N}. \tag{12.26}$$

In an s-wave superconductor with spin singlet pairing, both $\chi(0,0)$ and the corresponding Knight shift drop monotonically with decreasing temperature. In particular, in an s-wave superconductor, the Knight shift decays exponentially in low temperatures, similar to other thermodynamic quantities. In a d-wave superconductor, however, the Knight shift is proportional to the quasiparticle density of states and decays just linearly in low temperatures. If the magnetic field is applied along the c-axis, the magnetic susceptibility is determined by Eq. (12.11) in low temperatures and the corresponding Knight shift is approximately given by

$$K \approx \frac{(\ln 2)F(0)\gamma_e N_F}{2\gamma_N} \frac{k_B T}{\Delta_0}. \tag{12.27}$$

12.4 Spin-Lattice Relaxation

In an applied field, the energy levels of nuclear spins are split. Upon electromagnetic excitation, a nuclear spin can occupy a higher energy level state. At a finite temperature, a nuclear spin can be also excited from a lower energy state to a higher energy one by thermal fluctuations. The nuclear spin in an excited state is unstable and tends to relax back to the ground state. The relaxing process of nuclear spin from a non-equilibrium state to the thermal equilibrium state by exchanging energy with the environment is called the spin-lattice relaxation. Here the lattice is just the solid state environment surrounding the nuclei. During the relaxation process, the temperature of nuclear spins tends to approach the surrounding lattice temperature. At an equilibrium state, the energies gained and released by nuclear spins are balanced.

The spin-lattice relaxation rate T_1^{-1} is a characteristic quantity describing the process of a nuclear spin polarized along a particular direction relaxing to a random orientation through the scattering from surrounding electrons. The relaxation happens approximately in two steps: The first step happens with only nuclear spins, without exchanging energy with electrons. A nuclear spin tends to reach an instantaneous equilibrium state through exchanging energy with its internal degrees of freedom or with other nuclear spins. The probability of this nuclear spin at different configurations satisfies the statistical distribution of micro-canonical ensemble described by an effective nuclear spin temperature T_N. This temperature is generally higher than the temperature of surrounding electrons T_e. The second step is the relaxation of nuclear spins to the genuine equilibrium state by exchanging energy with electrons through the hyperfine interaction so that T_N is cooled down to the environment temperature T_e. The first step runs much faster than the second one. This is the reason why we can assume that the nucleus spins can reach an instantaneous micro-canonical equilibrium state. It can be also regarded as a basic assumption used in the analysis of spin-lattice relaxation. In real materials, the time scale of the first relaxation step is typically of order $10 \sim 100\mu s$, while the time scale of the second relaxation step is typically of order of ms. Hence the assumption of two-step relaxation is valid.

Under the assumption that the intra-nucleus relaxation is much faster than the spin-lattice relaxation, the nuclear spin relaxation is just the process of the nuclear spin temperature T_N approaching the equilibrium lattice temperature through the hyperfine interaction. To describe this process, let us consider a nuclear spin system in an external magnetic field H_0, whose energy level is given by

$$E_n = -\gamma_N \hbar n H_0 \qquad (n = -I, -I+1, ..., I). \qquad (12.28)$$

At the instantaneous temperature T_N, the probability of the nuclear spin in the state with energy E_n is

$$p_n = \frac{\exp(-\beta_N E_n)}{Z_N}, \tag{12.29}$$

where $\beta_N = 1/k_B T_N$ and

$$Z_N = \sum_n \exp(-\beta_N E_n)$$

is the partition function of the nuclear spin. The average energy of the nuclear spin is given by

$$E_N = \sum_n p_n E_n. \tag{12.30}$$

If W_{nm} is the transition rate from the state of E_n to that of E_m, then the time-derivative of p_n is determined by the formula

$$\frac{\mathrm{d}p_n}{\mathrm{d}t} = \sum_m (p_m W_{mn} - p_n W_{nm}). \tag{12.31}$$

The corresponding rate of change in the energy is

$$\frac{\mathrm{d}E_N}{\mathrm{d}t} = \sum_n E_n \frac{\mathrm{d}p_n}{\mathrm{d}t} = \frac{1}{2} \sum_{mn} (p_m W_{mn} - p_n W_{nm})(E_n - E_m). \tag{12.32}$$

When the system reaches the final equilibrium state, $T = T_e$, the probability of the nuclear spin at each energy level will no longer change with time, i.e. $\mathrm{d}p_n/\mathrm{d}t = 0$. In this case, W_{mn} satisfies the equation

$$\frac{W_{nm}}{W_{mn}} = e^{\beta_e(E_n - E_m)}, \tag{12.33}$$

where $\beta_e = 1/k_B T_e$. Substituting this equation into Eq. (12.32), we have

$$\frac{\mathrm{d}E_N}{\mathrm{d}t} = \frac{1}{2} \sum_{mn} p_m W_{mn} \left[1 - e^{(\beta_e - \beta_N)(E_n - E_m)}\right] (E_n - E_m). \tag{12.34}$$

In real NMR experiments, the energy of nuclear spins $\hbar\omega_0$ is usually 3 to 5 orders of magnitude smaller than the measurement temperatures, i.e. $\hbar\omega_0 \ll k_B T_e$. The right-hand side of Eq. (12.34) can be expanded using $(\beta_N - \beta_e)\hbar\omega_0$

as a small parameter. Correcting up to the first order of $(\beta_N - \beta_e)\hbar\omega_0$, we find that

$$\frac{\mathrm{d}E_N}{\mathrm{d}t} \approx -\frac{\beta_e - \beta_N}{2} \sum_{mn} p_m W_{mn}(E_n - E_m)^2. \tag{12.35}$$

From Eq. (12.32), we also have

$$\frac{\mathrm{d}E_N}{\mathrm{d}t} = \frac{\mathrm{d}\beta_N}{\mathrm{d}t} \sum_n E_n \frac{\mathrm{d}p_n}{\mathrm{d}\beta_N} = \frac{\mathrm{d}\beta_N}{\mathrm{d}t}\left(-\sum_n E_n^2 p_n + E_N^2\right). \tag{12.36}$$

Therefore,

$$\frac{\mathrm{d}\beta_N}{\mathrm{d}t} \approx \frac{\displaystyle\sum_{mn} p_m W_{mn}(E_n - E_m)^2}{2\left(\displaystyle\sum_n E_n^2 p_n - E_N^2\right)}(\beta_e - \beta_N). \tag{12.37}$$

By further expanding p_m up to the first order of β_N, we then obtain the equation for determining the spin-lattice relaxation rate:

$$\frac{\mathrm{d}\beta_N}{\mathrm{d}t} \approx \frac{\displaystyle\sum_{mn} W_{mn}(E_n - E_m)^2}{2\displaystyle\sum_n E_n^2}(\beta_e - \beta_N) = \frac{\beta_e - \beta_N}{T_1}, \tag{12.38}$$

where

$$T_1^{-1} = \frac{\displaystyle\sum_{mn} W_{mn}(E_n - E_m)^2}{2\displaystyle\sum_n E_n^2} = \frac{\displaystyle\sum_{mn} W_{mn}(n - m)^2}{2\displaystyle\sum_n n^2} \tag{12.39}$$

is the nuclear spin-lattice relaxation rate. It describes the speed of β_N approaching β_e under the electron-nucleus interaction. The transition matrix elements W_{mn} between two energy levels m and n induced by the hyperfine interaction satisfy the selection rule: $m = n, n \pm 1$. Thus the above expression can be further simplified as

$$T_1^{-1} = \frac{\displaystyle\sum_n (W_{n,n+1} + W_{n+1,n})}{2\displaystyle\sum_n n^2}. \tag{12.40}$$

In a spin-lattice relaxation process, the nuclear spin undergoes a transition in which it either absorbs or releases energy. In order to conserve energy, the lattice must undergo a compensating change. Simultaneously, electrons also undergo a transition from one state to another by releasing or absorbing energy. According to the Fermi golden rule, , the transition rate of nuclear spin from state E_n to E_m for the hyperfine interaction defined in Eq. (12.21) is given by

$$W_{mn} = \frac{2\pi}{\hbar} \sum_{\alpha,\alpha'} \left| \langle n, \alpha | \sum_{q,\mu} F_\mu(q) I_\mu S_\mu(q) | m, \alpha' \rangle \right|^2 \delta(E_{\alpha'} - E_\alpha + \omega_{mn}), \quad (12.41)$$

where $\omega_{mn} = E_m - E_n$. $|\alpha\rangle$ and E_α are the eigen wavefunction and eigen energy of electron, respectively. This formula of W_{mn} can be also expressed as

$$W_{mn} = \sum_{q,\mu,\nu} A_{\mu\nu}(q, m, n) \sum_{\alpha,\alpha'} \langle \alpha | S_\mu(q) \alpha' \rangle \langle \alpha' | S_\nu(-q) \alpha \rangle \delta(E_{\alpha'} - E_\alpha + \omega_{mn}),$$
$$(12.42)$$

where

$$A_{\mu\nu}(q, m, n) = \frac{2\pi}{\hbar} |F(q)|^2 \langle n | I_\mu | m \rangle \langle m | I_\nu | n \rangle. \quad (12.43)$$

From the fluctuation-dissipation theorem, it can be shown that the right-hand side of Eq. (12.42) is proportional to the imaginary part of the magnetic susceptibility:

$$\sum_{\alpha,\alpha'} \langle \alpha | S_\mu(q) \alpha' \rangle \langle \alpha' | S_\nu(-q) \alpha \rangle \delta(E_{\alpha'} - E_\alpha + \omega) = \frac{\mathrm{Im}\chi_{\mu\nu}(q, \omega)}{\pi\gamma_e^2 \hbar^2 (1 - e^{-\beta\omega})}. \quad (12.44)$$

Therefore, we have

$$W_{mn} = \frac{1}{\pi\gamma_e^2 \hbar^2 (1 - e^{-\beta\omega_{mn}})} \sum_{q,\mu,\nu} A_{\mu\nu}(q, m, n) \mathrm{Im}\chi_{\mu\nu}(q, \omega_{mn})$$
$$\approx \frac{k_B T}{\pi\gamma_e^2 \hbar^2} \sum_{q,\mu,\nu} A_{\mu\nu}(q, m, n) \lim_{\omega \to 0} \frac{\mathrm{Im}\chi_{\mu\nu}(q, \omega)}{\omega}. \quad (12.45)$$

The spin-lattice relaxation rate is then obtained as

$$T_1^{-1} = k_B T \sum_{q,\mu,\nu} \tilde{A}_{\mu\nu}(q) \lim_{\omega \to 0} \frac{\mathrm{Im}\chi_{\mu\nu}(q, \omega)}{\omega}, \quad (12.46)$$

where the coefficient $\tilde{A}_{\mu\nu}(q)$ is defined by

$$\tilde{A}_{\mu\nu}(q) = \frac{|F(q)|^2}{\gamma_e^2 \hbar^3} \frac{\sum_{mn} \langle n | I_\mu | m \rangle \langle m | I_\nu | n \rangle (E_n - E_m)^2}{\sum_n E_n^2}. \quad (12.47)$$

In an isotropic system, the spin susceptibility is diagonal and direction independent:

$$\chi_{\mu\nu} = \chi_{zz}\delta_{\mu,\nu}. \tag{12.48}$$

In this case, T_1^{-1} can be written as

$$T_1^{-1} = \frac{2k_B T}{\gamma_e^2 \hbar^3} \sum_q |F(q)|^2 \lim_{\omega \to 0} \frac{\mathrm{Im}\chi_{zz}(q,\omega)}{\omega}. \tag{12.49}$$

In the above derivation, the following identity is employed,

$$\frac{\sum_{\mu mn} \langle n |I_\mu| m \rangle \langle m |I_\mu| n \rangle (E_n - E_m)^2}{\sum_n E_n^2} = 2.$$

Based on the expression of structure factors, we know that antiferromagnetic fluctuations affect the NMR results at copper and oxygen sites differently. The antiferromagnetic fluctuation can enhance the spin-lattice relaxation at the copper sites, but it has almost no effect on the spin-lattice relaxation at the oxygen sites. For high-T_c cuprates, from the experimental measurement, it was found that the relaxation rate T_1^{-1} of Cu is significantly larger than that of O or Y. Furthermore, in the normal state, T_1^{-1} at the copper sites does not satisfy the Korringa relation that generally holds in a Landau Fermi liquid system. It implies that the antiferromagnetic fluctuation does have a strong impact on the spin-lattice relaxation rate. Nevertheless, in the discussion of low-temperature spin-lattice relaxation of d-wave superconductors, it is the nodal quasiparticles excitations rather than the antiferromagnetic fluctuations that play a more important role. To the leading order approximation, one can neglect the contribution of antiferromagnetic fluctuations to the spin-lattice relaxation.

If $F(q) = F(0)$ does not depend on q, the spin-lattice relaxation rate becomes

$$T_1^{-1} = \frac{2k_B T F^2(0)}{\gamma_e^2 \hbar^3} \sum_q \lim_{\omega \to 0} \frac{\mathrm{Im}\chi_{zz}(q,\omega)}{\omega}. \tag{12.50}$$

This expression is commonly used in the analysis of experimental data. It catches the main feature of the spin-lattice relaxation and holds in most of cases. For more general cases, we need to know the expression of $F(q)$.

From Eq. (12.8), $\mathrm{Im}\chi_{zz}$ can be written using the single-particle spectra function as

$$\lim_{\omega\to 0}\frac{\mathrm{Im}\chi_{zz}(q,\omega)}{\omega} = -\frac{\gamma_e^2\hbar^2}{4\pi V}\sum_k\int d\omega\,\frac{\partial f(\omega)}{\partial\omega}\mathrm{TrIm}G\,(q+k,\omega)\,\mathrm{Im}G\,(k,\omega).$$
(12.51)

Based on this expression, it is simple to show that the following equation holds for d-wave superconductors:

$$T_1^{-1} = -\frac{\pi k_B T F^2(0)}{\hbar}\int d\omega\,\frac{\partial f(\omega)}{\partial\omega}\rho^2(\omega),$$
(12.52)

where $\rho(\omega)$ is the electron density of states.

For s-wave superconductors, Eq. (12.50) can be simplified as

$$T_1^{-1} = -\frac{\pi k_B T F^2(0)}{\hbar}\int d\omega\,\frac{\partial f(\omega)}{\partial\omega}\rho^2(\omega)\left(1+\frac{\Delta^2}{\omega^2}\right).$$
(12.53)

In the normal state, $\Delta = 0$ and $\rho(\omega)\approx N_F$,

$$T_1^{-1} = \frac{\pi k_B T F^2(0)N_F^2}{\hbar}.$$
(12.54)

Combining this equation with the expression of the Knight shift, Eq. (12.26), we obtain

$$k_B T K^2 T_1 = \frac{\hbar}{4\pi}\frac{\gamma_e^2}{\gamma_N^2},$$
(12.55)

which is also called Korringa relation. It shows that $k_B T K^2 T_1$ is universal in the normal state, depending on the ratio of γ_e/γ_N, but not on electronic and lattice structures.

In an ideal s-wave superconductor, $\rho(\omega)$ diverges at $\omega = \Delta$, and $T_1^{-1}(T)$ should also diverge. But in real materials, this divergence is smeared out by strong coupling, impurity scattering, and other effects, leaving only a peak at T_c as a residual character of s-wave superconductor. This characteristic peak is often called the coherent or Hebel-Slichter peak [224]. It was observed in most metal or alloy-based superconductors. But there are exceptions. For example, in the strong coupling s-wave superconductor $TlMo_6Se_{7.5}$, this coherence peak is not observed.

In a d-wave superconductor, the density of states also diverges at $\omega = \Delta$. But the divergence is just logarithmic, much weaker than the square-root divergence of the density of states in an s-wave superconductor. In this case, there is no diverges in the integral of Eq. (12.52). Nevertheless, T_1^{-1} still exhibits a small coherence peak just below T_c. This peak is not as robust as in the s-wave superconductor. It is rather fragile against antiferromagnetic fluctuation, strong coupling and other effects. Thus it is difficult to observe this peak in a d-wave superconductor. In low temperatures, the integral in Eq. (12.52) contributes mainly by low-lying excitations. In this case, $\rho(\omega)$ is proportional to ω and T_1^{-1} is approximately equal to

$$T_1^{-1}(T) \approx \frac{\pi^3 N_F^2 F^2(0) k_B^3 T^3}{3\hbar \Delta_0^2}, \qquad T \ll T_c. \qquad (12.56)$$

Combining this expression with Eq. (12.27), we find that the Korringa relation holds for d-wave superconductors in low temperatures:

$$k_B T K^2 T_1 = \frac{3\hbar \ln^2 2}{4\pi^3} \frac{\gamma_e^2}{\gamma_N^2}. \qquad (12.57)$$

Again $k_B T K^2 T_1$ is a universal constant proportional to γ_e^2/γ_N^2, but the coefficient is changed in comparison with the normal state.

12.5 Effect of Impurity Scattering

The spin susceptibility $\chi(0,0)$ is approximately proportional to the real part of the in-plane current-current correlation function of electrons in the low-energy long-wave length limit. The latter, on the other hand, is proportional to the paramagnetic contribution to the superfluid density:

$$\chi_{zz}(0,0) = -\frac{\gamma_e^2 \hbar^2}{2e^2 v_F^2} \Pi_{ab}(0,0). \qquad (12.58)$$

Because the diamagnetic contribution to the superfluid density is nearly temperature independent, the temperature dependence of $\chi(0,0)$ is similar to that of the in-plane superfluid density.

In the gapless regime, it is simple to show, following the derivation for Π_{ab} presented in Chapter 8, the low temperature static uniform spin susceptibility varies quadratically with temperature in d-wave superconductors:

$$\chi_{zz}(0,0) = \frac{\gamma_e^2 \hbar^2 \Gamma_0 N_F}{\pi a \Delta_0} \left(\ln \frac{2\Delta_0}{\Gamma_0} + \frac{a^2 k_B^2 T^2}{6\Gamma_0^2} + o(T^4) \right). \tag{12.59}$$

Therefore, the Knight shift in the same temperature range, correcting up to the order of T^2, is given by

$$K = \frac{\gamma_e \Gamma_0 F(0) N_F}{\pi a \gamma_N \Delta_0} \left(\ln \frac{2\Delta_0}{\Gamma_0} + \frac{a^2 k_B^2 T^2}{6\Gamma_0^2} + o(T^4) \right). \tag{12.60}$$

The Knight shift is finite at zero temperature because the density of states of quasiparticles at the Fermi level is finite in disordered d-wave superconductors. The quadratic temperature dependence of the Knight shift is a consequence of the Sommerfeld expansion.

The spin-lattice relaxation rate T^{-1} is determined by the imaginary part of the magnetic susceptibility. For a d-wave or other unconventional super-conductor whose gap function vanishes after averaging over the entire Fermi surface, the spin-lattice relaxation rate is determined purely by the quasiparticle density of states $\rho(\omega)$ if the structural factor is momentum independent, i.e. $F(q) = F(0)$,

$$T_1^{-1} = -\frac{\pi k_B T F^2(0)}{\hbar} \int d\omega \frac{\partial f(\omega)}{\partial \omega} \rho^2(\omega). \tag{12.61}$$

In a disordered system, the quasiparticle density of states is finite at the Fermi energy. In the gapless regime, the quasiparticle density of states in the unitary and Born scattering limits are given by Eq. (8.63) and Eq. (8.64), respectively. Substituting these equations into Eq. (12.61), the spin-lattice relaxation rate is found to be

$$T_1^{-1} = \frac{\pi k_B T F^2(0) \rho^2(0)}{\hbar} \left[1 + \frac{\pi^4 \Delta_0^2 k_B^2 T^2}{3\Gamma_0^2 \Gamma_N^2} + o(T^4) \right] \tag{12.62}$$

in the Born scattering limit, and

$$T_1^{-1} = \frac{\pi k_B T F^2(0) \rho^2(0)}{\hbar} \left[1 - \frac{\pi^2 k_B^2 T^2}{6\Gamma_0^2} + o(T^4) \right]$$

in the unitary scattering limit. In either limit, T_1^{-1} scales linearly with T in low temperatures, different than the T^3-behavior in a pure d-wave superconductor. The linear temperature term results from the finite density of states of electrons on the Fermi surface. The leading correction to the linear temperature dependence of T_1^{-1} is proportional to T^3. But this T^3-term behaves quite differently in these two scattering limits. The coefficient of the T^3-term is positive in the Born scattering limit, but negative in the unitary scattering limit. This difference results from the difference in the energy dependence of low-energy density of states. From this difference one can in principle determine in which limit the impurity scattering potential is.

12.6 Contribution of Impurity Resonance States

Both the Knight shift and the spin-lattice relaxation are sensitive to the magnetic structure surrounding a nucleus whose NMR spectroscopy is measured. This property of NMR can be used to probe the magnetic structure in the vicinity of an impurity.

Around zinc or other non-magnetic impurities in high-T_c superconductors, it was found from NMR measurements that the impurity contribution to the spin susceptibility is Curie-Weiss-like [225, 226, 227, 228, 229, 230, 231]. One possible explanation to this phenomenon is that in a system with strong antiferromagnetic fluctuation, a zinc or other non-magnetic impurity would induce certain unscreened magnetic moments around the impurity. These induced moments lead to the Curie-Weiss behavior of the spin susceptibility. But this interpretation is not consistent with other experimental observations. First, in the zinc-doped $YBa_2(Cu_{1-x}Zn_x)O_8$ sample, the μSR measurement did not find any evidence of induced local moments [232]. Second, if the Curie-Weiss behavior is indeed induced by the unscreened local moments, then the spin-lattice relaxation rate should increase monotonically with decreasing temperature. However, in high-T_c superconductors the impurity contribution to $(T_1T)^{-1}$ decays exponentially in low temperatures. This exponential decay indicates that the induced magnetic moment (if exists) is frozen, although the mechanism of this frozen effect is unclear [229]. Third, in overdoped high-T_c cuprates, the magnetic correlations are strongly suppressed and the chance to create magnetic moments by a non-magnetic impurity is very slim if not completely impossible [233]. These facts rules out the possibility that the Curie-Weiss behavior observed in the NMR experiments is truly due to the contribution of induced magnetic moments.

In the analysis of NMR experimental data, an important but often overlooked point is the contribution of non-magnetic resonance states generated by non-magnetic impurities in high-T_c superconductors. From the discussion given in Chapter 7, we know that a zinc or other non-magnetic impurity can create a sharp low energy resonance state in the superconducting state of d-wave superconductors [234, 235]. In the absence of an external magnetic field, this low energy resonance state is not magnetically polarized. But its contribution to the magnetic susceptibility is finite. In fact, as will be shown below, if temperature is higher than the resonance energy, the contribution of the resonance state to the spin susceptibility is Curie-Weiss like. Thus as far as the zero-field spin susceptibility is concerned, the non-magnetic resonance state behaves like a local magnetic impurity. On the other hand, if temperature is lower than the resonance energy, it is difficult to excite an electron to the resonance state and the contribution of this state to the spin-lattice relaxation or the Knight shift is very small, exhibiting an activated behavior. This explains naturally the exponential behavior of the $(T_1 T)^{-1}$ at low temperatures.

In a system without translation invariance, the spin-lattice relaxation rate on site r can be similarly derived as for Eq. (12.49). The result is given by

$$\frac{1}{T_1(r)T} = \frac{2k_B}{\gamma_e^2 \hbar^3} \sum_{j,l} F_{j,r} F_{l,r} \lim_{\omega \to 0} \frac{\mathrm{Im}\chi_{zz}(j,l,\omega)}{\omega}, \qquad (12.63)$$

where j or l is the coordinate of r or any of its four nearest neighboring sites. $F_{j,r} = A$ is the structure factor of the hyperfine interaction at site $j = r$. $F_{j,r} = B$ is the indirect hyperfine interaction induced by the exchange interaction of copper spins if $j \neq r$.

In the limit of $\omega \to 0$, the magnetic susceptibility $\mathrm{Im}\chi_{zz}(j,l,\omega)$ can be expressed using the electron Green's function as

$$\lim_{\omega \to 0} \frac{\mathrm{Im}\chi_{zz}(j,l,\omega)}{\omega} = -\frac{\gamma_e^2 \hbar^2}{2\pi} \int_{-\infty}^{\infty} d\varepsilon A(j,l,\varepsilon) \frac{\partial f(\varepsilon)}{\partial \varepsilon}, \tag{12.64}$$

where

$$A(j,j',\varepsilon) = [\mathrm{Im}G_{11}(j,j';\varepsilon)]^2 + [\mathrm{Im}G_{12}(j,j';\varepsilon)]^2.$$

In a single-impurity system, the electron Green's function is given by Eq. (7.8). In the unitary scattering limit, there are poles in $G(r,r';\omega)$. These poles correspond to the impurity induced resonance states. In high-T_c superconductors, the phase shift induced by the zinc-impurity scattering potential is $\delta_0 \approx 0.48\pi$ and the scattering parameter $c \approx 0.0629$. The corresponding resonance frequency Ω' with its imaginary part Ω'' is significantly smaller than the superconducting gap, i.e. $(\Omega', \Omega'') \ll \Delta_0$.

Substituting Eq. (12.64) into Eq. (12.63), $1/T_1 T$ can be rewritten as

$$\frac{1}{T_1(r)T} = -\frac{k_B}{\pi\hbar} \int d\varepsilon \frac{\partial f(\varepsilon)}{\partial \varepsilon} \sum_{jl} F_{j,r} F_{l,r} A(j,l;\varepsilon). \tag{12.65}$$

In the low energy limit, the Green's function of electrons without considering the correction from the impurity scattering, $G^0(r,\omega)$, is a smooth function of ω. Its imaginary part $\mathrm{Im}G^0(r,\omega)$ approaches to zero as $\omega \to 0$. Hence, at $r \neq 0$, the imaginary part of $G^0(r,\Omega')$ is much smaller than its real part. In this case, we can neglect the imaginary part of $G^0(r,\omega)$, and express the correction to the Green's function as

$$\delta G(r,r'\omega) \approx \mathrm{Re}G^0(r,0)T(\omega)\mathrm{Re}G^0(-r',0). \tag{12.66}$$

In low temperatures, the spin-lattice relaxation rate is mainly determined by the resonant state. If the system is particle-hole symmetric, then the contribution to the spin-lattice relaxation rate from the resonance state can be expressed using Eq. (12.66) as

$$\delta \left[T_1(r)T\right]^{-1} \approx -\frac{k_B}{\pi\hbar} \int d\varepsilon \frac{\partial f(\varepsilon)}{\partial \varepsilon} Z^2(r,\varepsilon), \tag{12.67}$$

$$Z(r,\varepsilon) = \sum_j F_{j,r} \left([\mathrm{Re}G_{11}^0(j,0)]^2 + [\mathrm{Re}G_{12}^0(j,0)]^2 \right) T_{11}''(\varepsilon).$$

If the temperature is much larger than the resonance energy but far smaller than T_c, i.e. $k_B T_c \gg k_B T \gg \Omega'$, the integral in Eq. (12.67) contributes mainly by the pole of $T_{11}(\varepsilon)$. In this case, $\partial f(\varepsilon)/\partial \varepsilon|_{\varepsilon=\Omega'} \propto 1/T$, and the spin-lattice relaxation

$$\delta \left[T_1(r) T \right]^{-1} \propto \frac{1}{T} \tag{12.68}$$

has the standard Curie-Weiss form, similar as in a system of magnetic impurities. This $1/T$-behavior of $(T_1 T)^{-1}$ agrees well with the experimental result in the superconduting state of zinc-doped $YBa_2Cu_4O_8$ and $YBa_2Cu_3O_{6.7}$[236].

On the other hand, if temperature is much lower than the resonance frequency, i.e. $k_B T \ll \Omega'$, $\partial f(\varepsilon)/\partial \varepsilon$ decays exponentially with temperature, the impurity contribution to the spin-lattice relaxation also drops exponentially. As previously mentioned, an exponential decay of $(T_1 T)^{-1}$ is often regarded as a signature of spin frozen [229]. However, in this case, this exponential decay is purely due to the fact that the resonance energy is higher than the temperature and it is difficult to activate electrons to the impurity resonance state by thermal excitations.

Therefore, $\delta[T_1(r)T]^{-1}$ varies non-monotonically with temperature. By lowering temperature $\delta[T_1(r)T]^{-1}$ increases at the beginning, and then drops exponentially after reaching a maximum at a temperature close to the resonance energy. In the limit $c \rightarrow 0$, the peak temperature of $\delta[T_1(r)T]^{-1}$ is approximately located at $k_B T_f \approx 0.65\Omega'$. In the zinc-substituted BSCCO, the energy of the impurity resonance state is approximately $\Omega' \approx 17K$[237]. The peak temperature of $[T_1(r)T]^{-1}$ induced by the resonance state is estimated to be $T_f \approx 11K$, close to the experiment value of 10K for YBCO [229].

Fig. 12.1 shows the temperature dependence of the impurity contribution to the spin-lattice relaxation rate $\delta(1/T_1 T)$ on one of the nearest neighboring sites of the impurity. Clearly, the impurity contribution to $\delta(1/T_1 T)$ is very sensitive to the value of phase shift δ_0. In the unitary limit, $\delta_0 = \pi/2$, $\delta(T_1 T)^{-1}$ increases monotonically with decreasing temperature. But the peak value of $\delta(T_1 T)^{-1}$ decreases quickly with decreasing δ_0.

The Knight shift is determined by the real part of the magnetic suscepti-

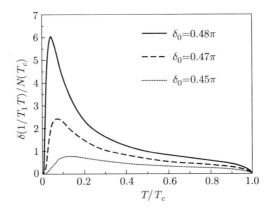

Figure 12.1   Temperature dependence of the impurity correction to the spin-lattice relaxation rate on the neighboring sites of the impurity. The normalization constant $N(T) = (T_1 T)^{-1}|_{T=T_c}$ is the spin-lattice relaxation rate at $T = T_c$.

bility $\mathrm{Re}\chi_{zz}$. At site r, it can be expressed as

$$K(r) = \frac{1}{\gamma_e \gamma_n \hbar^2} \sum_j F_{j,r} \mathrm{Re}\chi_{zz}(j), \qquad (12.69)$$

where

$$\mathrm{Re}\chi_{zz}(j) = \frac{\mu_B^2}{\pi} \int d\varepsilon \frac{\partial f(\varepsilon)}{\partial \varepsilon} \mathrm{Tr}\, \mathrm{Im}G\,(j,j,\varepsilon)\,. \qquad (12.70)$$

The contribution of the impurity resonance state to $\mathrm{Re}\chi$ is approximately given by

$$\delta\mathrm{Re}\chi_{zz}(j) \approx \frac{\mu_B^2 \, \mathrm{Tr}\mathrm{Re}G^0(j,0)\mathrm{Re}G^0(j,0)}{\pi} \int d\varepsilon T_{11}''(\varepsilon) \frac{\partial f(\varepsilon)}{\partial \varepsilon}. \qquad (12.71)$$

Similar to $\mathrm{Im}\chi$, the variation of $\delta\mathrm{Re}\chi(j)$ with temperature is determined mainly by the resonance pole:

$$\delta\mathrm{Re}\chi_{zz}(j) \approx \left.\frac{\partial f(\varepsilon)}{\partial \varepsilon}\right|_{\varepsilon=\Omega'}. \qquad (12.72)$$

In the limit $k_B T_c \gg k_B T \gg \Omega'$, $\partial f(\Omega')/\partial \omega \approx 1/T$. From Eq. (12.70) we have

$$\delta K(r) \approx \frac{1}{T}. \qquad (12.73)$$

Thus in this temperature range the resonance contribution to the Knight shift is Curie-Weiss like. In the low temperature limit, $k_B T \ll \Omega'$, the resonance contribution is negligibly small. $K(T)$ decays to zero exponentially, again different from the contribution of free magnetic moments.

Fig. 12.2 shows the impurity correction to the Knight shift as a function of temperature. It indicates that if temperature is not too low, the NMR Knight shift is approximately Curie-Weiss like, similar to the contribution of localized magnetic moments. From this Curie-Weiss behavior of the magnetic susceptibility, one can define an effective magnetic moment μ_{eff} corresponding to this resonance state:

$$\frac{\mu_{eff}^2}{3k_B T} = \sum_j \delta \mathrm{Re} \chi_{zz}(j), \tag{12.74}$$

where j is to sum over all four nearest neighboring sites of the impurity. Fig. 12.3 shows the effective magnetic moment obtained with this equation. In obtaining the result shown in this figure, the energy-momentum dispersion relation of electron proposed by Norman *et. al.* is used [238].

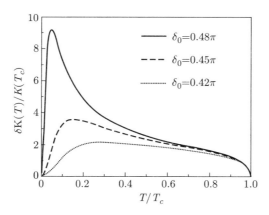

Figure 12.2   The impurity contribution to the Knight shift on the neighboring sites. $K(T_c)$ is the Knight shift at $T = T_c$.

The result in Fig. 12.3 shows that in the temperature range which is neither too low, nor too close to the superconducting transition temperature, the effective moment corresponding to the non-magnetic impurity is approximately

$0.3\mu_B$. This value is close to the effective moment estimated from the NMR data obtained in the slightly overdoped YBCO with Zn impurities. It shows that at least in this material, the Curie-Weiss behavior of the NMR spectroscopy is mainly the contribution of impurity induced resonance states.

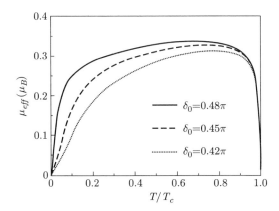

Figure 12.3   Temperature dependence of the effective magnetic moment corresponding to a resonance state induced by a single non-magnetic impurity.

12.7 Experimental Results of High-T_c Superconductors

NMR has good spatial and frequency resolution, which can be used to detect spatial magnetic fluctuations of electron spins around different nuclei. This is important to a comprehensive understanding of magnetic interactions, especially the property of local spin fluctuations. The nuclear spin-lattice relaxation rate T_1^{-1} in an intrinsic d-wave superconductor shows a weak temperature dependence in low temperatures. It scales as T^3. In order to distinguish this T^3-behavior from the exponential behavior in an s-wave superconductor, the measurement resolution needs to be high. Theoretical calculations for NMR are often obtained in the limit of zero magnetic field, but experimental measurements are usually done in a strong external magnetic field. In the normal state, the applied magnetic field affects weakly the electronic structure, and what the NMR measures is the intrinsic property of electrons. This situation changes in the superconducting state. In particular, an external magnetic field may suppress the critical temperature and generate magnetic flux lines, which can alter NMR spectra. In fact, the NMR spectra of high-T_c cuprates are changed by varying magnetic fields. For example, the ^{63}Cu nuclear spin-relaxation rate in YBa$_2$Cu$_3$O$_7$ is four times larger in an external magnetic field of $8.31T$ than in the zero field at $T = 0.2T_c$ [239]. Moreover, high-T_c cuprates are layered materials, the NMR results depend on whether the magnetic field is applied along the c-axis or parallel to the ab-plane [239]. Therefore, in the comparison of experimental results with theoretical calculations, effects of applied magnetic fields on the superconducting state need to be carefully considered. The experiment data of NMR for high-T_c cuprate superconductors were acquired mainly from copper, oxygen and yttrium atoms.

Below T_c, the spin-lattice relaxation rate T_1^{-1} drop monotonically with lowering temperatures in almost all nuclei, including copper, oxygen and yttrium. However, T_1^{-1} on the copper site is one order of magnitude larger than that on the oxygen site. It implies that the antiferromagnetic fluctuation is very strong even in the superconducting state [240, 241, 242]. This is a strong evidence supporting the antiferromagnetic fluctuation mechanism of high-T_c superconductivity.

Just below T_c, T_1^{-1} decays quickly and the Hebel-Slichter coherent peak is not observed in all high-T_c compounds [240, 243, 244, 245, 246, 247]. The Hebel-Slichter peak appears in most of s-wave superconductors. Nevertheless, the absence of this coherence peak does not rule out the possibility of s-wave superconducting pairing because the strong coupling and other physical effects can smear out this peak even in s-wave superconductors. But the absence of the Hebel-Slichter peak does mean that the density of states of quasiparticles in high-T_c superconductors is not strongly divergent or even not divergent at $\omega = \Delta$. In this sense, the NMR results favor more the d-wave scenario of high-T_c superconductivity, since the density of states of the d-wave superconductor diverges just logarithmically, weaker than the square-root divergence of the s-wave superconductor.

The spin-lattice relaxation rate T_1^{-1} of ^{63}Cu or ^{17}O in YBa$_2$Cu$_3$O$_{7-\delta}$ scales as T^3 in an intermediate temperature range with $T < T_c/2$. At very low temperatures, $T \ll T_c$, T_1^{-1} drops more slowly than T^3 in YBa$_2$Cu$_3$O$_{7-\delta}$ [248, 243, 249], and shows approximately a linear temperature dependence in Bi$_2$Sr$_2$CaCu$_2$O$_8$ [244, 247], overdoped La$_{2-x}$Sr$_x$CuO$_{4-x}$ [245], and Tl$_2$Ba$_2$Ca$_2$ Cu$_3$O$_{10}$ [242]. Both T^3 at relatively high temperatures and the linear low temperature dependence of T_1^{-1} agree with the behavior of disordered *d*-wave superconductors. In the underdoped La$_{2-x}$Sr$_x$CuO$_{4-x}$ [245], T_1^{-1} of ^{63}Cu tends to saturate at very low temperature. This saturation cannot be explained by the effect of non-magnetic impurity scattering in *d*-wave superconductors. Whether it is due to the experimental background or other physical effects, such as antiferromagnetic fluctuations, needs further clarification.

The Knight shift of nuclear spins K decreases monotonically as temperature is decreased in the superconducting phase. Its value at zero temperature, $K(0)$, is small but finite [32, 243, 244, 247, 250, 251, 252]. The monotonic decay of the Knight shift in the superconducting phase is a strong indication of singlet pairing. If superconducting electrons are spin-triplet paired, the Knight shift does not change much across the superconducting critical point, in particular, it will not drop steeply with temperature in the superconducting state. The shift in the NMR resonance frequency resulting from the coupling of nuclear spins with the orbital moments of electrons, K_{orb}, is usually temperature independent. It is also insensitive to the electron or hole doping level. In normal metals, K_{orb} can be obtained by measuring the Knight shift in the corresponding diamagnetic insulator in which electron spins are completely quenched. However, high-T_c superconductors are doped antiferromagnetic insulators and electron spins have a large contribution to the Knight shift, thus careful analysis of experimental data should be done in order to determine K_{orb}. For the *s* or intrinsic *d*-wave superconductor, the spin susceptibility vanishes at zero temperature, and $K(0) = K_{orb}$ contributes completely from the orbital angular momentum. However, in disordered *d*-wave superconductors, the quasiparticle density of states is finite at the Fermi surface, and its contribution to $K(0)$ is also finite.

Chapter 13

The Mixed State

13.1 The Semi-Classical Approximation

A type-II superconductor is in a mixed state with quantized magnetic flux lines which penetrate the material, turning a region of the superconductor normal, if the applied field is higher than the lower critical field but lower than the upper critical field, i.e. $H_{c1} \ll H \ll H_{c2}$. The flux lines are also the vortex lines of the superconducting order parameter. The presence of vortices changes the spectrum of quasiparticle excitations. It also changes thermodynamic and dynamic properties of the superconducting state. Quasiparticle excitations around vortex cores behave quite differently in the s- and d-wave superconductors. This leads to the difference in the field dependence of the specific heat and other physical quantities in these two kinds of superconductors.

Similar as in the s-wave superconductor, the vortex structure of the d-wave superconductor is determined by the superconducting coherence length ξ and the magnetic penetration depth λ. The radius of the vortex core equals roughly the coherence length ξ. The pairing order parameter is suppressed inside the vortex core and vanishes right at the core center. The external magnetic field passes through the vortex core as well as its vicinity within the characteristic length scale of the penetration depth λ.

High-T_c cuprates are typical type-II superconductors. The lower critical field is of the order of a few hundred Gauss, but the upper critical field is generally above 50T. Along the CuO_2 planes, the penetration depth is about 10^3Å, but the coherence length of Cooper pairs is just around $15 \sim 20$Å. The ratio between the magnetic penetration depth and the coherence length, $\lambda/\xi_0 \approx 10^2$. Thus the size of a vortex core of high-T_c superconductors is very small, and the spatial distribution of the magnetic field is very broad. The magnetic field can be approximately regarded as uniformly distributed in the whole system excluding the vortex cores if the average inter-vortex distance R lies between the above two length scales, i.e. $\xi \ll R \ll \lambda$.

In the mixed state, there exist two kinds of low energy excitations. The first is the fermionic core states localized mainly inside the vortex cores. In 1964, Caroli, de Gennes, and Matricon [253] showed that there are quasiparticle bound states inside a vortex core by solving the Bogoliubov-de Gennes equation for the isotropic *s*-wave superconductor. In 1989, Hess *et. al.* observed, for the first time, this kind of vortex bound states in the superconducting state of $NbSe_2$ through the STM experiments [254, 255], verifying the theoretical prediction. In the *d*-wave superconductor, there is not any vortex bound state because the gap function vanishes along the nodal directions. But there are sharp resonance states around each vortex. These resonance states behave similarly as the *s*-wave core bound states. It is difficult to distinguish a resonance state from a bound state if the energy resolution is not sufficiently high.

The second is the quasiparticle excitations induced by the applied magnetic field outside the cores. In the isotropic *s*-wave superconductor, the number of this kind of excitations is suppressed by the superconducting energy gap and its contribution is very small in low temperatures. However, it is different in the *d*-wave superconductor. Due to the existence of gap nodes, quasiparticles are relatively easy to excite outside vortex cores. Since the volume in which these quasiparticle excitations are populated is significantly larger than the size of the vortex cores, their contribution to thermodynamic quantities is also much larger than that of the vortex core states. Thus in the mixed state of the *d*-wave superconductor, as first pointed out by Volovik[256], the low-energy physics is predominately governed by the quasiparticle excitations induced by the applied field outside the vortex cores. In fact, this is also a basic assumption made in the analysis of physical properties of the mixed state of *d*-wave superconductors.

In addition, the vortex lines always form a lattice, either regular or irregular, in the mixed state. This vortex lattice can scatter quasiparticles and affect thermodynamic and dynamic properties of superconductors. This scattering effect is weak and negligible if the applied field is not very high and the inter-vortex distance is large. However, if the applied field is very high, namely close to the upper critical field, the scattering becomes strong and this effect should be more seriously considered.

It is difficult to study comprehensively properties of quasiparticle excitations in the mixed state of d-wave superconductors. There are two reasons for this. First, the vortex line in a d-wave superconductor is not rotation invariant, different than in its s-wave counterpart. In particular, the effective coherence length of Cooper pairs diverges along the nodal direction. This implies that the quasiparticle excitation inside the core is not perfectly confined, hence not forming a bound state in the d-wave superconductor, and the wavefunction of a vortex core state can escape along the gap nodal directions. It is impossible to find a rigorous solution for the vortex core states by solving the Bogoliubov-De Gennes equation for the d-wave superconductor. Second, the scattering process of superconducting quasiparticles by the vortex cores is complicated and lack of systematic study. Certain approximations have to be used in order to calculate the microscopic structures of vortex lines and related quasiparticle excitation spectra from quantum theory.

In the discussions below, we use the d_{xy}-wave superconductor as an example to discuss the properties of quasiparticle excitations in the mixed state. The results can be generalized to the $d_{x^2-y^2}$-wave superconductor straightforwardly.

In the mixed state, the dynamics of quasiparticles is governed by the Hamil-tonian defined in Eq. (3.53). For the d_{xy} superconductor, Eq. (3.53) becomes

$$
\hat{H} = \begin{pmatrix} \dfrac{1}{2m}\left(p - \dfrac{e}{c}A\right)^2 + U(r) - \varepsilon_F & \dfrac{1}{4p_F^2}\{p_x, \{p_y, \Delta_0(r)\}\} \\[3mm] \dfrac{1}{4p_F^2}\{p_x, \{p_y, \Delta_0(r)\}\} & -\dfrac{1}{2m}\left(p + \dfrac{e}{c}A\right)^2 - U(r) + \varepsilon_F \end{pmatrix}. \tag{13.1}
$$

In the limit $\xi \ll R \ll \lambda$, the magnetic field is uniformly distributed except inside or in the close vicinity of the vortex cores. The amplitude of the gap function $\Delta_0(r)$ is also approximately coordinate independent. But the phase of $\Delta_0(r)$ varies in space. It winds by 2π around a close loop enclosing a vortex flux. Hence, outside the vortex, we can take the approximation

$$\Delta_0(r) \approx \Delta_0 e^{i\phi(r)},$$

and assume Δ_0 to be r independent.

The phase of $\Delta_0(r)$ can be gauged out by a unitary transformation and replaced by an effective vector potential acting on quasiparticles. The corresponding gauge transformation is defined by

$$\hat{H} \to U^{-1}\hat{H}U, \quad U = \begin{pmatrix} e^{i\phi_e(r)} & 0 \\ 0 & e^{-i[\phi(r)-\phi_e(r)]} \end{pmatrix}, \tag{13.2}$$

where $\phi_e(r)$ is an arbitrary phase function. To ensure that the transformation matrix U is single-valued by winding a vortex line, we usually set $\phi_e(r) = 0$ or $\phi_e(r) = \phi(r)$. For a single vortex line, these are the only two values $\phi_e(r)$ can be taken. However, in a system with many vortex lines, $\phi_e(r)$ can also take other expressions.

It is usually convenient to use a single-valued U to solve the Hamiltonian. However, to understand qualitatively the physical property of the mixed state, sometimes it is more convenient to use a non-single-valued transformation. For example, if we take $\phi_e(r) = \phi(r)/2$, then the Hamiltonian Eq. (13.1) can be greatly simplified [257]. The resulting Hamiltonian reads

$$\hat{H} = \begin{pmatrix} \dfrac{1}{2m}(p+mv_s)^2 + U(r) - \varepsilon_F & \dfrac{\Delta_0}{p_F^2}\left(p_x p_y + \dfrac{i\hbar^2}{2}\phi''_{xy}\right) \\[2ex] \dfrac{\Delta_0}{p_F^2}\left(p_x p_y - \dfrac{i\hbar^2}{2}\phi''_{xy}\right) & -\dfrac{1}{2m}(p-mv_s)^2 - U(r) + \varepsilon_F \end{pmatrix}, \tag{13.3}$$

where

$$v_s = \frac{1}{m}\left(\frac{\hbar}{2}\nabla\phi - \frac{e}{c}A\right) \tag{13.4}$$

is the velocity of supercurrent.

In the study of quasiparticle properties in the mixed state, a frequently used approximation is to treat the velocity of supercurrent as a static field rather than a dynamic variable. This is a semi-classical approximation. Under this approximation, one can even to solve the Hamiltonian defined by Eq. (13.3) to obtain an analytic solution.

In Eq. (13.3), if v_s is coordinate independent, and both the disorder potential $U(r)$ and the spatial variation of the phase ϕ vanish, the above Hamiltonian can be readily diagonalized. The quasiparticle excitation spectrum such obtained is

$$E_k = \sqrt{\left(\frac{\hbar^2}{2m}k^2 + \frac{1}{2}mv_s^2 - \varepsilon_F\right)^2 + \Delta_k^2} + \hbar k \cdot v_s,$$

where $\Delta_k = \Delta_0 \hat{k}_x \hat{k}_y$. The role of the v_s^2 term is to change the Fermi energy. This term can be absorbed in the redefinition of ε_F. Then the above expression becomes

$$E_k = \sqrt{\xi_k^2 + \Delta_k^2} + \hbar k \cdot v_s, \tag{13.5}$$

in which

$$\xi_k = \frac{\hbar^2}{2m}k^2 - \varepsilon_F.$$

The first term on the right-hand side of Eq. (13.5) is the quasiparticle dispersion in the absence of the supercurrent. The second term is the correction from the supercurrent which is usually dubbed as the Doppler shift. If the supercurrent velocity does not vary in space, the expression of the Doppler shift $\delta\varepsilon(k) = \hbar k \cdot v$ is exact. This shift results from the finite center-of-mass momentum of Cooper pairs in the presence of supercurrent.

In the s-wave superconductor, the correction to the quasiparticle spectrum from the Doppler shift is too small to alter qualitatively the gap structure. Hence the Doppler shift does not affect much low energy properties of the s-wave superconductor. However, for the d-wave superconductor, the Doppler shift can change significantly the gap structure of quasiparticles near the nodal lines by lifting the chemical potential. Now the Fermi surface is no longer just a point. Instead, the volume of the Fermi surface and the corresponding zero energy density of states becomes finite, proportional to the energy scale of the Doppler shift.

In realistic superconductors, the supercurrent velocity v_s is spatially dependent. In particular, it varies significantly around the vortex cores. If the variance is small in comparison with the coherence length of Cooper pairs, i.e. $|\nabla v_s|\xi_0 \ll |v_s|$, Eq. (13.5) holds approximately. E_k defined by Eq. (13.5) can be approximately taken as the quasiparticle energy dispersion at the point where v_s is defined.

Under the semi-classical approximation, the supercurrent velocity v_s in the mixed state can be determined from the supercurrent density and the classical equations of electromagnetic fields:

$$\frac{4\pi\lambda^2}{c}\nabla \times j_s + H = \hat{z}\Phi_0 \sum_i \delta(r - R_i), \tag{13.6}$$

$$\nabla \times H = \frac{4\pi}{c}j_s. \tag{13.7}$$

Eq. (13.6) is the London equation in the presence of vortex lines. It holds approximately in the limit $\xi \ll R \ll \lambda$. R_i is the coordinate of the magnetic vortex core center. The solution to the above equations is

$$j_s = \frac{c\Phi_0}{4\pi} \int \frac{d^2k}{4\pi^2} \frac{ik \cdot \hat{z}}{1 + \lambda^2 k^2} e^{ik \cdot r}. \tag{13.8}$$

By further using the definition $j_s = en_s v_s$, $\Phi_0 = hc/(2e)$, and the relation between superfluid density n_s and λ, i.e. $n_s = mc^2/(4\pi e^2 \lambda^2)$, the supercurrent velocity is found to be

$$v_s = \frac{\hbar}{4\pi m} \sum_i \int d^2k \frac{ik \times \hat{z}}{k^2 + \lambda^{-2}} e^{ik \cdot (r - R_i)}. \tag{13.9}$$

In the limit $\lambda \to \infty$, the above integral can be solved analytically. It gives

$$v_s = \sum_i \frac{\hbar}{2m|r_i|} \hat{z} \times \hat{r}_i, \tag{13.10}$$

where $r_i \equiv r - R_i$. It shows that the change of the supercurrent velocity v_s is indeed small in comparison with the coherence length and the semi-classical approximation is valid away from the vortex cores. Therefore, the semi-classical approximation can be used in the calculation of physical quantities that are predominately determined by the quasiparticle excitations outside the vortex cores.

13.2 Low Energy Density of States

In the mixed state, there exist two types of quasiparticle excitations, residing inside and outside the vortex core, respectively. Below we discuss their contributions to the low energy density of states.

Inside the vortex core, the superconducting order parameter is suppressed, but there are circulating screening currents. In order to evaluate the contribution of quasiparticle excitations inside the cores, each vortex core can be regarded as a potential well with height Δ_0^2/ε_F and radius ξ_0. In the s-wave superconductor, the scattering potential is isotropic and the quasiparticles can form a few bound states inside the core. In the d-wave superconductor, there are no bound states due to the existence of gap nodes. Instead, the low-lying states of quasiparticles are resonance states. The eigenfunctions or energies of these resonant or bound states are determined purely by the intrinsic parameters of the superconductor in the limit $\lambda \gg \xi_0$, independent of the strength of applied magnetic field. Thus the contribution of these resonant or bound states to the low energy density of states by each vortex core is also independent of the external magnetic field. Since the density of vortex lines is proportional to the external magnetic field H, this means that the vortex contribution to the low energy density of states is also proportional to H, i.e.

$$\rho_{core} \approx H, \tag{13.11}$$

irrespective of the pairing symmetry.

Away from the vortex cores, the correction to the quasiparticle spectrum by the Doppler shift behaves differently in the s- and d-wave superconductors. In the s-wave superconductor, the correction is negligibly small compared to the quasiparticle excitation gap. Thus the low energy quasiparticle density of states contributes mainly by the core excitations in low temperatures, and is proportional to H as given by Eq. (13.11). For the d-wave superconductor, the Doppler shift correction to the excitation spectrum is larger than the gap value near the nodal lines. The contribution from the quasiparticle excitations outside the core cannot be neglected. In fact, its contribution is larger than the core excitations.

Under the semi-classical approximation, both the supercurrent velocity and the quasiparticle density of states are functions of coordinates. On average, the quasiparticle density of states contributed by each magnetic flux line is determined by the formula

$$\rho_{out}(\omega, H) = \int d\varepsilon \rho_0(\omega + \varepsilon) P(\varepsilon, H), \qquad (13.12)$$

where

$$P(\varepsilon, H) = \frac{1}{A} \int d^2 r \delta \left(\varepsilon - \hbar v_s(r) \cdot k \right) \qquad (13.13)$$

is the Doppler distribution function. It measures the average distribution of the Doppler shift in space. The domain of integration in Eq. (13.13) is the region of one flux line, and A is the corresponding area.

When the magnetic field is changed, the coordinates of the vortex core centers R_i are also changed. The density of vortices increases with the external magnetic field. The field dependence of R_i is determined, on average, by the following formula

$$R_i(H) = x^{-1} R_i(H_0),$$

where $x = \sqrt{H/H_0}$ and H_0 is a reference magnetic field. Here it is implicitly assumed that the system behaves qualitatively similarly in the two fields, H and H_0. Using Eq. (13.9), it can be shown that the supercurrent velocity satisfies the following scaling relation:

$$v_s(r, \lambda, H) = x v_s(xr, x\lambda, H_0). \qquad (13.14)$$

v_s depends on the penetration depth λ, which determines the characteristic length scale of v_s at long distance. For the system with $R \ll \lambda$, the effect of λ on v_s is very small on the length scale discussed here. Thus λ can be approximately taken as infinity. In this case,

$$v_s(r, H) \approx x v_s(xr, H_0). \qquad (13.15)$$

Substituting this equation into Eq. (13.13), and considering the fact that the area per flux line scales as $A \to x^2 A$ under the change of magnetic field from H to H_0, we find the following expression for the distribution function $P(\varepsilon, H)$:

$$P(\varepsilon, H) = x^{-1} P(x^{-1}\varepsilon, H_0). \qquad (13.16)$$

By further substituting it into Eq. (13.12), we then obtain the following equation

$$\rho_{out}(\omega, H) = \int d\varepsilon \rho_0(\omega + x\varepsilon) P(\varepsilon, H_0). \tag{13.17}$$

If $\omega = 0$ and the correction of the Doppler shift to the energy is small compared to the maximal gap, $\rho_0(x\varepsilon)$ is approximately a linear function of ε, $\rho_0(x\varepsilon) \approx x N_F \varepsilon / \Delta_0$. Thus at zero frequency, the average density of states contributed by the quasiparticle excitations outside the vortex core is approximately given by

$$\rho_{out}(0, H) = \alpha\sqrt{H}, \tag{13.18}$$

where the coefficient α is determined purely by the system parameters, independent of H:

$$\alpha = \frac{N_F}{\Delta_0\sqrt{H_0}} \int d\varepsilon\varepsilon P(\varepsilon, H_0). \tag{13.19}$$

Comparing Eq. (13.18) with Eq. (13.11), it is clear that the quasiparticle excitations outside the vortex cores contribute more to the low energy density of states than the vortex core states at low magnetic fields, i.e. $\rho_{core}/\rho_{out} \ll 1$. Thus to the leading order approximation, the density of states at the Fermi surface scales

$$\rho(H) \approx \sqrt{H}. \tag{13.20}$$

This is an important result for the d-wave superconductor that was first obtained by Volovik [256]. It implies that one can neglect the vortex core excitations in the study of thermodynamic and dynamic properties of d-wave superconductors in the mixed state. As only the quasiparticles outside the vortex cores need to be considered, this greatly simplifies the calculation for the excitation spectra. The coefficient of the $\sqrt{H}$ term in the density of states depends on the distribution function $P(\varepsilon, H)$. $P(\varepsilon, H)$, on the other hand, depends on the distribution of vortices. A detailed discussion on the expressions of $P(\varepsilon, H)$ at different vortex distributions is given in Ref. [258]. In general, $P(\varepsilon, H)$ is obtained by numerical calculations.

In low temperatures, the contribution of quasiparticle excitations to the specific heat coefficient C_v/T is proportional to the low-energy density of states.

Thus the low temperature specific heat of electrons in the mixed state is proportional to $\sqrt{H}$ [256]:

$$C_v \propto T\sqrt{H}. \qquad (13.21)$$

This square-root field dependence of the specific heat is a characteristic property of *d*-wave superconductors. In the *s*-wave superconductor, the low energy density of states is dominated by the quasiparticle excitations inside the cores. It is proportional to the magnetic field H, so is the low temperature specific heat coefficient C_v/T.

The specific heat contains the contribution from both electrons and phonons. Generally it is difficult to separate the electron contribution from the phonon one. This is the major obstacle in the analysis of experimental data of specific heat. However, phonons do not couple to the magnetic field. Their contribution to the specific heat does not depend on the applied field. This means that the difference of specific heat at different magnetic fields is purely the contribution of electrons. This property can be used to test the $\sqrt{H}$ scaling behavior of *d*-wave superconductors in a finite magnetic field.

The $\sqrt{H}$ scaling behavior of low temperature specific heat in the *d*-wave superconductor was first verified experimentally by Moler *et. al.* [259] for YBCO superconductors. Fig. 13.1 shows the field dependence of the specific heat coefficient C_v/T they obtained from the measurement data in the low temperature limit. The experimental results agree with the theoretical prediction. It shows that the low energy excitations in the *d*-wave mixed states indeed contribute mainly by the quasiparticle excitations outside the vortex cores. Later on, the $\sqrt{H}$ behaivor of the specific heat coefficient was further confirmed in YBCO[260, 261], $Y_{0.8}Ca_{0.2}Ba_2Cu_3O_{6+x}$[262], and $La_{2-x}Sr_xCuO_4$[263]. Wen *et. al.* [263] also found that the doping dependence of the maximal gap they obtained from the specific heat experiment is consistent with that obtained by the thermal conductance measurement [209]. However, in the underdoped $Y_{0.8}Ca_{0.2}Ba_2Cu_3O_{6+x}$[262], the specific heat varies almost linearly with H. This linear field dependence of the specific heat may result from the impurity scattering. It may also arise from the fact that the measurement temperature is still not low enough and the specific heat contains a significant contribution from the quasiparticles far away from the gap nodes. In the dirty scattering limit, Kubert and Hirschfeld found that the quasiparticle contribution to the specific heat scales as $H \ln H$, significantly different from the $\sqrt{H}$ behavior [264].

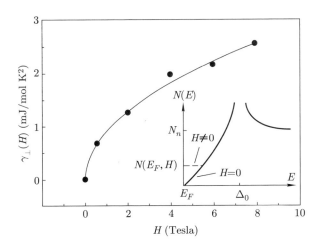

Figure 13.1   The specific heat coefficient $\gamma_\perp(H) = C_v(H)/T$ as a function of H in the low temperature limit for YBa$_2$Cu$_3$O$_{6.95}$. The solid line is the fitting curve for the experiment results with the formula $\gamma_\perp(H) = A\sqrt{H}$, and $A = 0.91$mJ/mol K^{2}T$^{1/2}$. The inset shows the density of states both at zero field and at finite field. Reprinted with permission from [259]. Copyright (1994) by the American Physical Society.

13.3   Universal Scaling Laws

Around the gap nodes, the quasiparticles of d-wave superconductors are Dirac-like. Their energy varies linearly with momentum. Thus the energy and momentum have the same scaling dimension under the scaling transformation. As a result of this, various thermodynamic quantities exhibit strong scaling behaviors [265]. Below, we take the d_{xy}-superconductor as an example to derive the scaling laws for several different thermodynamic quantities in the mixed states under the linear approximation of the energy dispersion.

There are four nodes in the quasiparticle spectra of the d_{xy}-superconductor.

If the scattering among these four nodes is neglected, the contributions to thermodynamic quantities from these four nodes are independent. The total contribution is simply the sum of the contribution from each node.

Around the gap node $k = (k_F, 0)$, the Hamiltonian of quasiparticles, Eq. (13.3), can be linearized according to the method introduced in Sect. 3.5. The linearized Hamiltonian corresponding to Eq. (3.69) is given by

$$\hat{H}_0(r, H) = \begin{pmatrix} v_F(p_x + mv_{s,x}) + U(r) & \dfrac{\Delta_0}{p_F}p_y \\ \dfrac{\Delta_0}{p_F}p_y & -v_F(p_x - mv_{s,x}) - U(r) \end{pmatrix}, \quad (13.22)$$

which is valid at $T \ll \Delta_0^2/\varepsilon_F$. This is the full Hamiltonian for describing the quasiparticle excitations outside the vortex cores.

Under the scaling transformation, $r \to xr$, if we assume both the number of vortices and the ratio between the average volumn of each vortex and the sample size are invariant, and the disorder potentials are uncorrelated, then it can be shown that the above Hamiltonian satisfies the following scaling equation

$$\hat{H}_0(r, H) = x\hat{H}_0(xr, H_0). \quad (13.23)$$

Of course in order to keep the total number of vortices unchanged under the scaling transformation, the total area of the system should scale with the magnetic field as

$$S_A(H) = x^{-2}S_A(H_0), \quad H = x^2 H_0.$$

Eq. (13.23) can be verified by analyzing the scaling behavior of each individual term in $\hat{H}_0$ under the transformation $r \to xr$. The variation of the momentum operator under the scaling scaling transformation is simple. From the definition of the momentum operator, we have

$$-i\hbar\partial_r = x\left(-i\hbar\partial_{(xr)}\right). \quad (13.24)$$

Generally the random potential is uncorrelated and its average is zero, $\langle U(r) \rangle = 0$. The spatial correlation of the random potential is a δ-function:

$$\langle U(r)U(r') \rangle = U_0\delta(r - r').$$

This short-ranged random potential has no characteristic length scale. In a two-dimensional system, $\delta(xr) = x^{-2}\delta(r)$. Thus $U(r)$ should satisfy the following relation under the scaling transformation:

$$U(r) = xU(xr). \tag{13.25}$$

Substituting the above equations and the scaling formula of the supercurrent velocity, Eq. (13.15), into (13.22), we then obtain Eq. (13.23).

The scaling relation revealed by Eq. (13.23) results from the linear approximation. This approximation is not valid in the strong impurity scattering limit, because the low energy density of states is changed by the random impurity potential and is no longer zero at the Fermi level.

From Eq. (13.23), it can be shown that the eigenvalue E_n and the corresponding eigenfunction $\tilde{\psi}_n$ of $\hat{H}_0$ transform under the scaling transformation as

$$\tilde{\psi}_n(r, H) = \tilde{\psi}_n(xr, H_0), \tag{13.26}$$

$$E_n(H) = xE_n(H_0). \tag{13.27}$$

This gives the equation that the internal energy satisfies at different magnetic fields:

$$U(T, H) = \sum_n E_n(H)f\left(\frac{E_n(H)}{T}\right) = xU(x^{-1}T, H_0), \tag{13.28}$$

where f is the Fermi distribution function. From the scaling behavior of the system size $S_A(H) = x^{-2}S_A(H_0)$, we then obtain the scaling law of the internal energy density

$$u(T, H) = H^{3/2}F_U(T/\sqrt{H}), \tag{13.29}$$

in which F_U is an unknown scaling function. The specific heat per unit area is determined by the derivative of $u(T, H)$ with respect to temperature:

$$C_v(T, H) = HF'_U(T/\sqrt{H}) = T\sqrt{H}F_C(T/\sqrt{H}), \tag{13.30}$$

where $F_C(T/\sqrt{H})$ is a universal scaling function of $T/\sqrt{H}$.

By integrating the specific heat with respect to temperature, we obtain the scaling formula of the entropy:

$$S = \int dT T C_v(T, H) = \sqrt{H} \int dT T^2 F_C(T/\sqrt{H}) = H^2 F_S(T/\sqrt{H}), \quad (13.31)$$

where F_S is a universal scaling function of the entropy.

The free energy is defined by $F = U - TS$. Its scaling law is given by the formula

$$F(H) = H^{3/2} F_F(T/\sqrt{H}). \tag{13.32}$$

The magnetic susceptibility is proportional to the second order derivative of the free energy with respect to the magnetic field. It satisfies the following scaling law:

$$\chi(T, H) = \frac{\partial^2 F}{\partial H^2} = \frac{T^2}{4H^{3/2}} F_F''(T/\sqrt{H}) = T^{-1} F_\chi(T/\sqrt{H}). \tag{13.33}$$

F_F and F_χ are the scaling functions for the free energy and the magnetic susceptibility, respectively.

These scaling laws of thermodynamic properties are obtained under the linear approximation. They were verified through the specific heat measurements in high-T_c superconductors. For YBCO superconductors, it was found that the specific heat indeed scales as $T/\sqrt{H}$, consistent with the theoretical prediction (Fig. 13.2) [266, 260, 261, 186]. Similar scaling behavior of the specific heat with $T/\sqrt{H}$ was also found in LSCO superconductors [263]. These experimental results gave a strong support to the scaling theory of thermodynamics quantitites in the mixed states of *d*-wave superconductors.

In addition to these thermodynamic quantities, Simon and Lee [265] found that the optical and thermal conductivity tensor determined by quasiparticle excitations also exhibits approximate scaling behaivor as a function of $T/\sqrt{H}$. In particular, they found that up to the leading order approximation, the thermal Hall conductance κ_{xy} satisfies the following scaling law:

$$\kappa \propto T^2 F_{xy}(T/\sqrt{H}). \tag{13.34}$$

This result agrees with the experiment results for YBCO superconductors.

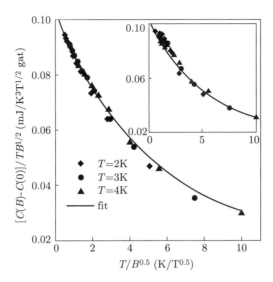

Figure 13.2   The scaling behavior of the specific heat versus the magnetic field B in the mixed state of high quality $YBa_2Cu_3O_7$ superconductor. The scaling variable is $T/B^{1/2}$. The inset is the scaling behavior before subtracting the Schottky impurity term. Reprinted with permission from [186]. Copyright (2001) by the American Physical Society.

Appendix A
Bogoliubov Transformation

The Bogoliubov transformation is used to diagonalize a bilinear Hamiltonian of fermions or bosons. The simplest bilinear Hamiltonian that can be diagonalized by the Bogoliubov transformation has the form

$$H = \lambda(a^\dagger a + b^\dagger b) + \left(\gamma a^\dagger b^\dagger + h.c.\right),\tag{A.1}$$

where (a, b) is a pair of fermion or boson operators. This kind of Hamiltonian does not conserve the particle number. It is widely used in the mean-field study of many-body physics.

The Bogoliubov transformation is canonical. It maintains the commutation rules of the creation and annihilation operators. In fermi systems, the Bogoliubov transformation is a unitary transformation because the fermion creation and annihilation operators can be transformed to each other by taking a particle-hole transformation. However, in bose systems, Bogoliubov transformation is no longer unitary. Instead, it is sympletic.

Below we discuss the Bogoliubov transformation for the fermion and boson systems separately. For simplicity, we assume γ is real. It is straightforward to generalize the results to the case of complex γ.

A.1 Fermi Systems

For fermions, Eq. (A.1) can be written in the matrix form as

$$H = \begin{pmatrix} a^\dagger & b \end{pmatrix} \begin{pmatrix} \lambda & \gamma \\ \gamma & -\lambda \end{pmatrix} \begin{pmatrix} a \\ b^\dagger \end{pmatrix} + \lambda.\tag{A.2}$$

The corresponding Bogoliubov transformation is defined by

$$\begin{pmatrix} a \\ b^\dagger \end{pmatrix} = \begin{pmatrix} u^* & v^* \\ -v & u \end{pmatrix} \begin{pmatrix} \alpha \\ \beta^\dagger \end{pmatrix}.\tag{A.3}$$

The inverse transformation is

$$\begin{pmatrix} \alpha \\ \beta^\dagger \end{pmatrix} = \begin{pmatrix} u & -v^* \\ v & u^* \end{pmatrix} \begin{pmatrix} a \\ b^\dagger \end{pmatrix}. \tag{A.4}$$

α and β are fermion operators, satisfying the anti-commutation relations. In order to maintain the Fermi-Dirac statistics of these operators, the transformation matrix must satisfy the equation

$$u^2 + v^2 = 1. \tag{A.5}$$

This is also the condition that the transformation matrix is unitary. Both u and v are real if γ is real.

After the transformation, the Hamiltonian becomes

$$H = \begin{pmatrix} \alpha^\dagger & \beta \end{pmatrix} \begin{pmatrix} \lambda\left(u^2 - v^2\right) - 2\gamma uv & 2\lambda uv + \gamma\left(u^2 - v^2\right) \\ 2\lambda uv + \gamma\left(u^2 - v^2\right) & -\lambda\left(u^2 + v^2\right) + 2\gamma uv \end{pmatrix} \begin{pmatrix} \alpha \\ \beta^\dagger \end{pmatrix} + \lambda. \tag{A.6}$$

This Hamiltonian is diagonalized if u and v also satisfy

$$\gamma(u^2 - v^2) + 2uv\lambda = 0. \tag{A.7}$$

By solving Eqs. (A.5) and (A.7), we find that

$$u = \sqrt{\frac{1}{2} + \frac{\lambda}{2\omega}}, \tag{A.8}$$

$$v = -\mathrm{sgn}(\gamma)\sqrt{\frac{1}{2} - \frac{\lambda}{2\omega}}, \tag{A.9}$$

where $\mathrm{sgn}(\gamma) = 1$ if $\gamma \geqslant 0$ or -1 otherwise, and

$$\omega = \sqrt{\lambda^2 + \gamma^2}. \tag{A.10}$$

After the diagonalization, the Hamiltonian becomes

$$H = \begin{pmatrix} \alpha^\dagger & \beta \end{pmatrix} \begin{pmatrix} \omega & 0 \\ 0 & -\omega \end{pmatrix} \begin{pmatrix} \alpha \\ \beta^\dagger \end{pmatrix} + \lambda$$
$$= \omega(\alpha^\dagger \alpha + \beta^\dagger \beta) - \omega + \lambda. \tag{A.11}$$

A.2 Bose Systems

Again we rewrite Eq. (A.1) in the matrix form as

$$H = \begin{pmatrix} a^\dagger & b \end{pmatrix} \begin{pmatrix} \lambda & \gamma \\ \gamma & \lambda \end{pmatrix} \begin{pmatrix} a \\ b^\dagger \end{pmatrix} - \lambda. \tag{A.12}$$

The Bogoliubov transformation is now defined as

$$\begin{pmatrix} a \\ b^\dagger \end{pmatrix} = \begin{pmatrix} u & v \\ v & u \end{pmatrix} \begin{pmatrix} \alpha \\ \beta^\dagger \end{pmatrix}. \tag{A.13}$$

Similar as in the fermion systems, u and v are not independent. They satisfy

$$u^2 - v^2 = 1, \tag{A.14}$$

if α and β are boson operators.

Substituting Eq. (A.13) into Eq. (A.12), we have

$$H = \begin{pmatrix} \alpha^\dagger & \beta \end{pmatrix} \begin{pmatrix} \lambda \left(u^2 + v^2 \right) + 2\gamma uv & 2\lambda uv + \gamma \left(u^2 + v^2 \right) \\ 2\lambda uv + \gamma \left(u^2 + v^2 \right) & \lambda \left(u^2 + v^2 \right) + 2\gamma uv \end{pmatrix} \begin{pmatrix} \alpha \\ \beta^\dagger \end{pmatrix} - \lambda. \tag{A.15}$$

To set the off-diagonal terms to zero, we obtain another equation that u and v satisfy

$$2uv\lambda + \gamma(u^2 + v^2) = 0. \tag{A.16}$$

By solving Eqs.(A.14) and (A.16), we find that

$$u = \sqrt{\frac{1}{2} + \frac{\lambda}{2\omega}}, \tag{A.17}$$

$$v = -\text{sgn}(\gamma)\sqrt{-\frac{1}{2} + \frac{\lambda}{2\omega}}, \tag{A.18}$$

where

$$\omega = \sqrt{\lambda^2 - \gamma^2}. \tag{A.19}$$

This solution is valid when $\lambda \geqslant |\gamma|$. Otherwise, the system described by Hamiltonian Eq. (A.13) is unstable. The diagonalized Hamiltonian then becomes

$$H = \omega(\alpha^\dagger \alpha + \beta^\dagger \beta) + \omega - \lambda. \tag{A.20}$$

The inverse transformation of Eq. (A.13) is

$$
\begin{pmatrix} \alpha \\ \beta^{\dagger} \end{pmatrix} = \begin{pmatrix} u & -v \\ -v & u \end{pmatrix} \begin{pmatrix} a \\ b^{\dagger} \end{pmatrix}.
\tag{A.21}
$$

Appendix B
Hohenberg Theorem

In 1967, Hohenberg proved an important theorem on the superfluid or super-conducting orders. It states that in both one and two dimensions, there is no superfluid long-range order in bosonic systems and no superconducting long-range order in electronic systems at any finite temperature. This theorem shows that there is no BCS-type superconductor in pure one- or two-dimensional materials. But it does not rule out the possibility of non-BCS-type superconducting phase transition, for example the Kosterlitz-Thouless (KT) transition, in low dimensions. The Hohenberg theorem puts strong constraint on the pairing mechanism and serves as an important guiding principle in the study of superconductivity. Below we give an introduction to the key steps and formulas used in the proof of this theorem.

B.1 Bogoliubov Inequality

The proof of the Hohenberg theorem uses the Bogoliubov inequality defined below. Given a Hamiltonian H, the Bogoliubov inequality reads

$$\langle \{A, A^\dagger\} \rangle \langle [[C, H], C^\dagger] \rangle \geqslant 2k_B T \, |\langle [C, A] \rangle|^2 , \tag{B.1}$$

where A and C are two arbitrary operators, and $\langle X \rangle$ represents the thermodynamic average of X defined by

$$\langle X \rangle = \frac{\mathrm{Tr} X \exp(-\beta H)}{\mathrm{Tr} \exp(-\beta H)},$$

with $\beta = 1/k_B T$. This inequality was also used by Mermin and Wagner ([267]) to prove the absence of ferromagnetic and antiferromagnetic long-range orders in one- or two-dimensional Heisenberg spin models at any finite temperatures.

There are many ways to prove this inequality. A relatively simple one was given by Mermin and Wagner in Ref. [267]. Below we briefly introduce their method.

We first define the inner product of A and B as

$$(A, B) = P \sum_{ij} A_{ij}^* B_{ij} \frac{W_i - W_j}{E_j - E_i + i0^\dagger}, \tag{B.2}$$

where $A_{ij} = \langle i \,|A|\, j\rangle$. P is to take the principal value for the expression behind. $W_i = \exp(-\beta E_i)/\mathrm{Tr}\exp(-\beta H)$ is the Boltzmann weight of the i'th eigenstate of H. Using the inequality

$$|\tanh x| \leqslant |x|,$$

it is simple to show that the following inequality is valid:

$$0 < \frac{W_i - W_j}{E_j - E_i} \leqslant \frac{1}{2}\beta(W_i + W_j).$$

We then obtain the inequality

$$(A, A) \leqslant \frac{1}{2}\beta\langle\{A, A^\dagger\}\rangle. \tag{B.3}$$

Similarly, using

$$A_{ij}^* A_{ij} B_{kl}^* B_{kl} + A_{kl}^* A_{kl} B_{ij}^* B_{ij} \geqslant A_{ij}^* B_{ij} B_{kl}^* A_{kl} + B_{ij}^* A_{ij} A_{kl}^* B_{kl},$$

it can be shown that (A, B) satisfies the Schwartz inequality:

$$(A, A)(B, B) \geqslant |(A, B)|^2. \tag{B.4}$$

Taking $B = [C^\dagger, H]$, from the definition we find that

$$(A, B) = \langle[C^\dagger, A^\dagger]\rangle,$$
$$(B, B) = \langle[C^\dagger, [H, C]]\rangle.$$

Substituting these expressions into Eq. (B.4) and using the inequality Eq. (B.3), we then obtain the Bogoliubov inequality Eq. (B.1).

B.2 Physical Meaning of the Bogoliubov Inequality

For a comprehensive understanding of the Bogoliubov inequality, let us analyze
the physical meaning of each term in Eq. (B.1). We start by introducting the
time-dependent correlation function of operators A and B

$$\tau_{AB}(t - t') = \langle [A^{\dagger}(t), B(t')] \rangle,$$

and the corresponding spectral function

$$\tau_{AB}(\omega) = \int \mathrm{d}t e^{i\omega t} \tau_{AB}(t) = \sum_{ij} 2\pi \delta(\omega + E_i - E_j) A_{ji}^* B_{ji} (W_i - W_j). \quad \text{(B.5)}$$

The response function of A and B is defined by

$$\chi_{AB}(\omega) = \int \frac{\mathrm{d}\omega'}{2\pi} \frac{\tau_{AB}(\omega')}{\omega' - \omega} = \sum_{ij} A_{ij}^* B_{ij} \frac{W_j - W_i}{E_i - E_j - \omega}. \quad \text{(B.6)}$$

Comparing this expression with Eq. (B.2), we find that the inner product of A
and B is just the zero-frequency response function

$$(A, B) = P\chi_{AB}(\omega = 0) \equiv \chi_{AB}^s.$$

Therefore, the Schwartz inequality (B.4) can be also expressed as

$$\chi_{AA}^s \chi_{BB}^s \geqslant |\chi_{AB}^s|^2. \quad \text{(B.7)}$$

This inequality reveals the relation between different response functions.

From the definition, it can be also shown that the expectation value of the
anticommutator of $A^{\dagger}$ and B and the spectral function of A and B satisfy the
following equation:

$$\langle \{A^{\dagger}, B\} \rangle = \int \frac{\mathrm{d}\omega}{2\pi} \tau_{AB}(\omega) \coth \frac{\beta\omega}{2}. \quad \text{(B.8)}$$

This equation associates the fluctuation (left hand side) with the dissipation
(right hand side), and is commonly known as the fluctuation-dissipation theorem. Hence the Bogoliubov inequality is just a constraint between fluctuations
and correlations.

B.3 Bose Systems

Now we use the Bogoliubov inequality to prove that there is no superfluid long-range order in one- or two-dimensional bosonic systems at any finite temperature by *Reductio ad absurdum*. We first assume that the system has a superfluid long-range order with the order parameter defined by

$$\langle a_k \rangle = \sqrt{V n_0}\delta(k), \tag{B.9}$$

where a_k is the boson operator, and V is the volume of the system.

According to the previous discussion, we know that $\langle \{A, A^\dagger\} \rangle$ in the Bogoliubov inequality (B.1) describes the fluctuation of the system. In the superfluid state, the fluctuation arises from the bosonic excitations at finite momenta, which have destructive effect on superfluidity. To describe this effect, it is natural to set

$$A = a_k, \tag{B.10}$$

$$C = \rho_k = \sum_q a_{q+k}^\dagger a_q. \tag{B.11}$$

We then have

$$\langle [C, A] \rangle = -\langle a_{q=0} \rangle = -\sqrt{n_0}, \tag{B.12}$$

$$\langle \{A, A^\dagger\} \rangle = 2\langle a_k^\dagger a_k \rangle + 1. \tag{B.13}$$

To evaluate the commutator between C and H, we assume

$$H = \sum_q \varepsilon_q a_q^\dagger a_q + H_I,$$

and the density operator C commutes with the interaction term H_I. Under this assumption, the continuity equation of electric charges holds,

$$\frac{\partial \rho}{\partial t} + \nabla \cdot j = 0,$$

and

$$[[C, H], C^\dagger] = \sum_q \left(\varepsilon_{k+q} + \varepsilon_{q-k} - 2\varepsilon_q \right) a_q^\dagger a_q. \tag{B.14}$$

If we further assume that the dispersion relation of free bosons is given by

$$\varepsilon_k = \frac{\hbar^2 k^2}{2m},$$

then

$$[[C, H], C^\dagger] = \frac{\hbar^2 k^2}{m} \sum_q a_q^\dagger a_q. \qquad (B.15)$$

Substituting the above results into the inequality Eq. (B.2), we obtain

$$\langle a_k^\dagger a_k \rangle \geqslant -\frac{1}{2} + \frac{k_B T m}{\hbar^2 k^2} \frac{n_0}{n}, \qquad (B.16)$$

where n is the density of bosons. The right hand side diverges quadratically as $k \to 0$, and its integral over momentum also diverges in both one and two dimensions. Clearly, this infrared divergence will invalidate the following sum rule

$$\frac{1}{V} \sum_{k \neq 0} \langle a_k^\dagger a_k \rangle = n - n_0 \qquad (B.17)$$

at any finite temperature $(T \neq 0)$. It indicates that the assumption made in Eq.(B.9) is invalid. Therefore, there is no superfluid long-range order in one and two-dimensional bose systems at finite temperatures.

B.4 Fermi Systems

Similar to the proof for the bose system, we assume that there is a superconducting long-range order in a fermi system. The order parameter is defined by

$$\Delta = \frac{1}{V} \sum_q \gamma_q \langle c_{q\uparrow} c_{-q\downarrow} \rangle, \qquad (B.18)$$

which is assumed to be finite and the pairing function γ_q is non-singular. Similar to Eqs. (B.10) and (B.11), we define

$$A = \frac{1}{V} \sum_q \gamma_q c_{k+q\uparrow} c_{-q\downarrow}, \qquad (B.19)$$

$$C = \rho_k = \sum_{q\sigma} c_{q+k\sigma}^\dagger c_{q\sigma}, \qquad (B.20)$$

where A and C are the Fourier components of the pairing and density operators at momentum k, respectively. The commutator between the above two operators is

$$\langle [A, C] \rangle = \Delta + \eta_k, \tag{B.21}$$

where

$$\eta_k = \frac{1}{V} \sum_q \gamma_{q-k} \langle c_{q\uparrow} c_{-q\downarrow} \rangle.$$

η_k is a function of k. In the limit $k \to 0$,

$$\lim_{k \to 0} \eta_k = \Delta.$$

Similarly, we define the Hamiltonian as

$$H = \sum_{q\sigma} \varepsilon_q c_{q\sigma}^\dagger c_{q\sigma} + H_I.$$

If the electric charge is conserved and C commutes with the interaction H_I, we have

$$\langle [[C, H], C^\dagger] \rangle = \sum_{q\sigma} (\varepsilon_{k+q} + \varepsilon_{q-k} - 2\varepsilon_q) \langle c_{q\sigma}^\dagger c_{q\sigma} \rangle = \frac{\hbar^2 k^2 n V}{m}. \tag{B.22}$$

In obtaining this equation, the energy dispersion of fermions is assumed to have the form $\varepsilon_k = \hbar^2 k^2 / 2m$.

The average value of the anticommutator of A and $A^\dagger$ is given by

$$\langle \{A, A^\dagger\} \rangle = \frac{1}{V} [F(k) + R(k)], \tag{B.23}$$

where

$$F(k) = \frac{1}{V} \sum_{qq'} \gamma_q \gamma_{q'}^* \langle c_{-q'\downarrow}^\dagger c_{k+q'\uparrow}^\dagger c_{k+q\uparrow} c_{-q\downarrow} \rangle, \tag{B.24}$$

$$R(k) = \frac{1}{V} \sum_q |\gamma_q|^2 \left(1 - \langle c_{q\downarrow}^\dagger c_{q\downarrow} \rangle - \langle c_{q+k\uparrow}^\dagger c_{q+k\uparrow} \rangle \right). \tag{B.25}$$

Since γ_q is non-singular and $0 \leqslant \langle c_{q\sigma}^\dagger c_{q\sigma} \rangle \leqslant 1$, $R(k)$ is always finite. The integral of $F(k)$ with respect to k equals

$$\frac{1}{V} \sum_k F(k) = \int dr_1 dr_2 \gamma(r - r_2) \gamma^*(r - r_1) \langle c_{r_1\downarrow}^\dagger c_{r\uparrow}^\dagger c_{r\uparrow} c_{r_2\downarrow} \rangle, \tag{B.26}$$

where

$$\gamma(r) = \frac{1}{V} \sum_q \gamma_q e^{iq \cdot r}.$$

Physically, $\langle a^\dagger b \rangle$ can be considered as an inner product between operators a and b. It is simple to show that they satisfy the definition of inner products as well as the Schwartz inequality

$$|\langle a^\dagger b \rangle|^2 \leqslant \langle a^\dagger a \rangle \langle b^\dagger b \rangle.$$

To apply this expression to $\langle c_{r_1\downarrow}^\dagger c_{r\uparrow}^\dagger c_{r\uparrow} c_{r_2\downarrow} \rangle$, we obtain the following inequality

$$\frac{1}{V} \sum_k F(k) < \left| \int dr' |\gamma(r - r')| \langle \rho_\downarrow(r')\rho_\uparrow(r) \rangle^{1/2} \right|^2 \equiv f_0. \tag{B.27}$$

f_0 is finite because the density-density correlation function is non-singular. Since $F(k = 0)$ is positive-definite, we obtain the following inequality for $F(k)$:

$$\frac{1}{V} \sum_{k \neq 0} F(k) = \frac{1}{V} \sum_k F(k) - \frac{F(k = 0)}{V} < f_0. \tag{B.28}$$

In addition, according to the Bogoliubov inequality, we find that $F(k)$ also satisfies the inequality

$$F(k) \geqslant \frac{2k_B T m |\Delta + \eta_k|^2}{\hbar^2 k^2 n} - R(k). \tag{B.29}$$

The momentum integration of the right-hand side is infrared divergent in both one and two dimensions. This apparently conflicts with Eq. (B.28). It implies that there is no superconducting long-range order in one- and two-dimensional Fermi systems at any finite temperature.

Appendix C
Degenerate Perturbation Theory

The degenerate perturbation theory is a useful tool for studying low energy physics in strongly correlated systems. It is widely used to derive low energy effective models of strongly correlated systems. The theory starts by assuming that the Hamiltonian H of a quantum system is a sum of two terms, $H = H_0 + H_I$, with H_0 the unperturbed Hamiltonian whose ground states are degenerate and can be diagonalized analytically, and H_I the perturbation which is small compared to H_0. The goal of the theory is to find systematically the corrections of H_I to the eigenvalues and eigenstates of H_0 by perturbation expansions. It is particularly useful when the energy scale of the problem is much smaller than the energy difference between the degenerate ground states and the first-excited states. In this case, the perturbation can be done to transform H into an effective Hamiltonian H_{eff} which acts only on the ground state subspace of H_0. It is sufficient to use this effective Hamiltonian to investigate low energy physics of the system. Both the t-J model and the Kondo lattice model are these kinds of effective Hamiltonians. The former is the effective low-energy model of the single-band or the three-band Hubbard model. The latter is the effective low energy model of the periodic Anderson model.

Let us consider the Schördinger equation of eigenstates:

$$(H_0 + H_I)|\Psi\rangle = E|\Psi\rangle. \tag{C.1}$$

After a simple transformation, this equation can be reexpressed as

$$
\begin{aligned}
|\Psi\rangle &= \frac{1}{E - H_0} H_I |\Psi\rangle \\
&= \frac{1}{E - H_0} P H_I |\Psi\rangle + \frac{1}{E - H_0} (1 - P) H_I |\Psi\rangle \\
&= \sum_\alpha a_\alpha |\alpha\rangle + \frac{1}{E - H_0} (1 - P) H_I |\Psi\rangle, \tag{C.2}
\end{aligned}
$$

where $\{|\alpha\rangle\}$ are the degenerate ground states of H_0, and P is the corresponding projection operator:

$$a_\alpha = \frac{\langle \alpha | H_I | \Phi \rangle}{E - E_0},$$

$$P = \sum_\alpha |\alpha\rangle\langle\alpha|.$$

To substitute iteratively the expression of $|\Phi\rangle$ in Eq. (C.2) into the right hand side of this equation itself, we have

$$|\Psi\rangle = \frac{1}{1-A} \sum_\alpha a_\alpha |\alpha\rangle = \left(1 + \frac{1}{1-A}A\right) \sum_\alpha a_\alpha |\alpha\rangle, \tag{C.3}$$

where

$$A \equiv \frac{1}{E - H_0}(1 - P)H_I, \tag{C.4}$$

whose projection onto the ground states of H_0 is zero, i.e. $PAP = 0$.

Using Eqs. (C.2) and (C.3), we find that

$$(E - E_0) \sum_\alpha a_\alpha |\alpha\rangle = \left[H_I \frac{1}{1-A} - (E - H_0)\frac{1}{1-A}A\right] \sum_\alpha a_\alpha |\alpha\rangle. \tag{C.5}$$

It indicates that the eigenvalues and eigenstates of H are determined by the following effective Hamiltonian

$$H_{\text{eff}}(E) = \left[H_I \frac{1}{1-A} - (E - H_0)\frac{1}{1-A}A\right] P. \tag{C.6}$$

If only the correction to the ground states of H_0 is considered, the effective Hamiltonian can be simplified as

$$H_{\text{eff}} = P\left[H_I \frac{1}{1-A} - (E - H_0)\frac{1}{1-A}A\right] P = PH_I \frac{1}{1-A}P. \tag{C.7}$$

By expansion in the order of A, the above Hamiltonian becomes

$$H_{\text{eff}} = PH_I \frac{1}{1-A}P = PH_I \sum_{n=0} A^n P. \tag{C.8}$$

This is the formula that is commonly used in practical calculations.

Appendix D
Anderson Theorem

In conventional s-wave superconductors, non-magnetic impurity scattering affects very little on the superconducting transition temperature as well as other physical quantities. This phenomenon was first noticed by Anderson. He gave a mathematical explanation to this phenomenon based on the self-consistent mean-field theory in 1959 [107], which is commonly referred to as the Anderson theorem.

The Anderson theorem results from the fact that non-magnetic impurity scattering does not break the time-reversal invariance of s-wave superconductors. Rigorously speaking, it is valid only when the superconducting correlation length ξ is much larger that the scattering mean free path l, i.e. $\xi \gg l$. In the opposite limit $l \gg \xi$, the time-reversal symmetry is still conserved, but the electronic band structures and the pairing interactions are strongly renormalized by the scattering potentials. This can affect significantly superconducting properties of s-wave superconductors and break the Anderson theorem.

In d-wave superconductors, as the gap function changes sign in momentum space, even non-magnetic impurities can affect strongly superconducting properties no matter whether the impurity potential is in the weak or strong scattering limit. These impurities interfere with the pairing phase and serve as pair-breakers. In particular, impurities can change significantly low energy or low temperature properties of d-wave superconductors. This is an important factor of d-wave superconductors that needs to be considered in the comparison of theoretical calculations with experimental results.

Two approximations are assumed in the proof of the Anderson theorem. First, the pairing gap function $\Delta(r)$ varies very little in space so that it can be replaced by its average value, $\Delta(r) = \Delta$. This approximation implies that the self-consistent mean-field equation of the energy gap is just a result of spatial average. Second, the scattering potential does not change the density of states around the Fermi surface of normal electrons. These two approximations are generally valid if the disorder scattering potential is not so strong. But the first approximation holds only when the correlation length ξ is much larger than the scattering mean-free path l. Under these approximations, the Bogoliubov-de-Gennes self-consistent field equation is given by

$$\begin{pmatrix} H_0(r) & \Delta \\ \Delta & -H_0(r) \end{pmatrix} \begin{pmatrix} u_n(r) \\ v_n(r) \end{pmatrix} = E_n \begin{pmatrix} u_n(r) \\ v_n(r) \end{pmatrix}, \qquad \text{(D.1)}$$

in which

$$H_0(r) = -\frac{\hbar^2}{2m}\nabla^2 + U(r) - \mu,$$

and $U(r)$ is the impurity scattering potential.

The gap function Δ does not depend on r. This greatly simplifies the calculation of the self-consistent gap equation. If $w_n(r)$ is the eigenstate of normal electrons,

$$H_0 w_n(r) = \xi_n w_n(r),$$

then $u_n(r)$ and $v_n(r)$ can be expressed using $w_n(r)$ as

$$u_n(r) = u_n w_n(r),$$
$$v_n(r) = v_n w_n(r).$$

Substituting these expressions into Eq. (D.1), we obtain the equation for determining the coefficients u_n and v_n:

$$\begin{pmatrix} \xi_n & \Delta \\ \Delta & -\xi_n \end{pmatrix} \begin{pmatrix} u_n \\ v_n \end{pmatrix} = E_n \begin{pmatrix} u_n \\ v_n \end{pmatrix}. \qquad \text{(D.2)}$$

This equation has exactly the same form as the standard BCS mean-field equation for a translation invariant system. The difference is that the momentum is not conserved and the basis states are now characterized by the quantum number n of H_0, instead of the momentum. By diagonalizing Eq. (D.2), we obtain the quasiparticle eigen energy

$$E_n = \sqrt{\xi_n^2 + \Delta^2},$$

and the corresponding eigenfunction

$$u_n = \sqrt{\frac{1}{2}\left(1 + \frac{\xi_n}{E_n}\right)},$$

$$v_n = -\sqrt{\frac{1}{2}\left(1 - \frac{\xi_n}{E_n}\right)}.$$

The energy gap is determined by the self-consistent equation

$$\Delta = -g\sum_n u_n(r)v_n(r)\tanh\frac{\beta E_n}{2}. \tag{D.3}$$

Substituting the above solutions into Eq. (D.3), we have

$$
\begin{aligned}
\Delta &= g\sum_n \langle w_n^2(r)\rangle \frac{\Delta}{2\sqrt{\xi_n^2 + \Delta^2}}\tanh\frac{\beta\sqrt{\xi_n^2 + \Delta^2}}{2} \\
&= g\int_{\omega_0}^{\omega_0} d\xi \rho(\omega)\frac{\Delta}{2\sqrt{\xi^2 + \Delta^2}}\tanh\frac{\beta\sqrt{\xi^2 + \Delta^2}}{2},
\end{aligned} \tag{D.4}
$$

where $\langle A\rangle$ is the spatial average of A, and

$$\rho(\xi) = \sum_n \delta(\xi - \xi_n)\langle w_n^2(r)\rangle$$

is the spatial average of electron density of states in the normal state. Since the impurity scattering does not change the density of states of normal electrons around the Fermi surface according to the previous assumption, Eq. (D.4) has exactly the same form as the gap equation for the impurity-free system with $U(r) = 0$. Thus the impurity scattering does not change the transition temperature T_c of the s-wave superconductor. This is the proof first given by Anderson. It is consistent with the experimental observations for conventional superconductors.

Appendix E

Sommerfeld Expansion

In the calculation of thermodynamic or dynamic quantities of electronic systems, we often encounter the following integral

$$I = \int_{-\infty}^{\infty} d\varepsilon g\left(\varepsilon\right) f\left(\varepsilon\right), \tag{E.1}$$

where $g(\varepsilon)$ is an arbitrary function of ε and $f\left(\varepsilon\right)$ is the Fermi distribution function

$$f\left(\varepsilon\right) = \frac{1}{e^{(\varepsilon-\mu)/k_B T} + 1}. \tag{E.2}$$

To ensure Eq. (E.1) integrable, $g(\varepsilon)$ is assumed to be at most exponentially divergent as $\varepsilon \to \infty$, and approach 0 as $\varepsilon \to -\infty$. It is impossible to rigorously solve this integral in most cases. But if we only want to know its low-temperature behavior, the Sommerfeld expansion can be used to obtain an approximate expression for this integral[268].

We first define a function

$$K(\varepsilon) = \int_{-\infty}^{\varepsilon} d\varepsilon g(\varepsilon), \tag{E.3}$$

whose derivative with respect to ε is just $g(\varepsilon)$. Integrating (E.1) by parts leads to the following expression

$$I = -\int_{-\infty}^{\infty} d\varepsilon K\left(\varepsilon\right) \frac{df\left(\varepsilon\right)}{d\varepsilon}. \tag{E.4}$$

If the deviation of the energy from the chemical potential is much larger than the temperature, $|\varepsilon - \mu| \gg k_B T$, $df(\varepsilon)/d\varepsilon$ decays exponentially with ε. Hence the integral in Eq. (E.4) is important only in the vicinity of the Fermi level. This implies that the integral I can be evaluated by performing the Taylor expansion for $K(\varepsilon)$ at $\varepsilon = \mu$ in low temperatures.

The Taylor expansion of $K(\varepsilon)$ is defined by

$$K(\varepsilon) = K(\mu) + \sum_{n=1}^{\infty} \frac{(\varepsilon - \mu)^n}{n!} \left(\frac{dK(\varepsilon)}{d\varepsilon} \right)_{\varepsilon = \mu}. \qquad (E.5)$$

Substituting it into Eq. (E.4) and integrating over ε, we then obtain the expression of the Sommerfeld expansion:

$$I = \int_{-\infty}^{\mu} d\varepsilon g(\varepsilon) + \sum_{n=1}^{\infty} a_n (k_B T)^{2n} \left[\frac{d^{2n-1}}{d\varepsilon^{2n-1}} g(\varepsilon) \right]_{\varepsilon = \mu}, \qquad (E.6)$$

where

$$a_n = -\frac{1}{(2n)!} \int dx x^{2n} \frac{d}{dx} \frac{1}{e^x + 1} = \frac{\left(2^{2n} - 2 \right) \pi^{2n}}{(2n)!} B_n, \qquad (E.7)$$

and B_n is the Bernoulli number. The first five Bernoulli numbers are

$$B_1 = \frac{1}{6}, \quad B_2 = \frac{1}{30}, \quad B_3 = \frac{1}{42}, \quad B_4 = \frac{1}{30}, \quad B_5 = \frac{5}{66}. \qquad (E.8)$$

Bibliography

[1] H. K. Onnes. The resistance of pure mercury at helium temperatures. *Commun. Phys. Lab. Univ. Leiden*, 120b, 1911.

[2] H. K. Onnes. The disappearance of the resistivity of mercury. *Commun. Phys. Lab. Univ. Leiden*, 122b, 1911.

[3] H. K. Onnes. On the sudden change in the rate at which the resistance of mercury disappears. *Commun. Phys. Lab. Univ. Leiden*, 124c, 1911.

[4] W. Meissner and R. Ochsenfeld. Ein neuer effekt bei eintritt der supraleitfähigkeit. *Naturwissenschaften*, 21(44): 787, 1933.

[5] C. J. Gorter and H. B. G. Casimir. On superconductivity I. *Physica*, 1: 30, 1934.

[6] F. London and H. London. The electromagnetic equations of the supraconductor. *Proceedings of the Royal Society of London. Series A-Mathematical and Physical Sciences*, 149(866): 71, 1935.

[7] L. N. Cooper. Bound electron pairs in a degenerate Fermi gas. *Phys. Rev.*, 104: 1189, 1956.

[8] J. Bardeen, L. N. Cooper, and J. R. Schrieffer. Theory of superconductivity. *Phys. Rev.*, 108: 1175, 1957.

[9] G. D. Mahan. *Many-Particle Physics*. Plenum Press, 2nd edition, 1981.

[10] P. G. De Gennes. *Superconductivity of Metals and Alloys (Advanced Book Classics)*. Addison-Wesley Publ. Company Inc, 1999.

[11] C. N. Yang. Concept of off-diagonal long-range order and the quantum phases of liquid He and of superconductors. *Rev. Mod. Phys.*, 34: 694, 1962.

[12] P. C. Hohenberg. Existence of long-range order in one and two dimensions. *Phys. Rev.*, 158: 383, 1967.

[13] P. W. Anderson. Coherent excited states in the theory of superconductivity: Gauge invariance and the Meissner effect. *Phys. Rev.*, 110: 827, 1958.

[14] V. L. Ginzburg and L. D. Landau. On the theory of superconductivity. *Zh. Eksperim. i. Teor. Fiz.*, 20: 1064, 1950.

[15]   L. P. Gor'kov. Microscopic derivation of the Ginzburg-Landau equations in the theory of superconductivity. *Sov. Phys. JETP*, 9(6): 1364, 1959.

[16]   V. J. Emery and S. A. Kivelson. Importance of phase fluctuations in superconductors with small superfluid density. *Nature*, 374(6521): 434, 1995.

[17]   Y. J. Uemura, G. M. Luke, B. J. Sternlieb, J. H. Brewer, J. F. Carolan, W. N. Hardy, R. Kadono, J. R. Kempton, R. F. Kiefl, S. R. Kreitzman, P. Mulhern, T. M. Riseman, D. L. Williams, B. X. Yang, S. Uchida, H. Takagi, J. Gopalakrishnan, A. W. Sleight, M. A. Subramanian, C. L. Chien, M. Z. Cieplak, G. Xiao, V. Y. Lee, B. W. Statt, C. E. Stronach, W. J. Kossler, and X. H. Yu. Universal correlations between T_c and $\dfrac{n_s}{m^*}$ (carrier density over effective mass) in high-T_c cuprate superconductors. *Phys. Rev. Lett.*, 62: 2317, 1989.

[18]   A. P. Mackenzie and Y. Maeno. The superconductivity of Sr_2RuO_4 and the physics of spin-triplet pairing. *Rev. Mod. Phys.*, 75: 657, 2003.

[19]   G. Kotliar and J. L. Liu. Superexchange mechanism and d-wave superconductivity. *Phys. Rev. B*, 38: 5142, 1988.

[20]   N. E. Bickers, D. J. Scalapino, and S. R. White. Conserving approximations for strongly correlated electron systems: Bethe-Salpeter equation and dynamics for the two-dimensional Hubbard model. *Phys. Rev. Lett.*, 62: 961, 1989.

[21]   T. Moriya, Y. Takahashi, and K. Ueda. Antiferromagnetic spin fluctuations and superconductivity in two-dimensional metals—a possible model for high T_c oxides. *J. Phys. Soc. Jpn.*, 59(8): 2905, 1990.

[22]   P. Monthoux, A. V. Balatsky, and D. Pines. Weak-coupling theory of high-temperature superconductivity in the antiferromagnetically correlated copper oxides. *Phys. Rev. B*, 46: 14803, 1992.

[23]   T. Yamashita, A. Kawakami, T. Nishihara, Y. Hirotsu, and M. Takata. AC Josephson effect in point-contacts of Ba-Y-Cu-O ceramics. *Jpn. J. Appl. Phys.*, 26(5A): L635, 1987.

[24]   T. Yamashita, A. Kawakami, T. Nishihara, M. Takata, and K. Kishio. Rf power dependence of AC Josephson current in point contacts of BaY(Tm)CuO ceramics. *Jpn. J. Appl. Phys.*, 26(5A): L671, 1987.

[25]   T. J. Witt. Accurate determination of $\dfrac{2e}{h}$ in Y-Ba-Cu-O Josephson junctions. *Phys. Rev. Lett.*, 61: 1423, 1988.

[26] H. F. C. Hoevers, P. J. M. Van Bentum, L. E. C. Van De Leemput, H. Van Kempen, A. J. G. Schellingerhout, and D. Van Der Marel. Determination of the energy gap in a thin $YBa_2Cu_3O_{7-x}$ film by Andreev reflection and by tunneling. *Physica C: Superconductivity*, 152(1): 105, 1988.

[27] P. J. M. Van Bentum, H. F. C. Hoevers, H. Van Kempen, L. E. C. Van De Leemput, M. J. M. F. de Nivelle, L. W. M. Schreurs, R. T. M. Smokers, and P. A. A. Teunissen. Determination of the energy gap in $YBa_2Cu_3O_{7-\delta}$ by tunneling, far infrared reflection and Andreev reflection. *Physica C: Superconductivity*, 153: 1718, 1988.

[28] C. E. Gough, M. S. Colclough, E. M. Forgan, R. G. Jordan, and M. Keene. Flux quantization in a high-T_c superconductor. *Nature*, 326 (6116): 855, 1987.

[29] R. H. Koch, C. P. Umbach, G. J. Clark, P. Chaudhari, and R. B. Laibowitz. Quantum interference devices made from superconducting oxide thin films. *Appl. Phys. Lett.*, 51(3): 200, 1987.

[30] P. L. Gammel, P. A. Polakos, C. E. Rice, L. R. Harriott, and D. J. Bishop. Little-parks oscillations of T_c in patterned microstructures of the oxide super-conductor $YBa_2Cu_3O_7$: Experimental limits on fractional-statistics-particle theories. *Phys. Rev. B*, 41: 2593, 1990.

[31] J. C. Campuzano, H. Ding, M. R. Norman, M. Randeira, A. F. Bellman, T. Yokoya, T. Takahashi, H. Katayama-Yoshida, T. Mochiku, and K. Kadowaki. Direct observation of particle-hole mixing in the superconducting state by angle-resolved photoemission. *Phys. Rev. B*, 53: R14737, 1996.

[32] M. Takigawa, P. C. Hammel, R. H. Heffner, and Z. Fisk. Spin susceptibility in superconducting $YBa_2Cu_3O_7$ from ^{63}Cu Knight shift. *Phys. Rev. B*, 39: 7371, 1989.

[33] S. E. Barrett, D. J. Durand, C. H. Pennington, C. P. Slichter, T. A. Friedmann, J. P. Rice, and D. M. Ginsberg. ^{63}Cu Knight shifts in the superconducting state of $YBa_2Cu_3O_{7-\delta}$ (T_c=90K). *Phys. Rev. B*, 41: 6283, 1990.

[34] T. Xiang. Physical properties of d-wave superconductors and pairing symmetry of high temperature superconducting electrons, in *Fundamental Research in High T_c Superconductivity*, Edited by W. Z. Zhou and W. Y. Liang. Shanghai Science and Technology Press, 1999.

[35] J. G. Bednorz and K. A. Müller. Possible high T_c superconductivity in the Ba-La-Cu-O system. *Z. Phys. B*, 64(2): 189, 1986.

[36] A. Damascelli, Z. Hussain, and Z. X. Shen. Angle-resolved photoemission studies of the cuprate superconductors. *Rev. Mod. Phys.*, 75: 473, 2003.

[37] P. W. Anderson. Hall effect in the two-dimensional Luttinger liquid. *Phys. Rev. Lett.*, 67: 2092, 1991.

[38] T. R. Chien, Z. Z. Wang, and N. P. Ong. Effect of Zn impurities on the normalstate Hall angle in single-crystal $YBa_2Cu_{3-x}Zn_xO_{7-\delta}$. *Phys. Rev. Lett.*, 67: 2088, 1991.

[39] T. Timusk and B. Statt. The pseudogap in high-temperature superconductors: an experimental survey. *Rep. Prog. Phys.*, 62(1): 61, 1999.

[40] F. Zhou, P. H. Hor, X. L. Dong, and Z. X. Zhao. Anomalies at magic charge densities in under-doped $La_{2-x}Sr_xCuO_4$ superconductor crystals prepared by floating-zone method. *Sci. Tech. Adv. Mater.*, 6(7): 873, 2005.

[41] Z. A. Xu, N. P. Ong, Y. Y. Wang, T. Kakeshita, and S. Uchida. Vortex-like excitations and the onset of superconducting phase fluctuation in underdoped $La_{2-x}Sr_xCuO_4$. *Nature*, 406(6795): 486, 2000.

[42] J. M. Tranquada, B. J. Sternlieb, J. D. Axe, Y. Nakamura, and S. Uchida. Evidence for stripe correlations of spins and holes in copper oxide superconductors. *Nature*, 375(6532): 561, 1995.

[43] J. L. Tallon and J. W. Loram. The doping dependence of T^*—what is the real high-T_c phase diagram? *Physica C*, 349(1): 53, 2001.

[44] C. Panagopoulos, J. L. Tallon, B. D. Rainford, T. Xiang, J. R. Cooper, and C. A. Scott. Evidence for a generic quantum transition in high-T_c cuprates. *Phys. Rev. B*, 66: 064501, 2002.

[45] V. J. Emery. Theory of high-T_c superconductivity in oxides. *Phys. Rev. Lett.*, 58: 2794, 1987.

[46] P. W Anderson. The resonating valence bond state in La_2CuO_4 and superconductivity. *Science*, 235(4793): 1196, 1987.

[47] F. C. Zhang and T. M. Rice. Effective Hamiltonian for the superconducting Cu oxides. *Phys. Rev. B*, 37: 3759, 1988.

[48] E. H. Lieb and F. Y. Wu. Absence of Mott transition in an exact solution of the short-range, one-band model in one dimension. *Phys. Rev. Lett.*, 20: 1445, 1968.

[49] T. Xiang and J. M. Wheatley. c-axis superfluid response of copper oxide superconductors. *Phys. Rev. Lett.*, 77: 4632, 1996.

[50] T. Xiang, C. Panagopoulos, and J. R. Cooper. Low temperature superfluid response of high-T_c superconductors. *International Journal of Modern Physics B*, 12(10): 1007, 1998.

[51] D. L. Feng, N. P. Armitage, D. H. Lu, A. Damascelli, J. P. Hu, P. Bogdanov, A. Lanzara, F. Ronning, K. M. Shen, H. Eisaki, C. Kim, Z. X. Shen, J. I. Shimoyama, and K. Kishio. Bilayer splitting in the electronic structure of heavily overdoped $Bi_2Sr_2CaCu_2O_{8+\delta}$. *Phys. Rev. Lett.*, 86: 5550, 2001.

[52] N. E. Hussey, M. Abdel-Jawad, A. Carrington, A. P. Mackenzie, and L. Balicas. A coherent three-dimensional Fermi surface in a high-transition-temperature superconductor. *Nature*, 425(6960): 814, 2003.

[53] T. Xiang, Y. H. Su, C. Panagopoulos, Z. B. Su, and L. Yu. Microscopic Hamiltonian for Zn- or Ni-substituted high-temperature cuprate superconductors. *Phys. Rev. B*, 66: 174504, 2002.

[54] J. W. Loram, K. A. Mirza, and J. R. Cooper. Properties of the supercondcuting condensate and the normal state pseudogap in high T_c cuprates derived from the electronic specific heat. *Research Review 1998, IRC in Superconductivity, University of Cambridge*: 77, 1998.

[55] N. Momono and M. Ido. Evidence for nodes in the superconducting gap of $La_{2-x}Sr_xCuO_4$. T^2 dependence of electronic specific heat and impurity effects. *Physica C*, 264(3): 311, 1996.

[56] G. D. Mahan. Theory of photoemission in simple metals. *Phys. Rev. B*, 2: 4334, 1970.

[57] W. L. Schaich and N. W. Ashcroft. Model calculations in the theory of photoemission. *Phys. Rev. B*, 3: 2452, 1971.

[58] C. N. Berglund and W. E. Spicer. Photoemission studies of copper and silver: Theory. *Phys. Rev.*, 136: A1030, 1964.

[59] P. J. Feibelman and D. E. Eastman. Photoemission spectroscopy-correspondence between quantum theory and experimental phenomenology. *Phys. Rev. B*, 10: 4932, 1974.

[60] J. C. Campuzano, M. R. Norman, and M. Randeira. *Physics of Superconductors*, volume II. Springer, 2003.

[61] T. Cuk, D. H. Lu, X. J. Zhou, Z. X Shen, T. P. Devereaux, and N. Nagaosa. A review of electron–phonon coupling seen in the high-T_c superconductors by

angle-resolved photoemission studies (arpes). *Phys. Stat. Sol. B*, 242(1): 11, 2005.

[62] J. M. Luttinger and J. C.Ward. Ground-state energy of a many-fermion system. II. *Phys. Rev.*, 118: 1417, 1960.

[63] A. A. Abrikosov, L. P. Gor'kov, and I. E. Dzyaloshinski. *Methods of Quantum Field Theory in Statistical Physics.* Dover Publications, New York, 1963.

[64] D. S. Marshall, D. S. Dessau, A. G. Loeser, C. H. Park, A. Y. Matsuura, J. N. Eckstein, I. Bozovic, P. Fournier, A. Kapitulnik, W. E. Spicer, and Z. X. Shen. Unconventional electronic structure evolution with hole doping in $Bi_2Sr_2CaCu_2O_{8+\delta}$ angle-resolved photoemission results. *Phys. Rev. Lett.*, 76: 4841, 1996.

[65] M. R. Norman, H. Ding, M. Randeria, J. C. Campuzano, T. Yokoya, T. Takeuchi, T. Takahashi, T. Mochiku, K. Kadowaki, P. Guptasarma, and D. G. Hinks. Destruction of the Fermi surface in underdoped high-T_c superconductors. *Nature*, 392(6672): 157, 1998.

[66] Z. X. Shen, D. S. Dessau, B. O. Wells, D. M. King, W. E. Spicer, A. J. Arko, D. Marshall, L. W. Lombardo, A. Kapitulnik, P. Dickinson, S. Doniach, J. Di-Carlo, T. Loeser, and C. H. Park. Anomalously large gap anisotropy in the a-b plane of $Bi_2Sr_2CaCu_2O_{8+x}$. *Phys. Rev. Lett.*, 70: 1553, 1993.

[67] H. Ding, M. R. Norman, J. C. Campuzano, M. Randeria, A. F. Bellman, T. Yokoya, T. Takahashi, T. Mochiku, and K. Kadowaki. Angle-resolved photoemission spectroscopy study of the superconducting gap anisotropy in $Bi_2Sr_2CaCu_2O_{8+x}$. *Phys. Rev. B*, 54: R9678, 1996.

[68] M. L. Titov, A. G. Yashenkin, and D. N. Aristov. Quasiparticle damping in two-dimensional superconductors with unconventional pairing. *Phys. Rev. B*, 52: 10626, 1995.

[69] M. B. Walker and M. F. Smith. Quasiparticle-quasiparticle scattering in high-T_c superconductors. *Phys. Rev. B*, 61: 11285, 2000.

[70] T. Valla, T. E. Kidd, J. D. Rameau, H. J. Noh, G. D. Gu, P. D. Johnson, H. B. Yang, and H. Ding. Fine details of the nodal electronic excitations in $Bi_2Sr_2CaCu_2O_{8+\delta}$. *Phys. Rev. B*, 73: 184518, 2006.

[71] D. Duffy, P. J. Hirschfeld, and D. J. Scalapino. Quasiparticle lifetimes in a $d_{x^2-y^2}$ superconductor. *Phys. Rev. B*, 64: 224522, 2001.

[72] T. Dahm, P. J. Hirschfeld, D. J. Scalapino, and L. Zhu. Nodal quasiparticle lifetimes in cuprate superconductors. *Phys. Rev. B*, 72: 214512, 2005.

[73] A. Hosseini, R. Harris, S. Kamal, P. Dosanjh, J. Preston, R. X. Liang, W. N. Hardy, and D. A. Bonn. Microwave spectroscopy of thermally excited quasi-particles in $YBa_2Cu_3O_{6.99}$. *Phys. Rev. B*, 60: 1349, 1999.

[74] A. F. Andreev. The thermal conductivity of the intermediate state in super-conductors. *Sov. Phys. JETP*, 19: 1228, 1964.

[75] C. R. Hu. Midgap surface states as a novel signature for $d_{x_a^2-x_b^2}$-wave super-conductivity. *Phys. Rev. Lett.*, 72: 1526, 1994.

[76] Y. Tanaka and S. Kashiwaya. Theory of tunneling spectroscopy of *d*-wave superconductors. *Phys. Rev. Lett.*, 74: 3451, 1995.

[77] S. Kashiwaya, Y. Tanaka, M. Koyanagi, and K. Kajimura. Theory for tun-neling spectroscopy of anisotropic superconductors. *Phys. Rev. B*, 53: 2667, 1996.

[78] G. E. Blonder, M. Tinkham, and T. M. Klapwijk. Transition from metallic to tunneling regimes in superconducting microconstrictions: Excess current, charge imbalance, and supercurrent conversion. *Phys. Rev. B*, 25: 4515, 1982.

[79] S. Sinha and K. W. Ng. Zero bias conductance peak enhancement in Bi_2Sr_2Ca Cu_2O_8/Pb tunneling junctions. *Phys. Rev. Lett.*, 80: 1296, 1998.

[80] H. Aubin, L. H. Greene, S. Jian, and D. G. Hinks. Andreev bound states at the onset of phase coherence in $Bi_2Sr_2CaCu_2O_8$. *Phys. Rev. Lett.*, 89: 177001, 2002.

[81] M. Covington, M. Aprili, E. Paraoanu, L. H. Greene, F. Xu, J. Zhu, and C. A. Mirkin. Observation of surface-induced broken time-reversal symmetry in $YBa_2Cu_3O_7$ tunnel junctions. *Phys. Rev. Lett.*, 79: 277, 1997.

[82] J. Y. T. Wei, N. C. Yeh, D. F. Garrigus, and M. Strasik. Directional tunneling and Andreev reflection on $YBa_2Cu_3O_{7-\delta}$ single crystals: Predominance of *d*-wave pairing symmetry verified with the generalized Blonder, Tinkham, and Klapwijk theory. *Phys. Rev. Lett.*, 81: 2542, 1998.

[83] M. Aprili, E. Badica, and L. H. Greene. Doppler shift of the Andreev bound states at the YBCO surface. *Phys. Rev. Lett.*, 83: 4630, 1999.

[84] R. Krupke and G. Deutscher. Anisotropic magnetic field dependence of the zero-bias anomaly on in-plane oriented [100] $Y_1Ba_2Cu_3O_{7-x}$/In tunnel junc-tions. *Phys. Rev. Lett.*, 83: 4634, 1999.

[85] J. Bardeen. Tunnelling from a many-particle point of view. *Phys. Rev. Lett.*, 6: 57, 1961.

[86] W. A. Harrison. Tunneling from an independent-particle point of view. *Phys. Rev.*, 123: 85, 1961.

[87] M. Franz and A. J. Millis. Phase fluctuations and spectral properties of underdoped cuprates. *Phys. Rev. B*, 58: 14572, 1998.

[88] M. B. Walker and J. Luettmer-Strathmann. Josephson tunneling in high-T_c superconductors. *Phys. Rev. B*, 54: 588, 1996.

[89] M. Sigrist and T. M. Rice. Paramagnetic effect in high T_c superconductors—a hint for d-wave superconductivity. *J. Phys. Soc. Jpn.*, 61(12): 4283, 1992.

[90] C. C. Tsuei, J. R. Kirtley, C. C. Chi, L. S. Yu-Jahnes, A. Gupta, T. Shaw, J. Z. Sun, and M. B. Ketchen. Pairing symmetry and flux quantization in a tricrystal superconducting ring of $YBa_2Cu_3O_{7-\delta}$. *Phys. Rev. Lett.*, 73: 593, 1994.

[91] D. J. Van Harlingen. Phase-sensitive tests of the symmetry of the pairing state in the high-temperature superconductors-evidence for $d_{x^2-y^2}$ symmetry. *Rev. Mod. Phys.*, 67: 515, 1995.

[92] C. C. Tsuei and J. R. Kirtley. Pairing symmetry in cuprate superconductors. *Rev. Mod. Phys.*, 72: 969, 2000.

[93] D. A. Wollman, D. J. Van Harlingen, W. C. Lee, D. M. Ginsberg, and A. J. Leggett. Experimental determination of the superconducting pairing state in YBCO from the phase coherence of YBCO-Pb DC squids. *Phys. Rev. Lett.*, 71: 2134, 1993.

[94] D. A. Wollman, D. J. Van Harlingen, J. Giapintzakis, and D. M. Ginsberg. Evidence for $d_{x^2-y^2}$ pairing from the magnetic field modulation of $YBa_2Cu_3O_7$-Pb Josephson junctions. *Phys. Rev. Lett.*, 74: 797, 1995.

[95] C. C. Tsuei and J. R. Kirtley. d-wave pairing symmetry in cuprate superconductors. *Physica C*, 341: 1625, 2000.

[96] C. Tsuei, J. R. Kirtley, M. Rupp, J. Z. Sun, A. Gupta, M. B. Ketchen, C. A. Wang, Z. F. Ren, J. H. Wang, and M. Bhushan. Pairing symmetry in single-layer tetragonal Tl_2Ba_2CuO superconductors. *Science*, 271: 329, 1996.

[97] C. C. Tsuei, J. R. Kirtley, Z. F. Ren, J. H. Wang, H. Ray, and Z. Z. Li. Pure $d_{x^2-y^2}$ order parameter symmetry in the tetragonal superconductor $Ti_2Ba_2Cu O_{6+\delta}$. *Nature*, 387(6632): 481, 1997.

[98]   C. C. Tsuei and J. R. Kirtley. Pure *d*-wave pairing symmetry in high-T_c cuprate superconductors. *J. Phys. Chem. Solids*, 59: 2045, 1998.

[99]   C. C. Tsuei and J. R. Kirtley. Phase-sensitive evidence for *d*-wave pairing symmetry in electron-doped cuprate superconductors. *Phys. Rev. Lett.*, 85: 182, 2000.

[100] C. C. Tsuei, J. R. Kirtley, G. Hammerl, J. Mannhart, H. Ray, and Z. Z. Li. Robust $d_{x^2-y^2}$ pairing symmetry in hole-doped cuprate superconductors. *Phys. Rev. Lett.*, 93: 187004, 2004.

[101] M. Sigrist and T. M. Rice. Unusual paramagnetic phenomena in granular high-temperature superconductors—a consequence of *d*-wave pairing? *Rev. Mod. Phys.*, 67: 503, 1995.

[102] W. Braunisch, N. Knauf, V. Kataev, S. Neuhausen, A. Grütz, A. Kock, B. Roden, D. Khomskii, and D. Wohlleben. Paramagnetic Meissner effect in Bi high-temperature superconductors. *Phys. Rev. Lett.*, 68: 1908, 1992.

[103] P. Svedlindh, K. Niskanen, P. Norling, P. Nordblad, L. Lundgren, B. Lönnberg, and T Lundström. Anti-Meissner effect in the BiSrCaCuO-system. *Physica C*, 162: 1365, 1989.

[104] D. J. Thompson, M. S. M. Minhaj, L. E. Wenger, and J. T. Chen. Observation of paramagnetic Meissner effect in niobium disks. *Phys. Rev. Lett.*, 75: 529, 1995.

[105] A. K. Geim, S. V. Dubonos, J. G. S. Lok, M Henini, and J. C. Maan. Paramagnetic Meissner effect in small superconductors. *Nature*, 396 (6707): 144, 1998.

[106] T. M. Rice and M. Sigrist. Comment on "Paramagnetic Meissner effect in Nb". *Phys. Rev. B*, 55: 14647, 1997.

[107] P. W. Anderson. Theory of dirty superconductors. *J. Phys. Chem. Solids*, 11(1): 26, 1959.

[108] T. Xiang and J. M. Wheatley. Nonmagnetic impurities in two-dimensional superconductors. *Phys. Rev. B*, 51: 11721, 1995.

[109] S. H. Pan, E. W. Hudson, K. M. Lang, H. Eisaki, S. Uchida, and J. C. Davis. Imaging the effects of individual zinc impurity atoms on superconductivity in $Bi_2Sr_2CaCu_2O_{8+\delta}$. *Nature*, 403(6771): 746, 2000.

[110] J. W. Loram, J. Luo, J. R. Cooper, W. Y. Liang, and J. L. Tallon. Evidence

on the pseudogap and condensate from the electronic specific heat. *Journal of Physics and Chemistry of Solids*, 62(1): 59, 2001.

[111] J. R. Schrieffer. *Theory of Supercondcutivity.* Benjamin/Cummings, 1964.

[112] C. J. Wu, T. Xiang, and Z. B. Su. Absence of the zero bias peak in vortex tunneling spectra of high-temperature superconductors. *Phys. Rev. B*, 62: 14427, 2000.

[113] I. Martin, A. V. Balatsky, and J. Zaanen. Impurity states and interlayer tunneling in high temperature superconductors. *Phys. Rev. Lett.*, 88: 097003, 2002.

[114] L. Yu. Bound state in superconductors with paramagnetic impurities. *Acta Phys. Sinica*, 21: 75, 1965.

[115] H. Shiba. Classical spins in superconductors. *Prog. Theor. Phys.*, 40(3): 435, 1968.

[116] J. Kondo. Resistance minimum in dilute magnetic alloys. *Prog. Theor. Phys.*, 32(1): 37, 1964.

[117] P. W. Anderson, G. Yuval, and D. R. Hamann. Exact results in the Kondo problem. II. scaling theory, qualitatively correct solution, and some new results on one-dimensional classical statistical models. *Phys. Rev. B*, 1: 4464, 1970.

[118] K. G. Wilson. The renormalization group: Critical phenomena and the Kondo problem. *Rev. Mod. Phys.*, 47: 773, 1975.

[119] A. A. Abrikosov and L. P. Gor'kov. Contribution to the theory of superconducting alloys with paramagnetic impurities. *Sov. Phys. JETP*, 12: 1243, 1961.

[120] L. S. Borkowski and P. J. Hirschfeld. Kondo effect in gapless superconductors. *Phys. Rev. B*, 46: 9274, 1992.

[121] C. R. Cassanello and E. Fradkin. Overscreening of magnetic impurities in $d_{x^2-y^2}$-wave superconductors. *Phys. Rev. B*, 56: 11246, 1997.

[122] G. M. Zhang, H. Hu, and L. Yu. Marginal Fermi liquid resonance induccd by a quantum magnetic impurity in d-wave superconductors. *Phys. Rev. Lett.*, 86: 704, 2001.

[123] A. Polkovnikov, S. Sachdev, and M. Vojta. Impurity in a d-wave superconductor: Kondo effect and STM spectra. *Phys. Rev. Lett.*, 86: 296, 2001.

[124] P. J. Hirschfeld, P. Wölfle, and D. Einzel. Consequences of resonant impurity scattering in anisotropic superconductors: Thermal and spin relaxation

properties. *Phys. Rev. B*, 37: 83, 1988.

[125] Y. Itoh, S. Adachi, T. Machi, Y. Ohashi, and N. Koshizuka. Ni-substituted sites and the effect on Cu electron spin dynamics of $YBa_2Cu_{3-x}Ni_xO_{7-\delta}$. *Phys. Rev. B*, 66: 134511, 2002.

[126] N. Schopohl and O. V. Dolgov. T dependence of the magnetic penetration depth in unconventional superconductors at low temperatures: Can it be linear? *Phys. Rev. Lett.*, 80: 4761, 1998.

[127] G. E. Volovik. Comment on "T dependence of the magnetic penetration depth in unconventional superconductors at low temperatures: Can it be linear?". *Phys. Rev. Lett.*, 81: 4023, 1998.

[128] D. L. Novikov and A. J. Freeman. Electronic structure and Fermi surface of the $HgBa_2CuO_{4+\delta}$ superconductor: Apparent importance of the role of Van Hove singularities on high T_c. *Physica C*, 212(1): 233, 1993.

[129] O. K. Andersen, A. I. Liechtenstein, O. Jepsen, and F. Paulsen. LDA energy bands, low-energy Hamiltonians, t', t'', t, and j perpendicular to k. *J. Phys. Chem. Solids*, 56(12): 1573, 1995.

[130] M. Tinkham. *Introduction to Supercondcutivity*. McGraw-Hill, 2nd edition, 1996.

[131] W. N. Hardy, D. A. Bonn, D. C. Morgan, R. X. Liang, and K. Zhang. Precision measurements of the temperature dependence of $YBa_2Cu_3O_{6.95}$: Strong evidence for nodes in the gap function. *Phys. Rev. Lett.*, 70: 3999, 1993.

[132] K. Zhang, D. A. Bonn, S. Kamal, R. X. Liang, D. J. Baar, W. N. Hardy, D. Basov, and T. Timusk. Measurement of the ab plane anisotropy of microwave surface impedance of untwinned $YBa_2Cu_3O_{6.95}$ single crystals. *Phys. Rev. Lett.*, 73: 2484, 1994.

[133] D. A. Bonn, S. Kamal, K. Zhang, R. Liang, and W. N. Hardy. The microwave surface impedance of $YBa_2Cu_3O_{7-\delta}$. *J. Chem. Phys. Solids*, 56(12): 1941, 1995.

[134] J. E. Sonier, R. F. Kiefl, J. H. Brewer, D. A. Bonn, J. F. Carolan, K. H. Chow, P. Dosanjh, W. N. Hardy, R. X. Liang, W. A. MacFarlane, P. Mendels, G. D. Morris, T. M. Riseman, and J. W. Schneider. New muon-spin-rotation measurement of the temperature dependence of the magnetic penetration depth in $YBa_2Cu_3O_{6.95}$. *Phys. Rev. Lett.*, 72: 744, 1994.

[135] J. Mao, D. H. Wu, J. L. Peng, R. L. Greene, and S. M. Anlage. Anisotropic surface impedance of YBa$_2$Cu$_3$O$_{7-\delta}$ single crystals. *Phys. Rev. B*, 51: 3316, 1995.

[136] L. A. deVaulchier, J. P. Vieren, Y. Guldner, N. Bontemps, R. Combescot, Y. Lemaitre, and J. C. Mage. Linear temperature variation of the penetration depth in YBa$_2$Cu$_3$O$_{7-\delta}$ thin films. *Europhys. Lett.*, 33(2): 153, 1996.

[137] T. Jacobs, S. Sridhar, Q. Li, G. D. Gu, and N. Koshizuka. In-plane and c-axis microwave penetration depth of Bi$_2$Sr$_2$Ca$_1$Cu$_2$O$_{8+\delta}$ crystals. *Phys. Rev. Lett.*, 75: 4516, 1995.

[138] S. F. Lee, D. C. Morgan, R. J. Ormeno, D. M. Broun, R. A. Doyle, J. R. Waldram, and K. Kadowaki. a-b plane microwave surface impedance of a high quality Bi$_2$Sr$_2$CaCu$_2$O$_8$ single crystal. *Phys. Rev. Lett.*, 77: 735, 1996.

[139] O. Waldmann, F. Steinmeyer, P. Müller, J. J. Neumeier, F. X. Regi, H. Savary, and J. Schneck. Temperature and doping dependence of the penetration depth in Bi$_2$Sr$_2$CaCu$_2$O$_{8+\delta}$. *Phys. Rev. B*, 53: 11825, 1996.

[140] C. Panagopoulos, J. R. Cooper, G. B. Peacock, I. Gameson, P. P. Edwards, W. Schmidbauer, and J. W. Hodby. Anisotropic magnetic penetration depth of grain-aligned HgBa$_2$Ca$_2$Cu$_3$O$_{8+\delta}$. *Phys. Rev. B*, 53: R2999, 1996.

[141] C. Panagopoulos, J. R. Cooper, T. Xiang, G. B. Peacock, I. Gameson, and P. P. Edwards. Probing the order parameter and the c-axis coupling of high-T_c cuprates by penetration depth measurements. *Phys. Rev. Lett.*, 79: 2320, 1997.

[142] C. Panagopoulos, J. R. Cooper, N. Athanassopoulou, and J. Chrosch. Effects of Zn doping on the anisotropic penetration depth of YBa$_2$Cu$_3$O$_7$. *Phys. Rev. B*, 54: R12721, 1996.

[143] C. Panagopoulos, J. R. Cooper, T. Xiang, G. B. Peacock, I. Gameson, P. P. Edwards, W. Schmidbauer, and J. W. Hodby. Anisotropic penetration depth measurements of high-T_c superconductors. *Physica C*, 282: 145, 1997.

[144] C. Panagopoulos, J. R. Cooper, and W. Lo. *unpublished.*

[145] C. Panagopoulos and T. Xiang. Relationship between the superconducting energy gap and the critical temperature in high-T_c superconductors. *Phys. Rev. Lett.*, 81: 2336, 1998.

[146] D. M. Broun, D. C. Morgan, R. J. Ormeno, S. F. Lee, A. W. Tyler, A. P. Mackenzie, and J. R. Waldram. In-plane microwave conductivity of the

single-layer cuprate $Tl_2Ba_2CuO_{6+\delta}$. *Phys. Rev. B*, 56: R11443, 1997.

[147] V. J. Emery and S. A. Kivelson. Superconductivity in bad metals. *Phys. Rev. Lett.*, 74: 3253, 1995.

[148] D. N. Basov, R. Liang, D. A. Bonn, W. N. Hardy, B. Dabrowski, M. Quijada, D. B. Tanner, J. P. Rice, D. M. Ginsberg, and T. Timusk. In-plane anisotropy of the penetration depth in $YBa_2Cu_3O_{7-x}$ and $YBa_2Cu_4O_8$ superconductors. *Phys. Rev. Lett.*, 74: 598, 1995.

[149] Z. Schlesinger, R. T. Collins, F. Holtzberg, C. Feild, S. H. Blanton, U. Welp, G. W. Crabtree, Y. Fang, and J. Z. Liu. Superconducting energy gap and normal-state conductivity of a single-domain $YBa_2Cu_3O_7$ crystal. *Phys. Rev. Lett.*, 65: 801, 1990.

[150] T. A. Friedmann, M. W. Rabin, J. Giapintzakis, J. P. Rice, and D. M. Ginsberg. Direct measurement of the anisotropy of the resistivity in the a-b plane of twinfree, single-crystal, superconducting $YBa_2Cu_3O_{7-\delta}$. *Phys. Rev. B*, 42: 6217, 1990.

[151] R. C. Yu, M. B. Salamon, J. P. Lu, and W. C. Lee. Thermal conductivity of an untwinned $YBa_2Cu_3O_{7-\delta}$ single crystal and a new interpretation of the superconducting state thermal transport. *Phys. Rev. Lett.*, 69: 1431, 1992.

[152] M. B. Gaifullin, Y. Matsuda, N. Chikumoto, J. Shimoyama, K. Kishio, and R. Yoshizaki. c-axis superfluid response and quasiparticle damping of underdoped Bi:2212 and Bi:2201. *Phys. Rev. Lett.*, 83: 3928, 1999.

[153] P. J. Hirschfeld and N. Goldenfeld. Effect of strong scattering on the low-temperature penetration depth of a *d*-wave superconductor. *Phys. Rev. B*, 48: 4219, 1993.

[154] D. A. Bonn, S. Kamal, K. Zhang, R. X. Liang, D. J. Baar, E. Klein, and W. N. Hardy. Comparison of the influence of Ni and Zn impurities on the electromagnetic properties of $YBa_2Cu_3O_{6.95}$. *Phys. Rev. B*, 50: 4051, 1994.

[155] J. Annett, N. Goldenfeld, and S. R. Renn. Interpretation of the temperature dependence of the electromagnetic penetration depth in $YBa_2Cu_3O_{7-\delta}$. *Phys. Rev. B*, 43: 2778, 1991.

[156] J. Y. Lee, K. M. Paget, T. R. Lemberger, S. R. Foltyn, and X. D. Wu. Crossover in temperature dependence of penetration depth $\lambda(t)$ in superconducting $YBa_2Cu_3O_{7-\delta}$ films. *Phys. Rev. B*, 50: 3337, 1994.

[157] J. Nagamatsu, N. Nakagawa, T. Muranaka, Y. Zenitani, and J. Akimitsu. Superconductivity at 39 K in magnesium diboride. *Nature*, 410 (6824): 63, 2001.

[158] N. Nakai, M. Ichioka, and K. Machida. Field dependence of electronic specific heat in two-band superconductors. *J. Phys. Soc. Jpn.*, 71(1): 23, 2002.

[159] I. I. Mazin, O. K. Andersen, O. Jepsen, O. V. Dolgov, J. Kortus, A. A. Golubov, A. B. Kuz'menko, and D. van der Marel. Superconductivity in MgB_2: Clean or dirty? *Phys. Rev. Lett.*, 89: 107002, 2002.

[160] T. Xiang and J. M. Wheatley. Superfluid anisotropy in YBCO: Evidence for pair tunneling superconductivity. *Phys. Rev. Lett.*, 76: 134, 1996.

[161] C. Panagopoulos, J. L. Tallon, and T. Xiang. Effects of the CuO chains on the anisotropic penetration depth of $YBa_2Cu_4O_8$. *Phys. Rev. B*, 59: R6635, 1999.

[162] H. G. Luo and T. Xiang. Superfluid response in electron-doped cuprate superconductors. *Phys. Rev. Lett.*, 94: 027001, 2005.

[163] T. Sato, T. Kamiyama, T. Takahashi, K. Kurahashi, and K. Yamada. Observation of $d_{x^2-y^2}$-like superconducting gap in an electron-doped high-temperature superconductor. *Science*, 291(5508): 1517, 2001.

[164] N. P. Armitage, D. H. Lu, D. L. Feng, C. Kim, A. Damascelli, K. M. Shen, F. Ronning, Z. X. Shen, Y. Onose, Y. Taguchi, and Y. Tokura. Superconducting gap anisotropy in $Nd_{1.85}Ce_{0.15}CuO_4$: Results from photoemission. *Phys. Rev. Lett.*, 86: 1126, 2001.

[165] G. Blumberg, A. Koitzsch, A. Gozar, B. S. Dennis, C. A. Kendziora, P. Fournier, and R. L. Greene. Nonmonotonic $d_{x^2-y^2}$ superconducting order parameter in $Nd_{2-x}Ce_xCuO_4$. *Phys. Rev. Lett.*, 88: 107002, 2002.

[166] B. Chesca, K. Ehrhardt, M. Mössle, R. Straub, D. Koelle, R. Kleiner, and A. Tsukada. Magnetic-field dependence of the maximum supercurrent of $La_{2-x}Ce_xCuO_{4-y}$ interferometers: Evidence for a predominant $d_{x^2-y^2}$ superconducting order parameter. *Phys. Rev. Lett.*, 90: 057004, 2003.

[167] L. Alff, S. Meyer, S. Kleefisch, U. Schoop, A. Marx, H. Sato, M. Naito, and R. Gross. Anomalous low temperature behavior of superconducting $Nd_{1.85}Ce_{0.15}CuO_{4-y}$. *Phys. Rev. Lett.*, 83: 2644, 1999.

[168] J. D. Kokales, P. Fournier, L. V. Mercaldo, V. V. Talanov, R. L. Greene, and

S. M. Anlage. Microwave electrodynamics of electron-doped cuprate super-conductors. *Phys. Rev. Lett.*, 85: 3696, 2000.

[169] R. Prozorov, R. W. Giannetta, P. Fournier, and R. L. Greene. Evidence for nodal quasiparticles in electron-doped cuprates from penetration depth measurements. *Phys. Rev. Lett.*, 85: 3700, 2000.

[170] A. Snezhko, R. Prozorov, D. D. Lawrie, R. W. Giannetta, J. Gauthier, J. Renaud, and P. Fournier. Nodal order parameter in electron-doped $Pr_{2-x}Ce_x$ $CuO_{4-\delta}$ superconducting films. *Phys. Rev. Lett.*, 92: 157005, 2004.

[171] M. S. Kim, J. A. Skinta, T. R. Lemberger, A. Tsukada, and M. Naito. Magnetic penetration depth measurements of $Pr_{2-x}Ce_xCuO_{4-\delta}$ films on buffered substrates: Evidence for a nodeless gap. *Phys. Rev. Lett.*, 91: 087001, 2003.

[172] J. A. Skinta, T. R. Lemberger, T. Greibe, and M. Naito. Evidence for a nodeless gap from the superfluid density of optimally doped $Pr_{1.855}Ce_{0.145}CuO_{4-y}$ films. *Phys. Rev. Lett.*, 88: 207003, 2002.

[173] J. A. Skinta, M. S. Kim, T. R. Lemberger, T. Greibe, and M. Naito. Evidence for a transition in the pairing symmetry of the electrondoped cuprates $La_{2-x}Ce_xCuO_{4-y}$ and $Pr_{2-x}Ce_xCuO_{4-y}$. *Phys. Rev. Lett.*, 88: 207005, 2002.

[174] A. V. Pronin, A. Pimenov, A. Loidl, A. Tsukada, and M. Naito. Doping dependence of the gap anisotropy in $La_{2-x}Ce_xCuO_4$ studied by millimeter-wave spectroscopy. *Phys. Rev. B*, 68: 054511, 2003.

[175] N. P. Armitage, F. Ronning, D. H. Lu, C. Kim, A. Damascelli, K. M. Shen, D. L. Feng, H. Eisaki, Z. X. Shen, P. K. Mang, N. Kaneko, M. Greven, Y. Onose, Y. Taguchi, and Y. Tokura. Doping dependence of an n-type cuprate superconductor investigated by angle-resolved photoemission spectroscopy. *Phys. Rev. Lett.*, 88: 257001, 2002.

[176] T. Yoshida, X. J. Zhou, T. Sasagawa, W. L. Yang, P. V. Bogdanov, A. Lanzara, Z. Hussain, T. Mizokawa, A. Fujimori, H. Eisaki, Z. X. Shen, T. Kakeshita, and S. Uchida. Metallic behavior of lightly doped $La_{2-x}Sr_xCuO_4$ with a Fermi surface forming an arc. *Phys. Rev. Lett.*, 91: 027001, 2003.

[177] T. Xiang and J. M. Wheatley. Quasiparticle energy dispersion in doped two-dimensional quantum antiferromagnets. *Phys. Rev. B*, 54: R12653, 1996.

[178] Z. Z. Wang, T. R. Chien, N. P. Ong, J. M. Tarascon, and E. Wang. Positive hall coefficient observed in single-crystal $Nd_{2-x}Ce_xCuO_{4-\delta}$ at low temperatures. *Phys. Rev. B*, 43: 3020, 1991.

[179] W. Jiang, S. N. Mao, X. X. Xi, X. G. Jiang, J. L. Peng, T. Venkatesan, C. J. Lobb, and R. L. Greene. Anomalous transport properties in superconducting $Nd_{1.85}Ce_{0.15}CuO_{4-\delta}$. *Phys. Rev. Lett.*, 73: 1291, 1994.

[180] P. Fournier, X. Jiang, W. Jiang, S. N. Mao, T. Venkatesan, C. J. Lobb, and R. L. Greene. Thermomagnetic transport properties of $Nd_{1.85}Ce_{0.15}CuO_{4+\delta}$ films: Evidence for two types of charge carriers. *Phys. Rev. B*, 56: 14149, 1997.

[181] C. Kusko, R. S. Markiewicz, M. Lindroos, and A. Bansil. Fermi surface evolution and collapse of the Mott pseudogap in $Nd_{2-x}Ce_xCuO_{4-\delta}$. *Phys. Rev. B*, 66: 140513, 2002.

[182] H. Matsui, K. Terashima, T. Sato, T. Takahashi, S. C. Wang, H. B. Yang, H. Ding, T. Uefuji, and K. Yamada. Angle-resolved photoemission spectroscopy of the antiferromagnetic superconductor $Nd_{1.87}Ce_{0.13}CuO_4$: Anisotropic spin-correlation gap, pseudogap, and the induced quasiparticle mass enhancement. *Phys. Rev. Lett.*, 94: 047005, 2005.

[183] Q. S. Yuan, Y. Chen, T. K. Lee, and C. S. Ting. Fermi surface evolution in the antiferromagnetic state for the electron-doped $t - t - t'' - J$ model. *Phys. Rev. B*, 69: 214523, 2004.

[184] S. K. Yip and J. A. Sauls. Nonlinear meissner effect in CuO superconductors. *Phys. Rev. Lett.*, 69: 2264, 1992.

[185] K. A. Moler, D. L. Sisson, J. S. Urbach, M. R. Beasley, A. Kapitulnik, D. J. Baar, R. X. Liang, and W. N. Hardy. Specific heat of $YBa_2Cu_3O_{7-\delta}$. *Phys. Rev. B*, 55: 3954, 1997.

[186] Y. X. Wang, B. Revaz, A. Erb, and A. Junod. Direct observation and anisotropy of the contribution of gap nodes in the low-temperature specific heat of $YBa_2Cu_3O_7$. *Phys. Rev. B*, 63: 094508, 2001.

[187] I. Kosztin and A. J. Leggett. Nonlocal effects on the magnetic penetration depth in d-wave superconductors. *Phys. Rev. Lett.*, 79: 135, 1997.

[188] M. R. Li, P. J. Hirschfeld, and P. Wölfle. Is the nonlinear Meissner effect unobservable? *Phys. Rev. Lett.*, 81: 5640, 1998.

[189] D. C. Mattis and J. Bardeen. Theory of the anomalous skin effect in normal and superconducting metals. *Phys. Rev.*, 111: 412, 1958.

[190] R. A. Ferrell and R. E. Glover. Conductivity of superconducting films: A sum rule. *Phys. Rev.*, 109: 1398, 1958.

[191] M. Tinkham and R. A. Ferrell. Determination of the superconducting skin depth from the energy gap and sum rule. *Phys. Rev. Lett.*, 2: 331, 1959.

[192] C. C. Homes, S. V. Dordevic, D. A. Bonn, R. X. Liang, and W. N. Hardy. Sum rules and energy scales in the high-temperature superconductor $YBa_2Cu_3O_{6+x}$. *Phys. Rev. B*, 69: 024514, 2004.

[193] D. N. Basov, S. I. Woods, A. S. Katz, E. J. Singley, R. C. Dynes, M. Xu, D. G. Hinks, C. C. Homes, and M. Strongin. Sum rules and interlayer conductivity of high-T_c cuprates. *Science*, 283(5398): 49, 1999.

[194] H. J. A. Molegraaf, C. Presura, D. Van Der Marel, P. H. Kes, and M. Li. Superconductivity-induced transfer of in-plane spectral weight in $Bi_2Sr_2CaCu_2O_{8+\delta}$. *Science*, 295(5563): 2239, 2002.

[195] L. B. Ioffe and A. J. Millis. Superconductivity and the c-axis spectral weight of high-T_c superconductors. *Science*, 285(5431): 1241, 1999.

[196] P. J. Hirschfeld, W. O. Putikka, and D. J. Scalapino. Microwave conductivity of *d*-wave superconductors. *Phys. Rev. Lett.*, 71: 3705, 1993.

[197] P. J. Hirschfeld, W. O. Putikka, and D. J. Scalapino. *d*-wave model for microwave response of high-T_c superconductors. *Phys. Rev. B*, 50: 10250, 1994.

[198] P. A. Lee. Localized states in a *d*-wave superconductor. *Phys. Rev. Lett.*, 71: 1887, 1993.

[199] A. Hosseini, S. Kamal, D. A. Bonn, R. X. Liang, and W. N. Hardy. c-axis electrodynamics of $YBa_2Cu_3O_{7-\delta}$. *Phys. Rev. Lett.*, 81: 1298, 1998.

[200] T. Xiang and W. N. Hardy. Universal c-axis conductivity of high-T_c oxides in the superconducting state. *Phys. Rev. B*, 63: 024506, 2000.

[201] T. Valla, A. V. Fedorov, P. D. Johnson, B. O. Wells, S. L. Hulbert, Q. Li, G. D. Gu, and N. Koshizuka. Evidence for quantum critical behavior in the optimally doped cuprate $Bi_2Sr_2CaCu_2O_{8+\delta}$. *Science*, 285(5436): 2110, 1999.

[202] T. Valla, A. V. Fedorov, P. D. Johnson, Q. Li, G. D. Gu, and N. Koshizuka. Temperature dependent scattering rates at the Fermi surface of optimally doped $Bi_2Sr_2CaCu_2O_{8+\delta}$. *Phys. Rev. Lett.*, 85: 828, 2000.

[203] L. B. Ioffe and A. J. Millis. Zone-diagonal-dominated transport in high-T_c cuprates. *Phys. Rev. B*, 58: 11631, 1998.

[204] Y. I. Latyshev, T. Yamashita, L. N. Bulaevskii, M. J. Graf, A. V. Balatsky, and M. P. Maley. Interlayer transport of quasiparticles and cooper pairs in $Bi_2Sr_2CaCu_2O_{8+\delta}$ superconductors. *Phys. Rev. Lett.*, 82: 5345, 1999.

[205] V. Ambegaokar and A. Griffin. Theory of the thermal conductivity of su-
       perconducting alloys with paramagnetic impurities. *Phys. Rev.*, 137: A1151,
       1965.

[206] A. C. Durst and P. A. Lee. Impurity-induced quasiparticle transport and
       universal-limit Wiedemann-Franz violation in *d*-wave superconductors. *Phys.
       Rev. B*, 62: 1270, 2000.

[207] L. Taillefer, B. Lussier, R. Gagnon, K. Behnia, and H. Aubin. Universal heat
       conduction in $YBa_2Cu_3O_{6.9}$. *Phys. Rev. Lett.*, 79: 483, 1997.

[208] S. Nakamae, K. Behnia, L. Balicas, F. Rullier-Albenque, H. Berger, and T.
       Tamegai. Effect of controlled disorder on quasiparticle thermal transport in
       $Bi_2Sr_2CaCu_2O_8$. *Phys. Rev. B*, 63: 184509, 2001.

[209] M. Sutherland, D. G. Hawthorn, R. W. Hill, F. Ronning, S. Wakimoto, H.
       Zhang, C. Proust, E. Boaknin, C. Lupien, L. Taillefer, R. X. Liang, D. A.
       Bonn, W. N. Hardy, R. Gagnon, N. E. Hussey, T. Kimura, M. Nohara, and H.
       Takagi. Thermal conductivity across the phase diagram of cuprates: Low-
       energy quasiparticles and doping dependence of the superconducting gap.
       Phys. Rev. B, 67: 174520, 2003.

[210] A. Abrikosov and V. M. Genkin. On the theory of Raman scattering of light
       in superconductors. *Zh. Eksp. Teor. Fiz.* 65: 842, 1973.

[211] P. B. Allen. Fermi-surface harmonics: A general method for nonspherical prob-
       lems. Application to Boltzmann and Eliashberg equations. *Phys. Rev. B*, 13:
       1416, 1976.

[212] M. V. Klein and S. B. Dierker. Theory of Raman scattering in superconduc-
       tors. *Phys. Rev. B*, 29: 4976, 1984.

[213] H. Monien and A. Zawadowski. Theory of Raman scattering with final-state
       interaction in high-T_c BCS superconductors: Collective modes. *Phys. Rev. B*,
       41: 8798, 1990.

[214] T. P. Devereaux and D. Einzel. Erratum: Electronic Raman scattering in
       superconductors as a probe of anisotropic electron pairing. *Phys. Rev. B*, 54:
       15547, 1996.

[215] T. Staufer, R. Nemetschek, R. Hackl, P. Müller, and H. Veith. Investigation of
       the superconducting order parameter in $Bi_2Sr_2CaCu_2O_8$ single crystals. *Phys.
       Rev. Lett.*, 68: 1069, 1992.

[216]   X. K. Chen, J. C. Irwin, H. J. Trodahl, T. Kimura, and K. Kishio. Investigation of the superconducting gap in $La_{2-x}Sr_xCuo_4$ by Raman spectroscopy. *Phys. Rev. Lett.*, 73: 3290, 1994.

[217]   T. P. Devereaux, D. Einzel, B. Stadlober, R. Hackl, D. H. Leach, and J. J. Neumeier. Electronic Raman scattering in high-T_c superconductors: A probe of $d_{x^2-y^2}$ pairing. *Phys. Rev. Lett.*, 72: 396, 1994.

[218]   C. Kendziora, R. J. Kelley, and M. Onellion. Superconducting gap anisotropy vs doping level in high-T_c cuprates. *Phys. Rev. Lett.*, 77: 727, 1996.

[219]   M. M. Qazilbash, A. Koitzsch, B. S. Dennis, A. Gozar, H. Balci, C. A. Kendziora, R. L. Greene, and G. Blumberg. Evolution of superconductivity in electron-doped cuprates: Magneto-Raman spectroscopy. *Phys. Rev. B*, 72: 214510, 2005.

[220]   G. Blumberg, M. M. Qazilbash, B. S. Dennis, and R. L. Greene. Evolution of coherence and superconductivity in electron-doped cuprates. AIP Conference Proceedings, edited by Y. Takano, S. P. Hershfield, P. J. Hirschfeld, and A. M. Goldman, 24th International Conference on Low Temperature Physics (LT24). *Low Temperature Physics, Pts. A and B*, 850: 525, 2006.

[221]   C. S. Liu, H. G. Luo, W. C. Wu, and T. Xiang. Two-band model of Raman scattering on electron-doped high-T_c superconductors. *Phys. Rev. B*, 73: 174517, 2006.

[222]   C. P. Slichter. *Principles of Magnetic Resonance.* Springer, 1996.

[223]   F. Mila and T. M. Rice. Spin dynamics of $YBa_2Cu_3O_{6+x}$ as revealed by NMR. *Phys. Rev. B*, 40: 11382, 1989.

[224]   L. C. Hebel and C. P. Slichter. Nuclear spin relaxation in normal and superconducting aluminum. *Phys. Rev.*, 113: 1504, 1959.

[225]   H. Alloul, P. Mendels, H. Casalta, J. F. Marucco, and J. Arabski. Correlations between magnetic and superconducting properties of Zn-substituted $YBa_2Cu_3O_{6+x}$. *Phys. Rev. Lett.*, 67: 3140, 1991.

[226]   P. Mendels, H. Alloul, G. Collin, N. Blanchard, J. F. Marucco, and J. Bobro. Macroscopic magnetic properties of Ni and Zn substituted $YBa_2Cu_3O_x$. *Physica C: Superconductivity*, 235: 1595, 1994.

[227]   A. V. Mahajan, H. Alloul, G. Collin, and J. F. Marucco. [89]Y NMR probe of Zn induced local moments in $YBa_2(Cu_{1-y}Zn_y)_3O_{6+x}$. *Phys. Rev. Lett.*, 72: 3100, 1994.

[228] A. V. Mahajan, H. Alloul, G. Collin, and J. F. F. Marucco. ^{89}Y NMR probe of Zn induced local magnetism in YBa$_2$(Cu$_{1-y}$ZnY)$_3$O$_{6+x}$. *The European Physical Journal B-Condensed Matter and Complex Systems*, 13(3): 457, 2000.

[229] M. H. Julien, T. Fehér, M. Horvatić, C. Berthier, O. N. Bakharev, P. Ségransan, G. Collin, and J. F. Marucco. ^{63}Cu NMR evidence for enhanced antiferromagnetic correlations around Zn impurities in YBa$_2$Cu$_3$O$_{6.7}$. *Phys. Rev. Lett.*, 84: 3422, 2000.

[230] K. Ishida, Y. Kitaoka, K. Yamazoe, K. Asayama, and Y. Yamada. Al NMR probe of local moments induced by an Al impurity in high-T_c cuprate La$_{1.85}$ Sr$_{0.15}$CuO$_4$. *Phys. Rev. Lett.*, 76: 531, 1996.

[231] J. Bobroff, W. A. MacFarlane, H. Alloul, P. Mendels, N. Blanchard, G. Collin, and J. F. Marucco. Spinless impurities in high-T_c cuprates: Kondo-like behavior. *Phys. Rev. Lett.*, 83: 4381, 1999.

[232] C. Bernhard, C. Niedermayer, T. Blasius, G. V. M. Williams, R. De Renzi, C. Bucci, and J. L. Tallon. Muon-spin-rotation study of Zn-induced magnetic moments in cuprate high-T_c superconductors. *Phys. Rev. B*, 58: R8937, 1998.

[233] J. L. Tallon, J. W. Loram, and G. V. M. Williams. Comment on "Spinless impurities in high-T_c cuprates: Kondo-like behavior". *Phys. Rev. Lett.*, 88: 059701, 2002.

[234] A. V. Balatsky, M. I. Salkola, and A. Rosengren. Impurity-induced virtual bound states in d-wave superconductors. *Phys. Rev. B*, 51: 15547, 1995.

[235] M. I. Salkola, A. V. Balatsky, and D. J. Scalapino. Theory of scanning tunneling microscopy probe of impurity states in a d-wave superconductor. *Phys. Rev. Lett.*, 77: 1841, 1996.

[236] G. V. M. Williams and S. Krämer. Localized behavior near the Zn impurity in YBa$_2$Cu$_4$O$_8$ as measured by nuclear quadrupole resonance. *Phys. Rev. B*, 64: 104506, 2001.

[237] S. H. Pan, J. P. O'neal, R. L. Badzey, C. Chamon, H. Ding, J. R. Engelbrecht, Z. Wang, H. Eisaki, S. Uchida, A. K. Gupta, et al. Microscopic electronic inhomogeneity in the high-T_c superconductor Bi$_2$Sr$_2$CaCu$_2$O$_{8+x}$. *Nature*, 413(6853): 282, 2001.

[238] M. R. Norman, M. Randeria, H. Ding, and J. C. Campuzano. Phenomenological models for the gap anisotropy of Bi$_2$Sr$_2$CaCu$_2$O$_8$ as measured by angle-resolved photoemission spectroscopy. *Phys. Rev. B*, 52: 615, 1995.

[239] J. A. Martindale, S. E. Barrett, C. A. Klug, K. E. O'Hara, S. M. DeSoto, C. P. Slichter, T. A. Friedmann, and D. M. Ginsberg. Anisotropy and magnetic eld dependence of the planar copper NMR spin-lattice relaxation rate in the superconducting state of $YBa_2Cu_3O_7$. *Phys. Rev. Lett.*, 68: 702, 1992.

[240] P. C. Hammel, M. Takigawa, R. H. Heffner, Z. Fisk, and K. C. Ott. Spin dynamics at oxygen sites in $YBa_2Cu_3O_7$. *Phys. Rev. Lett.*, 63: 1992, 1989.

[241] S. E. Barrett, J. A. Martindale, D. J. Durand, C. H. Pennington, C. P. Slichter, T. A. Friedmann, J. P. Rice, and D. M. Ginsberg. Anomalous behavior of nuclear spin-lattice relaxation rates in $YBa_2Cu_3O_7$ below T_c. *Phys. Rev. Lett.*, 66: 108, 1991.

[242] G. Q. Zheng, Y. Kitaoka, K. Asayama, K. Hamada, H. Yamauchi, and S. Tanaka. Characteristics of the spin fluctuation in $Tl_2Ba_2Ca_2Cu_3O_{10}$. *Journal of the Physical Society of Japan*, 64(9): 3184, 1995.

[243] T. Imai, T. Shimizu, H. Yasuoka, Y. Ueda, and K. Kosuge. Anomalous temperature dependence of Cu nuclear spin-lattice relaxation in $YBa_2Cu_3O_{6.91}$. *Journal of the Physical Society of Japan*, 57(7): 2280, 1988.

[244] K. Ishida, Y. Kitaoka, K. Asayama, K. Kadowaki, and T. Mochiku. Cu NMR study in single crystal $Bi_2Sr_2CaCu_2O_8$ cobservation of gapless superconductivity. *Journal of the Physical Society of Japan*, 63(3): 1104, 1994.

[245] S. Ohsugi, Y. Kitaoka, K. Ishida, G. Q. Zheng, and K. Asayama. Cu NMR and NQR studies of high-T_c superconductor $La_{2-x}Sr_xCuO_4$. *Journal of the Physical Society of Japan*, 63(2): 700, 1994.

[246] K. Magishi, Y. Kitaoka, G. Q. Zheng, K. Asayama, K. Tokiwa, A. Iyo, and H. Ihara. Spin correlation in high-T_c cuprate $HgBa_2Ca_2Cu_3O_{8+\delta}$ with T_c=133 K —an origin of T_c-enhancement evidenced by ^{63}Cu-NMR study. *Journal of the Physical Society of Japan*, 64(12): 4561, 1995.

[247] M. Takigawa and D. B. Mitzi. NMR studies of spin excitations in superconducting $Bi_2Sr_2CaCu_2O_{8+\delta}$ single crystals. *Phys. Rev. Lett.*, 73: 1287, 1994.

[248] Y. Kitaoka, K. Ishida, G. Zheng, S. Ohsugi, K. Yamazoe, and K. Asayama. NMR study of symmetry of the superconducting order parameter in high-T_c cuprate superconductor. *Physica C*, 1881: 235, 1994.

[249] J. A. Martindale, S. E. Barrett, K. E. O'Hara, C. P. Slichter, W. C. Lee, and D. M. Ginsberg. Magnetic-field dependence of planar copper and oxygen

spinlattice relaxation rates in the superconducting state of $YBa_2Cu_3O_7$. *Phys. Rev. B*, 47: 9155, 1993.

[250] M. Horvatic, T. Auler, C. Berthier, Y. Berthier, P. Butaud, W. G. Clark, J. A. Gillet, P. Segransan, and J. Y. Henry. NMR investigation of single-crystal $YBa_2Cu_3O_{6+x}$ from the underdoped to the overdoped regime. *Phys. Rev. B*, 47: 3461, 1993.

[251] Y. Q. Song, M. A. Kennard, K. R. Poeppelmeier, and W. P. Halperin. Spin susceptibility in the $La_{2-x}Sr_xCuO_4$ system from underdoped to overdoped regimes. *Phys. Rev. Lett.*, 70: 3131, 1993.

[252] Y. Kitaoka, K. Fujiwara, K. Ishida, K. Asayama, Y. Shimakawa, T. Manako, and Y. Kubo. Spin dynamics in heavily-doped high-T_c superconductors $Tl_2Ba_2CuO_{6+y}$ with a single CuO_2 layer studied by ^{63}Cu and ^{205}Tl NMR. *Physica C: Superconductivity*, 179(1C3): 107, 1991.

[253] C. Caroli, P. G. De Gennes, and J. Matricon. Bound fermion states on a vortex line in a type II superconductor. *Physics Letters*, 9(4): 307, 1964.

[254] H. F. Hess, R. B. Robinson, R. C. Dynes, J. M. Valles, and J. V. Waszczak. Scanning-tunneling-microscope observation of the Abrikosov flux lattice and the density of states near and inside a fluxoid. *Phys. Rev. Lett.*, 62: 214, 1989.

[255] H. F. Hess, R. B. Robinson, and J. V.Waszczak. Vortex-core structure observed with a scanning tunneling microscope. *Phys. Rev. Lett.*, 64: 2711, 1990.

[256] G. E. Volovik. Superconductivity with lines of gap nodes: Density of states in the vortex. *JETP Letters C*, 58: 469, 1993.

[257] M. Franz and Z. Tesanovic. Quasiparticles in the vortex lattice of unconventional superconductors: Bloch waves or Landau levels? *Phys. Rev. Lett.*, 84: 554, 2000.

[258] I. Vekhter, P. J. Hirschfeld, and E. J. Nicol. Thermodynamics of d-wave superconductors in a magnetic field. *Phys. Rev. B*, 64: 064513, 2001.

[259] K. A. Moler, D. J. Baar, J. S. Urbach, R. X. Liang, W. N. Hardy, and A. Kapitulnik. Magnetic field dependence of the density of states of $YBa_2Cu_3O_{6.95}$ as determined from the specific heat. *Phys. Rev. Lett.*, 73: 2744, 1994.

[260] B. Revaz, J. Y. Genoud, A. Junod, K. Neumaier, A. Erb, and E. Walker. d-wave scaling relations in the mixed-state specific heat of $YBa_2Cu_3O_7$. *Phys. Rev. Lett.*, 80: 3364, 1998.

[261]   D. A. Wright, J. P. Emerson, B. F. Woodfield, J. E. Gordon, R. A. Fisher, and N. E. Phillips. Low-temperature specific heat of $YBa_2Cu_3O_{7-\delta}$, $0 \leqslant \delta \leqslant 0.2$: Evidence for *d*-wave pairing. *Phys. Rev. Lett.*, 82: 1550, 1999.

[262]   J. L. Luo, J. W. Loram, T. Xiang, T. R. Cooper, and J. L. Tallon. The magnetic field dependence of the electronic specific heat of $Y_{0.8}Ca_{0.2}Ba_2$ Cu_3O_{6+x}. arXiv:cond-mat/0112065, 2001.

[263]   H. H. Wen, L. Shan, X. G. Wen, Y. Wang, H. Gao, Z. Y. Liu, F. Zhou, J. W. Xiong, and W. X. Ti. Pseudogap, superconducting energy scale, and Fermi arcs of underdoped cuprate superconductors. *Phys. Rev. B*, 72: 134507, 2005.

[264]   C. Kübert and P. J. Hirschfeld. Vortex contribution to specific heat of dirty *d*-wave superconductors: Breakdown of scaling. *Solid State Communications*, 105(7): 459, 1998.

[265]   S. H. Simon and P. A. Lee. Scaling of the quasiparticle spectrum for *d*-wave superconductors. *Phys. Rev. Lett.*, 78: 1548, 1997.

[266]   A. Junod, M. Roulin, B. Revaz, A. Mirmelstein, J. Y. Genoud, E. Walker, and A. Erb. Specific heat of high temperature superconductors in high magnetic fields. *Physica C: Superconductivity*, 282: 1399, 1997.

[267]   N. D. Mermin and H. Wagner. Absence of ferromagnetism or antiferromagnetism in one- or two-dimensional isotropic Heisenberg models. *Phys. Rev. Lett.*, 17: 1133, 1966.

[268]   N. W. Ashcroft and N. D. Mermin. *Solid State Physics*. Holt, Rinehart and Winston, 1976.